A Laboratory of Her Own

A Laboratory of Her Own

Women and Science in Spanish Culture

Edited by
VICTORIA L. KETZ,
DAWN SMITH-SHERWOOD, AND
DEBRA FASZER-MCMAHON

VANDERBILT UNIVERSITY PRESS
Nashville

Copyright 2021 Vanderbilt University Press
All rights reserved
First printing 2021

Library of Congress Cataloging-in-Publication Data

Library of Congress Cataloging-in-Publication Data
Names: Ketz, Victoria L., editor. | Smith-Sherwood, Dawn, 1968– editor. | Faszer-McMahon, Debra, 1974– editor.
Title: A laboratory of her own : women and science in Spanish culture / edited by Victoria L. Ketz, Dawn Smith-Sherwood, and Debra Faszer-McMahon.
Description: Nashville : Vanderbilt University Press, [2020] | Includes bibliographical references and index.
Identifiers: LCCN 2020034108 (print) | LCCN 2020034109 (ebook) | ISBN 9780826501288 (paperback) | ISBN 9780826501295 (hardback) | ISBN 9780826501301 (epub) | ISBN 9780826501318 (pdf)
Subjects: LCSH: Women in science—Spain. | Women in science—Spain—History. | Women in science—Social aspects—Spain.
Classification: LCC Q130 .L33 2020 (print) | LCC Q130 (ebook) | DDC 500.82/0946—dc23
LC record available at https://lccn.loc.gov/2020034108
LC ebook record available at https://lccn.loc.gov/2020034109

For young women (y *las chicas raras*) everywhere, that they might be welcomed into and challenged by both the STEM fields and the arts

Contents

Acknowledgments ix

Foreword by Roberta Johnson xiii

Introduction. The Story of Women and STEM in Spanish Culture   1
Victoria L. Ketz, Dawn Smith-Sherwood, and Debra Faszer-McMahon

PART I. ON ROLE MODELS: FEMALE SCIENTISTS AND SPANISH LETTERS

1. Las chicas raras de STEM: Recuperating #WomensPlace in Spanish Literary and Scientific Histories   33
Dawn Smith-Sherwood

2. "The Doctor Is In": Elena Arnedo Soriano (1941–2015), Women's Health, and the Cultural History of Gender and Medicine in Spain   52
Silvia Bermúdez

3. Gender and the Critique of "Ascientific Traditions": Science as Text and Intertext in Rosa Montero's *La ridícula idea de no volver a verte*   70
Ellen Mayock

4. From *la santidad de la escoba* to *la trinidad higiénica*: Rosario de Acuña (1851–1923) and a More Inclusive Vision of Spain's Public Health   89
Erika M. Sutherland

5. Science, History, and Gender: An Interview with María Jesús Santesmases   119
Victoria L. Ketz and Debra Faszer-McMahon

PART II. ON STE(A)M: INTEGRATING SCIENTIFIC INQUIRY INTO THE CULTURAL REALM

6. Science in the Works of Clara Janés: A Poetics of Theoretical (Meta)physics   141
Debra Faszer-McMahon

7. An Extension of Sympathy: Science and Posthumanism in the Paintings of Remedios Varo   166
Marta del Pozo Ortea

8. Subversive, Combative, Corrective: Carmen de Burgos's Interventionist Translation of Möbius's *Öber den physiologischen Schwachsinn des Weibes* (The mental inferiority of women) 187
Leslie Anne Merced

9. Contrasting Images of Women Scientists in the Early Postwar Period (1940–1945) and the Novel *María Elena, ingeniero de caminos* by Mercedes Ballesteros 207
Miguel Soler Gallo

10. Unorthodox Theories and Beings: Science, Technology, and Women in the Narratives of Rosa Montero 235
Maryanne L. Leone

PART III. ON GENDER: USING STEM TO CRITIQUE GENDERED ROLES

11. Biotech, Barceló, Bustelo: Reproduction, Motherhood, and Gendered Hierarchies in Spanish Science Fiction 265
Mirla González

12. Challenging Boundaries of Time, Science, and Gender: Einstein's Theory of Relativity in Marina Mayoral's "Admirados colegas" 291
Victoria L. Ketz

13. Technological Portrayals: Framing *Fernandinas* in the Colonial Context through Photography and Press during the Spanish Second Republic 315
Inés Plasencia

14. Punishing Narratives: The Challenges of Gender and Scientific Authority in Spanish Science Fiction Film 350
Raquel Vega-Durán

Appendix. List of Works by Genre Addressed in This Volume 371
Contributors 379
Index 385

Acknowledgments

Many of our female colleagues, including two of this collection's editors, initially focused on the sciences in their educational formation, only to abandon this line of inquiry to pursue a career in the humanities. While humanistic pursuits have provided us with fascinating and fulfilling work, this pattern of shifting disciplinary focus sparked an interest in exploring the factors dissuading many of our female colleagues from STEM fields. While such patterns have been shifting over time, the power of academic structures and gender norms still seems to hold strong sway, and we see this in our own classrooms and with our female students. As our collection demonstrates, women in science, in Spain and worldwide, have faced pervasive and systematic discrimination that has affected their interactions with STEM fields. Yet our story is one of hope, experimentation, and interdisciplinarity. Spanish female cultural producers have insisted since at least the nineteenth century on connecting science and art, and we have attempted to honor their work and their passion for interdisciplinary STE(A)M endeavors throughout this volume.

The rejection of disciplinary silos and the importance of diverse perspectives is the focus of Banu Subramaniam's *Ghost Stories for Darwin: The Science of Variation and the Politics of Diversity*.[1] She offers intriguing examples of how women scientists might challenge the boundaries of established scientific research in order to connect with humanistic pursuits. Her own personal story echoes "la mirada tuerta" posited by Peninsular novelist and journalist Montserrat Roig. In the preface, subtitled "On Interdisciplinarity," Subramaniam describes the challenges of looking in both directions simultaneously: "Traversing liminal spaces, traveling the hallways of academia, at the borderlands of disciplines. [. . .] Dare I speak? Almost there, but never quite. Almost a scientist, yet a feminist; almost a feminist, yet a scientist; almost a native, yet an alien; almost an alien, yet a native; almost an outsider, yet inside; almost an insider, yet outside . . . Almost there, but never quite" (vii). Her words communicate the frustration of attempting to live in the liminal spaces between disciplinary boundaries and gender norms. Yet those frustrations, in her work and in the work of many Spanish women writers, have led to incredible creative contributions. This project is an attempt to bridge the divides defining the "borderlands of disciplines" such that future women, from all parts of the world, will feel less alien, less outsider, less divided, and will dare to speak. We have not yet arrived, but this collection, with its focus on women and STE(A)M in Spain, is an effort within our own discipline to traverse those divides.

We would like to thank our contributors for sharing with us their provocative research and for being patient and responsive throughout the editing process. We hope that you have benefited from the experience as much as we have. We would also like to express our appreciation to our respective institutions (La Salle University, Indiana University of Pennsylvania, and Seton Hill University) for their research support. Although all institutions of higher learning are under increasing financial constraints, our communities have nonetheless assisted us generously as we pursued important research in our disciplines. Moreover, we would like to acknowledge the valuable role played by both the Mid-America Conference on Hispanic Literatures (MACHL) and the Kentucky Foreign Language Conference (KFLC), where our collaborators were able to present papers that would later become the essays contained in this volume. A special thanks to our favorite KFLC breakfast joint, Shakespeare and Company, where we often met to discuss ideas and plan the stages of this project. We are also grateful to our editor at Vanderbilt, Zachary Gresham, whose enthusiasm, encouragement, and patience have made working on this collection an enjoyable experience. This volume would not have been possible without the generous support of Vanderbilt University Press's entire editorial staff, who assisted us with every aspect of the publication process.

We would also like to acknowledge the Artists Rights Society (ARS) New York / VEGAP for allowing us to publish the images of Remedios Varos's paintings, included in Chapter 7 by Marta del Pozo Ortea. A note of thanks to the Archives at the Liverpool Records Office, to the Spanish

National Library, the Spanish Patrimonio Nacional – Real Biblioteca del Palacio, and to Spain's Archivo General de la Administración at the Ministerio de Educación, Cultura y Deporte for sharing with us the wonderful photographic images in Inés Plasencia's contribution on the Fernandinas in Chapter 13. We would also like to thank Laida Memba for graciously allowing us to reprint images from her personal archive in this collection.

Finally, we would like to express our sincerest gratitude to our families for their patience over the past several years. Thank you for supporting us through this process, for getting takeout and making meals, for putting up with our early morning Google Hangout sessions, and for reading and rereading our drafts (Patrick, Kenneth, and Christopher). The co-editing of this volume has been a rewarding experience for us as it has allowed us the opportunity to create a community that has often been denied to women. For this, we raise our Alicante Specials and toast the unsung heroines of research!

NOTE

1. Subramaniam, Banu. *Ghost Stories for Darwin: The Science of Variation and the Politics of Diversity*. U of Illinois P, 2014.

Foreword
ROBERTA JOHNSON

Where were the Spanish Madame Curies, Barbara McClintocks, and Alice Stewarts? They existed, only hidden from view, overshadowed in a masculinist society that did not recognize women's achievements in the public sphere, especially the scientific arena. The present volume goes some way toward rectifying this oversight. An impressive group of mostly younger scholars delves into this new and necessary topic. The volume provides a nearly panoramic view of Spanish women's engagement with science.

The approach via literature to Spanish women's participation in scientific endeavor is a unique feature of the volume, and, one hopes, a strategy that will make the information on Spanish women scientists more accessible to a readership that might eschew biography. There is plenty of biography here, but it is enhanced by fictional and poetical contexts that are particularly reader friendly. Perhaps the lack of a biographical woman scientist in María Martínez Sierra's novel *El amor catedrático* (mentioned briefly in the Introduction to the volume) has occasioned omitting that work here. In the novel, a fictional female student marries her geology professor and aids him with his scientific work during their married life (shades of María Martínez Sierra's own life as ghost writer of her husband Gregorio's vast literary *oeuvre*).

The final section of the volume, which explores literary genre—particularly science fiction as it relates to gender—is forward-looking (in a literal as well as a literary sense). The volume is thought provoking and surely will inspire other studies on Spanish women in professions usually associated with men. One can think, for example, of the women who in the 1960s during the Franco dictatorship came together to conduct social scientific research on matters relating to women and sexuality. The group, known as SESM (Seminario de Estudios Sociológicos sobre la Mujer), included the Countess Campo Alange, Lilí Álvarez, Consuelo de la Gándara, and María Salas, among several others. They signed their publications collectively, and their methodology followed the dictates of modern sociological research with questionnaires completed by representative populations of respondents.

In closing, I return to the volume at hand. Several of the contributions are offered by seasoned scholars, while others are written by emerging researchers, signaling the passing of the baton to a new generation of feminist scholars in Hispanism. If the field began in the 1980s with the groundbreaking work of scholars like Susan Kirkpatrick (*Las románticas*) and Maryellen Bieder (pioneer of feminist work on Emilia Pardo Bazán and Carmen de Burgos), the present volume assures us that feminist approaches are firmly entrenched as a subfield within modern and contemporary Hispanism and that there are still areas of research to keep the field interesting and vital. As my role in Spanish Feminist Studies comes to an end, I am heartened and encouraged to see so many bright new stars on the horizon. Thank you for the opportunity to have one last say.

INTRODUCTION

The Story of Women and STEM in Spanish Culture

VICTORIA L. KETZ,
DAWN SMITH-SHERWOOD, AND
DEBRA FASZER-MCMAHON

"Training for and pursuing a career in science can be treacherous for women; many more begin than ultimately complete at every stage. Characterizing this as a pipeline problem, however, leads to a focus on individual women instead of structural conditions."

ENOBONG HANNAH BRANCH, *Pathways, Potholes, and the Persistence of Women in Science*

Virginia Woolf's 1929 book-length essay *A Room of One's Own* offered a ground-breaking critique of the exclusion of women from the English literary canon, calling for the creation of new spaces and financial resources to support women writers. In the essay's opening pages, Woolf recounts how she had been given the task of producing an essay on women and fiction and had arrived surprisingly soon at the essay's conclusion, namely that "a woman must have money and a room of her own if she is to write fiction" (4). A place, funding, and time: these are the essential, if not sufficient, elements that afford women's participation in traditionally male-dominated fields, be they literary or scientific. This study, *A Laboratory of Her Own*, addresses how Spanish literary texts and cultural producers since before the turn of the twentieth century have been reflecting on women's exclusion not only from literary but also from STEM fields. The volume focuses on the diverse ways the arts and humanities provide avenues for deepening the conversation about how women have been involved in, excluded from, and represented within the scientific realm. While women's historic exclusion from STEM fields has been receiving increased scrutiny worldwide, within the Spanish context women have been perceived as perhaps even more peripheral, given the complex socio-cultural structures emanating from gender norms and political ideologies dominant in the nineteenth and twentieth centuries.[1] Nonetheless, Spanish female cultural producers have long engaged with science and technology, and their increasing access to the aesthetic realm has led to important and interesting reflections about their ongoing marginalization within scientific contexts.

A central concern for this volume is thus the role of gender and culture in the production of scientific knowledge in Spain. Contributors offer cultural analyses of how Spanish society has viewed female scientists, how those perceptions have evolved over time, and how cultural producers, particularly women, have attempted to influence those changing perceptions. The volume employs an inclusive notion of culture that emphasizes the importance of diverse genres and the varied spaces inhabited by female scientists and cultural producers. Contributions analyze not only specific women scientists in Spain and their interactions with scientific power structures, but also the gendered ways that scientific knowledge has been politically and ideologically represented. In "Feminism and Cultural Studies," Anne Balsamo notes that cultural analyses are "equally preoccupied with the construction of identity and subjectivity as well as the politics of representation" (56), and this volume critiques the cultural structures that organize scientific knowledge in Spain, addressing concerns about power codified in scientific discourse and enacted in its application. Although science has sometimes been held captive by a masculine cult of rationality that has excluded women and the arts, full participation in the scientific realm by men and women, as well as artists and cultural producers, is vital to the production of inclusive knowledge. Balsamo points out the importance of feminist cultural projects, like that undertaken in this volume, in order to analyze the cultural substructures driving women's exclusion from STEM fields:

1) science is a culturally determined discourse that organizes or narrates a particular worldview; 2) scientific knowledge is socially constructed and the practice, production, and organization of science is likewise structured by social relations; and 3) the manifestations of contemporary science, technology, and other institutionalized systems of rationality (medicine, for example) are multifaceted, multi-national, and radically dispersed and decentered, and therefore, require the development of numerous feminist projects that will engage, critique, and struggle over such sites of the organization of power and knowledge. (63)

Certainly, there is need for a cultural studies analysis of the representation of women and science in the Spanish context, as there are no other books, in either English or Spanish, that look specifically at the issue within the realm of cultural or literary production. This volume focuses on how women, as scientists, novelists, poets, painters, translators, photographers, and in a range of other roles, have attempted, through artistic works, to address the challenges faced by Spanish women in science. It begins with a foreword by Roberta Johnson, foremost scholar of Spanish women writers and Spanish feminisms and author of ground-breaking works like *Gender and Nation in the Spanish Modernist Novel* (2003), *Antología de pensamiento feminista* (co-edited with Maite Zubiaurre 2012), *Spanish Women Writers and Spain's Civil War* (co-edited with Maryellen Bieder 2017) and *A New History of Iberian Feminisms* (co-edited with Silvia Bermúdez 2018). The chapters that follow include contributions from established scholars within the fields of peninsular literary and cultural studies as well as fresh voices, including the voice of a contemporary Spanish woman scientist and scientific historian, María Jesús Santesmases. The complete work offers thirteen scholarly essays and one interview that cover a range of topics related to women, science, and cultural production in Spain. Chapters, detailed below, are organized following a theoretical framework inspired by the work of Evelyn Fox Keller in *Feminism in Twentieth-Century Science, Technology and Medicine*, and the structure highlights three broad categories: 1) women scientists and their contributions and presence in Spanish letters; 2) female artists and authors who integrate scientific inquiry into their works; and 3) authors using STEM to comment on female and gendered roles. As the varied contributions to this volume attest, while no other monographs have addressed the issue of women and science in Spanish cultural production, Spanish cultural producers have long been working to shift damaging perceptions that have inhibited women's full participation in STEM fields.

Donna Haraway, in her essay "A Manifesto for Cyborgs," calls for a cultural studies agenda that intervenes, infiltrates, and reconstructs. While this collection, in contrast to Haraway's work, focuses on the specific national context of Spain, the work of intervention and infiltration that Haraway advocates is clearly evident in the diverse genres and theoretical frameworks addressed by contributors, including film studies, literary studies, journalism, art, medical science, translation studies, feminist studies, and ethnography. Many of the works have not been analyzed before, and those that have adopt a new focus. The diverse sources, methodologies, and genres that inform this collection signal the interdisciplinarity of cultural studies and also the contemporary movement to integrate the arts with STEM fields in a movement often termed STE(A)M. STEAM infiltrates, intervenes, and reconstructs in ways that recall Haraway's groundbreaking work, pushing the boundaries and limitations that have long stymied women's involvement in scientific endeavors.

This collection thus highlights how humanistic texts challenge the barriers that have marginalized women in Spanish scientific contexts, and contributors argue that artistic endeavors have been, and continue to be, essential for changing the status quo. Recent surveys and data compiled in Spain attest to the ongoing need for change. Nuria López notes, citing UNESCO, that only 28 percent of scientific researchers worldwide are women, and that only 35 percent of female secondary school students in Spain are contemplating a career in a STEM field. In the European Union (EU), Spain occupies ninth place for the number of women in the scientific workforce, with only 31.6 percent compared to 47 percent for the EU as a whole ("Las mujeres 'asaltan'"). The 2019 "Encuesta de Percepción Social de la Ciencia" (Survey of the social perception of science), conducted by the Spanish Foundation of Science and Technology, registered an increase in Spanish women's desire to enter scientific fields, but little actual change in statistical involvement appears on the horizon (Biosca and Sánchez). There are a series of barriers that younger women face when they choose to pursue a career in the sciences, and stereotypes about women's aptitudes, as well as the lack of female role models and social support, continue to be significant challenges (Lavy and Sand; Bian, Leslie, and Cimpian).[2] In Spain, the two most famous female scientists are Margarita Salas (principal investigator at CSIC, the Consejo Superior de Investigaciones Científicas) and María Blasco (Director of CNIO, Centro Nacional de Investigaciones Oncológicas), whose contributions rightly deserve further attention and accolades.[3] And yet, there are many other important Spanish women scientists (as noted in Chapter 5 by the biochemist and scientific historian María Jesús Santesmases), who have done incredible work in the field and whose contributions have not been acknowledged or celebrated.

Historically in Spain there has been systematic underrepresentation of women in scientific disciplines. José Manuel Lechado's *Científicas: Una historia, muchas injusticias* (Women scientists: one history, many injustices) addresses the issue from a global history perspective, highlighting cultural and religious dynamics from as early as the sixteenth century through the Franco era that created a particularly hostile environment. Lechado argues that for many Spanish women scientists in the nineteenth and twentieth centuries, "De haber nacido en otro país le[s] habría ido mucho mejor" (181; If they had been born in another country, they would have been much better off). The only science-related fields that were traditionally considered acceptable for women were midwifery, nursing, and home economics (Wyer et al. xxiii). Perhaps for that reason, historical participation of women in the scientific realm was often labeled under the more societally palatable guise of social justice. Elena Serrano, in "Chemistry in the City: The Scientific Role of Female Societies in Late Eighteenth-Century Madrid," finds women acting as scientists in surprising yet socially acceptable contexts, performing public works of charity but in effect conducting chemical experiments. Serrano details the charitable activities of two female societies involved in directing infant nutrition in a foundling house and working to purify air for prisoners in Madrid jail cells. Serrano's study explores the ways such efforts, and such women, contributed surreptitiously to the scientific findings of the time: "In the Spanish context, for women to meet solely for their own philosophical education was unthinkable. It would have been perceived as either ridiculous or dangerous, or both. [...] Any engagement with natural philosophy that enhanced individual spiritual progress without the mediating role of the priest had to be handled carefully in Catholic Spain" (141). Under the guise of works of charity, however, these aristocratic women were able to "carv[e] a place for women in the public sphere" (142) and in the scientific realm. Literary genres were similarly employed as a safe space for women to be engaged in scientific analysis, especially in the nineteenth and early twentieth centuries. For example, Margot Versteeg and Susan Walter include a section in their edited collection *Approaches to Teaching the Writings of Emilia Pardo Bazán* that is focused on evolution, race, naturalism, technology, and science in Pardo Bazán's oeuvre (64–85). Nineteenth and early to mid-twentieth century writers like Concepción Arenal (1820–1893), Emilia Pardo Bazán (1851–1921), Rosario de Acuña (1851–1923), Carmen de Burgos (1867–1932), María Martínez Sierra (1874–1974), Rosa Chacel (1898–1994), and María Zambrano (1904–1991) demonstrated a keen interest in scientific discoveries, and, as contributions from Sutherland (Chapter 4) and Merced (Chapter 8) attest, they often used literary genres to explore STEM topics and themes.[4]

While less has been written about women and science related to Spanish literary or cultural studies, the exclusion of women from scientific contexts has been addressed in numerous sociological analyses. For example, Ana M. González Ramos's edited volume *Mujeres en la ciencia contemporánea: La aguja y el camello* (2018; Women in contemporary science: The needle and the camel) offers an in-depth sociological study of the factors contributing to the gender disparity faced by women scientists in Spain today. Contributors address issues like early career abandonment, professional tension, glass ceilings, and socialized archetypes of masculine and feminine skills, basing their findings on surveys and interviews with participants in STEM fields. As Jorge Sáinz González notes in the volume's prologue, the underrepresentation of women is simply not acceptable:

> No es suficiente. En España las estudiantes universitarias representan un 54,1% del total mientras que el profesorado femenino es solo del 40,5%, número que baja al 35,5% si consideramos solo a funcionarias. Este dato es todavía menos esperanzador cuando se analiza el número de profesoras que llegan a posiciones de cátedra, direcciones de departamento, decanatos o rectorados. [. . .] El porcentaje de mujeres con alta cualificación se concentra además en algunas áreas de conocimiento, siendo las de ciencias e ingenierías las que menor presencia femenina concentran. (9)

> It is not sufficient. In Spain female university students represent 54.1 percent of the total while female professors are only 40.5 percent, a number that lowers to 35.5 percent if we consider only public-school faculty. This statistic is even less hopeful when one analyzes the number of female professors who become full professors, department chairs, deans, or provosts. [. . .] The percentage of women with high qualifications is also concen-

trated in a few areas of knowledge, with the sciences and engineering having the lowest female presence.

Spanish women in science thus have a double disadvantage—they are already underrepresented in general in higher education, and in STEM fields the level of exclusion becomes increasingly pronounced. The prologue goes on to emphasize the "theme of maximum importance," which is the need to increase university women's access to and interest in STEM careers (Sáinz González 10). Other Spanish sociological analyses continue to highlight similar concerns. Milagros Sáinz, an expert on gender and STEM in Spain, edited the 2018 study ¿*Por qué no hay más mujeres STEM? Se buscan ingenieras, físicas y tecnólogas* (Why are there not more STEM women? Seeking women engineers, physicists, and technologists). The work offers chapters on sociological theories and data to explain the disparity in women's access to STEM fields and follows case studies of Spanish high school, college, and professional women and their interactions with STEM careers. One finding is that "las personas jóvenes toman decisiones respecto a qué estudiar y en qué trabajar basándose en ideas preconcebidas o estereotipos sobre la clase de personas que trabajan en un determinado ámbito y sobre el tipo de trabajo que estas personas desarrollan" (Sáinz 13; young people make decisions about what to study and in what fields to work based on preconceived notions or stereotypes about the type of people that work in a particular area and the kinds of work that such people do). Thus, in order to change the statistics, young women need to see models of other social formations more open to women's participation in STEM.[5] Such imagining of possible alternative futures fits well with literary pursuits and particularly with certain literary genres, such as science fiction. Several contributors to this volume, including Maryanne L. Leone (Chapter 10), Mirla González (Chapter 11), and Raquel Vega Durán (Chapter 14) address the alternative possible worlds being imagined for women scientists in contemporary Spanish science fiction and film.

In addition to a lack of social support, the lack of recognition received by the women who do choose to pursue STEM fields is also troubling, both globally and in Spain. After twelve decades of Nobel Prizes in science, for example, only twenty-three have been awarded to women, and more than half of those have been awarded in the past twenty years.[6] Of those, all but five of the recipients shared their Nobel Prizes with male colleagues. 2020 saw the first year a prize was shared by an all female team, comprised of Emmanuelle Charpentier and Jennifer A. Doudna (Nobel in Chemistry) for their work creating CRISPR-Cas9, a method of genome editing that is rapidly changing biomedicine.[7] For the Fields Medal in Mathematics, only one woman, Maryam Mirzakhani (in 2014), has won, out of sixty total recipients. In the Spanish context, after thirty-eight years of Premios Nacionales de Investigación (National Research Prizes), there have been only six female winners, and no woman has won in the following categories, all named after Spanish male scientists: Premio Gregorio Marañón in Medicine, Premio Enrique Moles in Science and Chemical Technologies, Premio Alejandro Malaspina in Sciences and Natural Resources Technology, Premio Julio Rey Pastor in Mathematics and Information and Communication Technologies, Premio Juan de la Cierva in Technology Transference, and Premio Blas Cabrera in Physical Sciences and Inorganic Materials.[8] As Victoria L. Ketz points out in Chapter 12, there is an inherent bias toward recognizing male scientists' achievements while ignoring the contributions of their female colleagues.

The absence of women scientists receiving top honors in Spain has caught the attention of the press. In a 2016 *El País* article titled "Las mujeres no existen para los premios científicos" (Women do not exist for scientific awards), Manuel Ansede reported on the frustration expressed by the Association of Women Researchers and Technologists (AMIT), a Spanish non-governmental organization dedicated to promoting women's participation in STEM. AMIT was concerned by the announcement of that year's *Fronteras* prizes, awarded by the Fundación BBVA in Spain, and Ansede shared their concerns: "Un año más, sólo han premiado a hombres, a varones, a investigadores del sexo masculino. Es bastante desmoralizador ver, año tras año, esas fotos de los premiados: todos ellos hombres" (Ansede; Once again they have only recognized men, males, investigators of the masculine sex. It is rather demoralizing to see, year after year, those pictures of the award winners: all of them men). The article notes that in the first eight years of the award's inception, sixty-one men and only three women received a *Fronteras* prize, with no women receiving prizes in 2013, 2014, 2015, or 2016.[9] Ansede and AMIT describe not only the lack of female representation among the awardees as unjust, but also the lack of female representation among the jurors as bad optics, even anti-aesthetic: "Para muchas personas empieza a ser antiestético ver, año tras año, esas fotografías de provectos y encorbatados varones del jurado premiando a otros varones algo más jóvenes" (For many people it starts to be in poor taste to see, year after year, those photographs of the old, tie-wearing males of the jury awarding prizes to other somewhat younger males). The description of this exclusion as "anti-aesthetic" or "in poor taste" recalls the important relationship between aesthetics and STEM, or STE(A)M, and the power of the arts to critique and change cultural norms that have reinforced women's exclusion from the scientific realm.

The disparity between men's and women's participation and acceptance in STEM fields is not only apparent in Spain but has been acknowledged and addressed internationally. On February 11, 2016, the United Nations (UN) celebrated its first *International Day of Women and Girls in Science* with the theme "Transforming the World: Parity in Science." This annual event resulted from the eighty-first plenary meeting of the UN General Assembly where resolution 70/212 was formally adopted, declaring each February 11 *International Day of Women and Girls in Science*. Just before the 2018 UN celebration, *El País* published a special feature focused on women in science, highlighting several Spanish women for their involvement in STEM fields, including Sara Borrell (1917–1999), Josefina Castellví (1935–), and Gabriela Morreale (1930–2017) (Valdés). As Smith-Sherwood points out in Chapter 1 and Vega-Durán highlights in Chapter 10, the bio-summaries offered in *El País* reveal certain patterns experienced by these women, including exclusions from desired fields of study and the need to travel abroad in order to pursue professional opportunities. Nevertheless, these scientific women persisted, and the *El País* special web feature demonstrates the growing public interest in the topic of women and STEM in Spain.

As evidence of this increasing attention, in 2019 Spain planned its own "11 de Febrero" UN-style celebration of women in science, and as Nuria López noted, it included over 2,200 activities all over the country, with nine hundred scientists (the majority women) giving 1,900 talks and workshops in over eight hundred Spanish educational centers, research centers, universities, museums, businesses, and libraries. Rocío Ibarra, who headed the Spanish initiative, says that the idea came to fruition in order to promote younger women: "Necesitamos que estudien y se dediquen a carreras y que se visibilice a la mujer científica" (López; We need them to study and work in scientific careers and become visible as scientific women). Besides this macro-initiative, the Foundation of Catalan Investigation and Innovation has established the project "100tifiques" which targets over ten thousand high school students via interactions with one hundred female scientists who promote scientific vocations and highlight women in science (López).

In addition to initiatives reported by national and international media sources, women's scientific organizations, like AMIT and Women in Physics, have been tracking developments, particularly since the Spanish government and parliament in 2005 and 2007, respectively, adopted several gender equality measures. According to a 2009 report from the international conference on Women in Physics, "[b]y law, women must make up at least 40 percent of evaluation panels to hire and promote scientific personnel" (López-Sancho et al. 171). Additionally, the report states, "[t]he law has also established the number of research projects to be led by women" (172). It is clear from the report that its writers expect the group to continue to participate in multiple acts of self-advocacy; the report concludes, for example, "[i]t is expected that the Women in Science Unit created in 2007 will play an important role in the application of the laws" (172). The 2009 report also outlines a series of initiatives undertaken to promote the advancement of female colleagues in the field. These initiatives include inviting women physicists to be plenary speakers at conferences, including conference sessions that address gender issues, creating a special issue of the *Revista Española de Física* (Spanish Journal of Physics) in which all articles are contributed by women, linking the "Spanish Women in Physics" website to the broader national association site and sharing relevant statistical data in appropriate forums (171).[10]

Other researchers have offered similarly practical initiatives. The Spanish Ministerio de Ciencia e Innovación (Ministry of Science and Innovation) published its definitive early study in 2011, *Libro blanco: Situación de las mujeres en la ciencia española* (White book: The situation of women in Spanish science), citing the need for a change in cultural structures in order to make meaningful statistical progress (Sánchez de Madariaga et al.). However, the situation for women in science, nearly ten years later, continues to be disturbingly imbalanced. While Spanish women have made great strides in employment over the past several decades, the statistics are uneven related to leadership roles, particularly in the sciences.[11] The 2019 data from the Encuesta de Población Activa (Spanish Labor Force Survey) provided by the Instituto Nacional de Estadísticas (Spanish Institute of Statistics) notes an overall increase in women's participation in the field of professional scientific and technical activities. However, the survey does not indicate the level or type of positions held. As María Jesús Santesmases indicates in Chapter 5 of this collection, women are often absent from leadership roles. Reinforcing this concern is the recently released study by CSIC, noting that 57 percent of the scientific researchers in training at the research council are women, but advancement in their careers seems uncertain, since only around 25 percent of women at CSIC have historically been named principal investigators (López). Biosca and Sánchez argue that one notable step forward in moving the needle for women in STEM in Spain is the recent appointment of female CEOs at several technology companies. The Spanish branches of Google, Facebook, Microsoft, HP, and IBM are all headed by women with STEM expertise. As Wyer et al. attest, offering new models for more equitable gender roles related to science and technology appears critical in the new global economy (xvii). Several chapters in this volume, including Chapter 1 by Dawn Smith-Sherwood and Chapter 11 by Mirla González, look specifically at digital and technological disruptions that might offer young women alternative visions.

Although national organizations and media sources have begun to pay more attention to the challenges related to women and STEM in Spain, the possible positive influence of arts and letters in enacting change has not been deeply explored. In María Teresa García Nieto's insightful edited collection *Mujeres, ciencia e información* (2015; Women, science, and information), contributors offer in-depth analyses of a range of cultural and sociological factors leading to the "desequilibrio preocupante" (worrisome imbalance) between men and women in Spanish scientific fields, including the social representation of women scientists, the image of women scientists among the general Spanish population, the representation of women scientists in the press, and the notoriety (or lack of recognition) of women scientists online (5–8). However, the exhaustive study, with statistical data, interviews, and analysis, does not look at any cultural production or the impact that film, literature, or fine arts might have on the conversation. In the prologue to the study, Asunción Bernárdez Rodal describes science as a "symbolic space of prestige" and notes that "in popular culture, science contains something very similar to what religions used to have," arguing that scientific spaces, like religious ones, have been very exclusionary toward women "por las leyes y las costumbres" (9; by law and customs). Undeniably, the symbolic space of prestige evoked by STEM fields is extremely powerful. Bernárdez Rodal compares it to the force of religious conviction, implying that shifting the status quo will require a cultural and emotive response, rather than something purely empirical or analytic. Scientific, historical, or sociological analyses alone do not employ the emotional tools necessary to change minds and shift hearts. The role of cultural studies and the arts must be considered as part of the broader critique of this troubling exclusion.

The present study seeks to address how cultural studies, particularly works emanating from the arts and humanities, provide important avenues for deepening the conversation about women's involvement in, exclusion from, and representation within the scientific realm in Spain. Despite long-standing STEM omissions, female cultural producers have been regularly engaged with science and technology as expressed in literature, art, film, and other areas. Contributors to this collection study representations of women and science beginning in the late nineteenth century and offer a particular focus on twentieth and twenty-first century cultural production. STEM topics include environmental issues, biodiversity, temporal and spatial theories, medicinal practice and reproductive rights, neuroscience, robotics, artificial intelligence, and quantum physics. These scientific themes and issues are analyzed within diverse forms of cultural production, including narrative, painting, poetry, photography, medical texts, translation, newswriting, film, and other forms. Three contributors address late nineteenth and early twentieth-century works, including translations (Merced), speeches and news articles (Sutherland), and photography (Plasencia) in Peninsular and colonial contexts. Four contributors address mid-twentieth century cultural production during the Franco dictatorship, including scientific and literary histories (Smith-Sherwood), early postwar propaganda (Soler), mid-century art (del Pozo Ortea), and late Francoist medical restrictions and challenges for women's health (Bermúdez). The seven remaining contributions focus on mid-to-late twentieth and twenty-first century cultural production related to women and science, including film (Vega-Durán), biotechnology (González), short stories (Ketz), poetry (Faszer-McMahon), novels (Leone and Mayock), and ethnography (Santesmases).

The collection also addresses diverse racial and ethnic dynamics related to women and science in Spanish culture. Contributions move beyond white European women's perspectives to include African viewpoints emanating from women's roles in Spain's former colonies, as well as the cultural, linguistic, and ethnic variation within the Iberian Peninsula itself, and futuristic visions of races and ethnicities beyond the human. For example, Chapter 13, by Inés Plasencia, focuses entirely on race and ethnicity in the context of colonial photography, with Fernandinas (free African women from Fernando Poo) as the central figures and actors. As Plasencia points out in her introduction, Chapter 13 "problematizes how colonialism dealt with gender and 'racial difference' in the context of an elite African social class that challenged the hegemonic colonial visual culture," and thus this volume addresses how early technologies like photography manipulated and constructed the colonial social and racial order.

While Chapter 13 provides the most in-depth study of non-white women in the volume, other chapters also address race and ethnicity. For example, racial dynamics in the colonial context are raised in Chapter 9 by Miguel Soler when analyzing the plot of an early Franco-era novel, *María Elena, ingeniero de caminos* (María Elena, civil engineer), which places its protagonist in the colonial context halfway through the work. The novel reveals the patriarchal systems that dominated the colonial space, and Soler argues that "the novel thus provides important material for analyzing Francoist colonialism, sexism, and even racism." Although racism is not mentioned specifically in the text, the focus on colonial interactions via the plot highlights the intersectionality between gendered discourse and other forms of oppression. In addition, Chapters 7, 10, and 11 also address issues of race and ethnicity, often related to cyborg protagonists, or those with non-human, non-gendered, or non-European ethnicity. For example, Chapter 11, by Mirla González, discusses race and eugenics in the context of science fiction dystopias and heteronormativity. The manipulation of genes in *Planeta hembra* (Female planet) raises concerns about eugenics, exclusionary practices, and oppressive control within the political and social order. Thus, issues of race and ethnicity are addressed in nearly every section of the volume. Nonetheless, contributors also acknowledge how the patterns of exclusion and racial homogeneity marking real-life women engaged in scientific pursuits in Spain in the nineteenth and twentieth centuries make analysis of the lack of racial parity, and imagined alternative futures, all the more pressing.

While the collection analyzes diverse cultural forms from the late nineteenth through the twenty-first centuries, the editors have chosen not to follow a chronological presentation of those works. Instead, the collection follows a thematic approach inspired by the theories of Evelyn Fox Keller in her article "Making a Difference: Feminist Movement and Feminist Critiques of Science." Fox Keller's study offers an organizational framework that helps clarify the complex dynamics surrounding studies of women and science, arguing that feminist critiques of science have focused on three main strands: "writing women [scientists] back into history," analyzing the "role of dominant ideologies of gender in scientific, technological, and medical history," and "changing cultural maps" related to sex, women, and the body (100–01).[12] Chapters in this volume have thus been organized to address those broad areas via the editors' own three-part cultural analysis: Part I: "On Role Models: Female Scientists and Spanish Letters"; Part II: "On STE(A)M: Integrating Scientific Inquiry into the Cultural Realm"; and Part III: "On Gender: Using STEM to Critique Gendered Roles." The volume thus employs Fox Keller's framework as a conceptual tool that offers a more nuanced organization of the diverse analyses offered by contributors than could be offered by a purely chronological approach.

In Part I, "On Role Models: Female Scientists and Spanish Letters," contributors address the lives and histories of real Spanish women involved in scientific pursuits and issues related to women scientists' reception, contributions, and legacies. In the first chapter, Dawn Smith-Sherwood compares contemporary Spanish women scientists' efforts to recognize the contributions of their female forebears with post-Franco era Spanish women humanists' efforts to recover lost women's voices for literary history. Her chapter, "Las chicas raras de STEM: Recuperating #WomensPlace in Spanish Literary and Scientific Histories," argues that practicing Spanish women scientists of the early twenty-first century are engaged in a project like that of practicing Spanish women writers of the late twentieth century. In addition to contributing original work in their respective scientific and literary fields, these women engage in the work of recuperating lost (grand)mothers for Spanish scientific and literary herstories. The chapter explores parallels to the ongoing project of recuperating historical memory in Spain, as well as to the intersection of personal literary and scientific histories in contemporary non-fiction.

In Chapter 2, titled "'The Doctor Is In': Elena Arnedo Soriano, Women's Health, and the Cultural History of Gender and Medicine in Spain," Silvia Bermúdez addresses the important contributions of Elena Arnedo (Madrid, 1941–2015), gynecologist, author, and activist. Arnedo has yet to receive the critical attention she deserves as a leading figure in the defense of sexual and reproductive rights in Spain since the early 1970s. This chapter furthers the cultural history of women scientists in Spain by discussing Arnedo's life as a medical practitioner, committed PSOE-socialist, and feminist activist. The study analyzes Arnedo's specific contributions in the area of women's health, particularly the reference volume *El gran libro de la mujer* (1997). The chapter concentrates on the medical aspects of Arnedo's writing in order to expose how gender and science are at the forefront of a feminist agenda in Spain.

Ellen Mayock's contribution in Chapter 3 continues this focus on women scientists and extends it more explicitly into the literary realm. Her study, "Gender and the Critique of 'Ascientific Traditions': Science as Text and Intertext in Rosa Montero's *La ridícula idea de no volver a verte*," analyzes the history of women and science in Spain and charts how Rosa Montero develops a textual relationship with scientists and scientific concepts, particularly the famous scientist Marie Curie. Montero's work is a captivating hybrid memoir that braids her own story of the loss of her husband with Marie Curie's short diary about love, death, and science. The chapter analyzes how Montero celebrates women scientists' engagement with STEM fields and highlights the astute blending of Montero's own personal and career trajectory with Marie Curie's roles as scientist, two-time Nobel Prize winner, wife, mother, and lover. Montero vindicates and celebrates women scientists' participation in physics and chemistry and maps biological, anatomical, and physiological concerns through both medicalization and memoir.

A similar focus on popular and public scientific endeavors continues in Chapter 4, titled "From *la santidad de la escoba* to *la trinidad higiénica*: Rosario de Acuña and a More Inclusive Vision of Spain's Public Health." The contribution shifts time periods to address late nineteenth and early twentieth-century approaches to training public health's front-line practitioners. In the sole chapter from Part I focused on fin de siècle Spain, Erika Sutherland analyzes how Rosario de Acuña (Madrid, 1851–Gijón, 1923) championed health and access for all women. Rejecting the prevailing discourse of public health as a top-down repressive device, Acuña declared health a universal right. She wrote numerous essays and articles just as the modern science of hygiene was developing, and she addressed a range of issues including the benefits of fresh air, clean water, and natural milk, as well as the health risks posed by certain regulations and lifestyles. Acuña's prolific contributions to popular, women's, and working-class periodicals point to her insistence on bringing the developing principles of hygiene directly to the broadest group of women on the front lines of public health.

The final contribution to Part I, titled "Science, History, and Gender: An Interview with María Jesús Santesmases," brings the contemporary voice of a Spanish woman scientist and scientific historian into the collection and provides a bookend to the important project of recuperating women's lost voices discussed in Chapter 1. María Jesús Santesmases is a member of the Department of Science, Technology, and Society at the Institute of Philosophy of the Spanish National Research Council (CSIC). Her publications, such as *Mujeres científicas en España 1940–1970: Profesionalización y modernización social* (2000), *The Circulation of Penicillin in Spain: Health, Wealth and Authority* (2018), *Gendered Drugs and Medicine: Historical and Socio-Cultural Perspectives* (co-authored with Teresa Ortiz 2014), and "Towards Denaturalization: Women Scientists and Academics in Twentieth-Century Spain" (2011) exemplify the ways in which Santesmases blends, through the lens of gender, her formal training as a biochemist with her professional concern to preserve scientific history. Additionally, Santesmases serves as an occasional contributor to *El País*, where her articles consider science and gender in the context of contemporary sociological, political, and economic issues. This interview provides her perspective on the contemporary context for women scientists in Spain and the way the realities for women scientists have been shifting, particularly since the transition toward democracy.

The five essays in Part I offer in-depth studies of actual women working in the scientific realm, highlighting how contributions from Elena Arnedo, María Jesús Santesmases, Marie Curie (in the hybrid work of Rosa Montero), and Rosario de Acuña have affected attitudes, access, and approaches to women and science in Spain. However, all these essays also link the important scientific contributions of actual women with contemporary cultural production and the ways Spanish women writers have employed and continue to employ the cultural realm in order to raise awareness, increase access, and highlight the work of women in science. The first section thus provides a natural segue to the second part of the collection, focused more explicitly on the representation of women scientists (as opposed to the work of scientists themselves) and the integration of scientific inquiry into Spanish cultural production.

In Part II, "On STE(A)M: Integrating Scientific Inquiry into the Cultural Realm," contributors address the diverse ways female artists and authors have been integrating science into their works. Chapter 6 offers an analysis of the integration of scientific topics in contemporary Spanish poetry via "Science in the Works of Clara Janés: A Poetics of Theoretical (Meta)physics." Debra Faszer-McMahon demonstrates how Janés (Barcelona 1940–), throughout her prolific career as a poet and translator, has consistently shown an interest in the poetics of scientific discourse, particularly via the combination of mystical poetry and theoretical physics. Janés's interrogations of human existence and mystical thought have been informed by many scientific leaders, including Nicolescu, Einstein, Schrodinger, Heisenberg, and Hawking. This chapter argues that for Janés, the creative poetic process replicates and challenges the quest for order, disruption, and meaning found in scientific thought, and Janés's works reveal the unstable boundaries between scientific inquiry and mystical poetic rumination. The chapter begins with background on Janés's larger corpus and how her interest in bridging scientific and poetic discourse connects with her larger poetic trajectory. It then analyzes three specific ways in which Janés's works bridge the scientific and poetic realms via close readings from several works. Finally, the study addresses why these efforts by Janés to bridge scientific and poetic worlds are important for Spain, for global women's issues, and for science today.

Just as contemporary poets like Janés have been incorporating specific scientific theories into their work, so other earlier female cultural producers have integrated concrete scientific ideas into their own unique cultural forms. Chapter 7, by Marta del Pozo Ortea, focuses on scientific representations in mid-twentieth-century art via the works of Remedios Varo (Anglès, Gerona, 1908–Ciudad de México, 1963). In "An Extension of Sympathy: Science and Posthumanism in the Paintings of Remedios Varo," del Pozo Ortea introduces Varo as a visual artist who professed that only science blended with art could respond to the ultimate meaning of reality. Her work represents diverse styles, including fantasy, surrealism, and symbolism, and in many works, a "humanoid" reacts to other living objects, animals, alchemical tools, or fantastic vehicles, frequently depicting "invisible" threads of relationships established between human and nonhuman figures. This contribution's analysis of Varo's work moves beyond existing art-historical studies by using hermeneutic tools from posthumanism, including agential realism, cybernetics, and object-oriented ontology, as well as concepts from quantum physics such as entanglement theory. The goal is to illuminate Remedios Varo's visual ecologies within a scenario that ontologically and epistemologically decenters the human and advocates, as science has been postulating in recent decades, for an eminently relational experience of reality.

The relational aspects of scientific discovery are also highlighted in Chapter 8, which shifts back in time to look at nineteenth-century translation and the work of Carmen de Burgos in the context of German medical treatises. Leslie Anne Merced's chapter, "Subversive, Combative, Corrective: Carmen de Burgos's Interventionist Translation of Möbius's *Über den physiologischen Schwachsinn des Weibes*" (The mental inferiority of women), analyzes how and why Burgos translated such a controversial and misogynistic text. Merced describes the late nineteenth and early twentieth-century fascination with phrenology and other pseudo-medical topics of the day. German perspectives on such issues became known in Spain, and Carmen de Burgos (Rodalquilar, 1863–Madrid, 1932) made the decision to publish a Spanish translation of Paul Julius Möbius's influential text. As Burgos states in the prologue, she embarked on the task of translating the German text with apprehension, and she assured her readers that she would work free from preconceived ideas that might impede the task. However, what follows reveals an approach by Burgos that gives as much voice to the author of the translated text as to the original. Such lack of translational "fluency" allows Burgos to demonstrate the lack of representation of the voiceless, in this case not only the female translator, but also women in science more generally. This study analyzes how Burgos offers a feminist approach to the translator's task and addresses issues of authorship, ideology, and rewriting in the context of women and scientific theories circulating in late nineteenth and early twentieth-century Spain.

Chapter 9 moves from the early twentieth century to analyze the representation of women and science after the Spanish Civil War, particularly within the context of Franco-era censorship. Miguel Soler Gallo's contribution, "Contrasting Images of Women Scientists in the Early Postwar Period (1940–1945) and the Novel *María Elena, ingeniero de caminos* by Mercedes Ballesteros," analyzes the representation of women scientists in literary and cultural magazines published by the Women's Section of the Falange. It begins by tracing a discourse aimed at narrowing female occupations whereby women scientists, engineers, chemists, and mathematicians became labeled as "viragos" (mannish women), "guarras" (dirty or slutty women), or "monsters," and publications often cast doubt on the intellectual capacity of women to exercise scientific professions. Women with aspirations to enter university were directed toward the humanities, implying an opposition between intellectual branches and reserving the scientific realm for men. The article focuses in particular on an analysis of Mercedes Ballesteros's (Madrid, 1913–1995) novel, which portrays a female civil engineer who, due to social pressure, takes on a male identity to compete and thrive in the workplace. The novel, published in 1940, offers a glimpse into early Franco-era discourse related to women and science and the problematic positioning of women from both within and outside the regime. Soler's article not only addresses the representation of women in

STEM fields in postwar Spain but also, like the study of colonial photography later in the collection, addresses the intersectionality between gender and race in the colonial context via the plot of Ballesteros's novel.

The tenth chapter and final contribution to Part II moves forward into the twenty-first century and also connects with earlier studies of female scientific achievement via the work of Rosa Montero analyzed in Part I. Maryanne L. Leone's "Unorthodox Theories and Beings: Science, Technology, and Women in the Narratives of Rosa Montero" analyzes several of Montero's recent novels, which engage directly with the involvement of women in, and the marginalization and exclusion of women from, STEM fields. In *Instrucciones para salvar el mundo* (2008), *Lágrimas en la lluvia* (2011), *La ridícula idea de no volver a verte* (2013), and *El peso del corazón* (2015), Montero voices concern for the damaging impact on individuals and the environment of consumer-oriented growth and scientific developments. Women scientists are often central characters, and a technohuman is the protagonist of *Lágrimas en la lluvia* and *El peso del corazón*. This contribution explores Montero's representation of female scientists, their exclusion from the scientific community, and their political and social contexts, which include turn-of-the-twentieth-century Paris, the Franco period, the contemporary era, and one hundred years into the future. The essay argues that Montero not only brings recognition to women in STEM fields but also interrogates the tensions between scientific inquiry and life sustainability. Montero's works suggest the desire to understand the individual self within a broader ecological co-dependence.

As described above, the contributions in Part II increasingly analyze specific scientific concepts, such as quantum theory, environmental studies, cybernetics, or phrenology, and contributors address the ways Spanish cultural producers have been incorporating those and other diverse scientific ideas into their works. Yet the authors and cultural productions studied also connect with the focus of Part III, namely the critique of gendered roles and the problematization of gender-based categories. Thus, in diverse periods, authors like Ballesteros have created protagonists who take on masculine identities, and artists like Varo depict humanoid figures that challenge gendered categories and question the binary gender divide. These challenges to traditional gendered categories are the focus of the final group of essays in the volume.

Part III of the collection, titled "On Gender: Using STEM to Critique Gendered Roles," focuses on the way cultural production related to science challenges traditional categories and boundaries, not only for women in

STEM fields but also in racialized, colonial, literary, and other varied contexts. In Chapter 11, Mirla González explores the gender-bending concepts in contemporary Spanish science fiction focused on biotechnologies. Her contribution, "Biotech, Barceló, Bustelo: Reproduction, Motherhood, and Gendered Hierarchies in Spanish Science Fiction," addresses one of the literary genres in Spain that has received the least attention due to a series of social, political, and economic factors. In an effort to highlight the contributions of Spanish women authors to the science-fiction genre, the study explores how utopias and dystopias have been used to invert gender roles in Gabriela Bustelo's *Planeta hembra* (2000), as well as how authors create equality between both sexes in works like Elia Barceló's *Consecuencias naturales* (1994; Natural consequences). The works analyzed deal with childbirth, women's reproductive rights, and various societal models regarding power relations between different genders and sexes. The study addresses how Spanish cultural productions engage with biotechnology, including natural and artificial reproduction, genetic engineering, and contraceptives. By drawing upon feminist thought and critical works regarding sex and gender, this chapter explores gender relations and matriarchal societies that offer an alternative to the patriarchal traditions of Spain.

The Spanish patriarchal tradition also comes under critique in Chapter 12 via Victoria L. Ketz's study "Challenging Boundaries of Time, Science, and Gender: Einstein's Theory of Relativity in Mayoral's 'Admirados colegas.'" Marina Mayoral (Mondoñedo, 1942–) is a professor of literature at the Complutense in Madrid, and she is thus well versed in literary history and Spanish literary tradition. Her short story "Admirados colegas" demonstrates an intriguing incorporation of scientific theories into contemporary fictional plots. The story highlights Einstein's Theory of Special Relativity when two female science students and their advisor are displaced through time to parallel temporal universes, interacting with Spanish literary giants such as Lope de Vega while simultaneously managing the scientific insights of space-time travel. This warping of narrative expectations serves to highlight gendered norms, and it challenges biased perceptions of females' intellectual abilities. By examining the structure, diction, and themes contained in the short story through a feminist lens, it becomes evident that Mayoral's narrative denounces the lack of opportunities available for women's advancement in the scientific realm.

In Chapter 13, "Technological Portrayals: Framing *Fernandinas* in the Colonial Context through Photography and Press during the Spanish Second Republic," the collection offers a global Hispanophone perspective on Women and STEM in Spanish Culture. Inés Plasencia contributes an intriguing study of the relationship between technologies of power (in this case photography), colonialism, and gendered representation within the Fernandino culture of Fernando Poo in Equatorial Guinea. The emerging technology of photography was employed in colonial contexts to forward Spanish territorial ambitions as well as to diminish the role and importance of the African elite, feminizing the men and sexualizing Fernandina women. However, the Fernandinos exercised unique agency, and Plasencia's essay highlights the ways in which the African elite represented themselves and challenged the gendered and racist representations of the period. The chapter analyzes photographic images found in British and Spanish archives, both public and private, and highlights the troubling exclusions evident in the Spanish colonial history of the African elite and the need to rectify that deliberate elision of the technological and historical record.

The notion of increasing visibility for marginalized groups in the context of gender norms continues to surface in Chapter 14, where Raquel Vega-Durán offers an analysis of visual representation via "Punishing Narratives: The Challenges of Gender and Scientific Authority in Spanish Science Fiction Film." While women have played an important role in the development of science and technology in Spain, it is only in the last decades of the twentieth century that women scientists have begun to gain visibility in Spanish film and television. Vega-Durán notes that while documentary film has offered Spanish audiences several examples of key women scientists, such as biologist Margarita Salas and oceanographer Josefina Castellví, television and film fiction have not produced many significant biopics. Perhaps surprisingly, science fiction, one of the least common film genres in Spain, has been most apt in recent years to include female scientists. Two notable examples include *Órbita 9* (Hatem Khraiche, dir., 2017) and *EVA* (Kike Maíllo, dir., 2011). However, Vega-Durán argues that while *Órbita 9* and *EVA* present audiences with women engaged in scientific breakthroughs—some as researchers and some as guinea pigs—these women's fates eventually become dependent on their male co-protagonists, thus challenging their scientific authority. This chapter reflects on how contemporary Spanish film relates to the conversations taking place in Spain today about encouraging women and girls to enter scientific fields.

The articles herein maintain that women scientists, though often marginalized by their male counterparts and societal norms more generally, have been supported by other females in the cultural realm and have made progress in STEM fields. The works of practicing female scientists, as well as poets, novelists, artists, and translators, reflect how diverse efforts are contributing to the evolving perception of female scientists in Spanish culture.[13] The cultural studies perspective offered in this collection aims to decipher and reinterpret the roles of science and gender in modern times. As can be seen by the works selected for analysis, this movement has been in place since at least the nineteenth century and has developed with increasing urgency. By examining the place of women in science and linking it to ideological, social, and political perspectives, this collection seeks to highlight the important contributions of the arts to the ongoing debate. In calling attention to the relationships between science and gender, power structures are revealed, and different futures can be imagined. Cultural studies, in the context of STE(A)M, can thus create spaces for innovative vision, rewriting the roles and expectations related to science and gender.

As Biosca and Sánchez state in their 2019 article on Spanish women leaders in technology, "Quedan muchas islas que conquistar. Muchos territorios para explorar. Muchos caminos que recorrer" (There are many islands left to conquer. Many territories to explore. Many roads to travel), indicating that tremendous work, in a range of fields and genres, remains for women in science in Spain. The article offers hope for the future, yet the language of conquest chosen by Biosca and Sánchez recalls a history of colonialist and gendered exclusions: "conquistar," "explorar," "recorrer." The work toward equity for women in science is in progress, and has been underway for some time, but there is clearly still much work to be done, including breaking down such gendered and colonial worldviews. As Dawn Smith-Sherwood, Ellen Mayock, and Maryanne L. Leone note in their literary studies in Chapters 1, 3, and 10, contemporary Spanish authors like Rosa Montero are helping to move the needle and simultaneously commenting on the speed of social change via intertextual hashtags like #LugarDeLaMujer, #Mutante, and #Raras. In order to generate a cultural shift that gives equity to women in the field of science, changes must be made not only in political, social, and economic contexts, but also in the cultural realm. Stereotypes and barriers cannot be eliminated without creating emotional reactions and disruptions that allow for new perspectives, and cultural productions allow for that kind of mental, emotional, and textured long-term impact.

In the conclusion to Fox Keller's essay "Making a Difference: Feminist Movement and Feminist Critiques of Science," she asserts that "Second-wave feminism has been one of the most powerful social movements of modern times" and emphasizes the important work yet to be done via third wave and developing feminisms (108). At the risk of returning to a 'first wave,' essentialist feminism, one cannot help but note the irony that while in Spanish, "science" or *la ciencia* is linguistically feminine, women in Spain continue to find themselves estranged from STEM fields. The still peripheral relationship of women to STEM, in Spain and around the globe, confirms the complex socio-economic-cultural structures that have also governed gender norms at different times and places, and in other disciplines, including literary and cultural studies. In 1929, Virginia Woolf imagined that Shakespeare might have had a talented sister, but without the space and financial backing to support her craft, she would have been erased from history. Each year, previously hidden or silenced literary and scientific figures are recuperated, and their contributions are acknowledged. Take, for instance, the case of María Teresa Toral, whose life is recounted in the 2012 biography *Una mujer silenciada* (A silenced woman) by Antonina Rodrigo. A student of pharmacy and chemistry under Enrique Moles, Toral escaped to Mexico, following persecution under the Franco regime, and there became known as an artist, in large part due to the efforts of another woman writer and journalist, Elena Poniatowska. The work of women writers and journalists, like Poniatowska and Rosa Montero, has been essential to the women-in-STEM cause.

In *Dime que me quieres aunque sea mentira* (Tell me you love me even if it's a lie), another Spanish woman writer and journalist, Montserrat Roig, presents women as both observers and observed, possessing what she terms "la mirada tuerta"—one eye looking in, one eye looking out—and the resultant "ya no, todavía no" (no longer, still not) historical perspective. Roig theorizes that, as a woman looks both within and without, she sees what has already been gained and what has yet to been attained. As this volume has considered the "ya no" of what has been achieved and the "todavía no" of what still needs to be done, perhaps the time has come not only to be concerned about the women who do not win the significant science prizes but also to highlight the important artistic and cultural representations showcasing those gaps and signaling others that might merit scrutiny. Should women scientists as well as cultural producers, in Spain and around the world, continue to play the game, participate in the construct, or disconnect and reconstruct?[14]

As several contributors to this volume note, female characters in recent works often dissociate themselves from the traditional system established by male predecessors in order to create, as contributor Ellen Mayock projects, "un tipo de laboratorio innovador, colectivo, tal vez igualitario y orgánico" (193; a type of innovative, collective laboratory, perhaps also egalitarian and organic). Women cannot ignore the negative ways that they are portrayed and perceived as scientists and scholars, and they should not have to forgo the possibilities of tenure and promotion and prestige. Nonetheless, perhaps it is possible to fight the system and simultaneously rise above, or as Mayock describes it, reject "cualquier noción de una postura de humildad y objetividad precisamente en su facilidad con su propia sujetividad" (191; whatever notion of a humble, objective posture precisely via an ease with one's own subjectivity). Many of the fictitious twenty-first century protagonists studied in this volume are the formally educated, socially and economically liberated descendants of the real life nineteenth-century amateur chemists conducting experiments in Madrid jails and foundling houses. They have the education and training, the social and economic freedom, and the technological know-how. If these futuristic possibilities become the collective project of both women writers and women scientists, then Woolf's ideal of women's space and financial support may finally be realized, or turned on its head, in a way not yet ("todavía no") imagined.

In the meantime, social media hashtag movements, from #DistractinglySexy to #MeToo, identify, recuperate, (re)mark, and (re)claim #WomensPlace and perhaps even #WomensTime, realizing the democratizing power of new technologies and obviating the perennial snubbing by official award agencies, both artistic and scientific, from Spanish Royal Academies to Nobel Prize boards. Two recent examples of women scientists in Spain and beyond being recognized and recuperated in their own time are encouraging. On October 2, 2018, physicist Donna Strickland became only the third woman to receive the Nobel Prize in physics. That same day, Wikipedia editors finally agreed that she met their "notability requirement," having previously rejected a May 2017 entry. As Dawn Bazely noted in the *Washington Post*, "Strickland's biography went up shortly after her award was announced. If you click on the 'history' tab to view the page's edits, you can replay the process of a woman scientist finally gaining widespread recognition, in real time."

In the Spanish context, *El País* recently highlighted the incredible story of María Wonenburger, the Galician mathematician whose story had all but disappeared until two female scientists at the University of A Coruña happened upon a brief anecdote regarding her life and work while attending an algebra conference (Moreno Iraola). While María Wonenburger was more than eighty years old before her STEM story came to light, the current pace of cultural change and the involvement of the arts in STE(A)M provide hope that many more Spanish women in science will be recognized, and many new Spanish women will choose to enter STEM fields with the space and financial support required for their success. #TimesUp.

NOTES

1. As evidence of the ongoing challenges unique to the Spanish context, the most recent data from the Unidad de Mujeres y Ciencia de España (The Spanish Unit on Women and Science) confirm that Spanish women scientists are surprisingly underrepresented, particularly in the context of the broader European Union: "Las investigadoras están infrarrepresentadas en todos los órganos unipersonales de gobierno y, además, en el caso de las universidades públicas, se ha observado un retroceso en la proporción de rectoras y vicerrectoras. Sorprende especialmente que en 2015 hubiera una rectora en las 50 universidades públicas españolas. En el conjunto de universidades públicas y privadas la proporción de rectoras asciende a 10% pero, en cualquier caso, está bastante por debajo del promedio de la Unión Europea (20% en 2014)" (Puy Rodríguez 13; Female scientists are underrepresented in all administrative governing bodies, and what is more, in the case of public universities, there has been a decline in the proportion of female deans and associate deans. It is particularly concerning that in 2015 there was only one female dean among fifty Spanish public universities. Counting public and private universities together, the proportion of female deans rose to 10 percent but, in any case, that is substantially below the average of the European Union [20 percent in 2014]).
2. Indeed, the cultural barriers for women are established in childhood and proliferate over time, often not allowing women to flourish in scientific fields as they mature. First, there is a stereotype that women do not perform well in mathematics and sciences. Lavy and Sand studied the unconscious gender bias of schoolteachers toward their students. The results indicated that when blindly assessed, the females performed better than the males, but when graded by their instructors, they received lower scores while those of their male colleagues were inflated. Bian, Leslie, and Cimpian's study noted that females began to shy away from sciences and math at the age of six because they did not feel that they were as bright as their male counterparts. Eisenhart and Holland have found that as women mature the challenges continue: women in higher education sometimes divest themselves from their school-related identities and invest more time in their romantic identities due to implicit pressure from peer groups. Wyer et al. studied many of the factors that dissuade women from pursuing careers in science, including the following: rejection by male peers, the societal pressure to follow traditional domestic and parental roles, the male domination of upper echelons of academia, and peer reviewers downgrading females' grant applications for science funding (71–72). Family support also matters. In their 2012 article "Romper la brecha digital de

género. Factores implicados en la opción por una carrera tecnológica" (Breaking the digital gender divide. Factors implicated in selecting a technological major), authors Naira Sánchez Vadillo et al. share the results of their research based on interviews conducted with young Spanish women pursuing a degree in Information Technologies and Communication. Primary among their findings is the importance of family attitudes toward or even encouragement of a non-gender normative interest in technology, access to technology, and early, deep, noncompetitive, and frequent use of video games.

3. Margarita Salas sadly passed away on November 7, 2019, after more than sixty years of scientific research and just a few months after the European Patent Office recognized her with a lifetime achievement award in June, 2019 (Viguera).

4. In the case of María Martínez Sierra, her work *El amor catedrático* (1955) offers a fictional example of what María Jesús Santesmases describes as an historical trope for Spanish women interested in science—serving as laboratory assistants to their husbands. Such roles offered socially acceptable forms for furthering women's STEM interests and pursuits. In the case of Martínez Sierra, such assistance mirrors her own work as a ghostwriter for her husband, Gregorio. Work to recuperate María Martínez Sierra's important literary role resonates with Dawn Smith-Sherwood's analysis in Chapter 1 of this volume regarding the parallels between Spanish women's literary and scientific recuperation.

5. Kevin Kumashiro explores the important role of disruptive knowledge in changing the persistence of oppression, particularly in educational contexts, in his study "Toward a Theory of Anti-Oppressive Education." He argues that "changing oppression requires disruptive knowledge, not simply more knowledge" (34) and states that diverse types of stories must be told in order to change the status quo, which is over-determined by repeated tropes and biases that permeate across culture. Such disruption is precisely what this volume posits Spanish cultural producers have been attempting to provide via artistic, literary, and other cultural texts.

6. They are: Marie Curie in Physics (1903) and Chemistry (1911), Irène Joliot Curie (Chemistry, 1935), Gerty Cori (Physiology, 1947), Maria Goeppert-Mayer (Physics, 1963), Dorothy Crowfoot Hodgkin (Chemistry, 1964), Rosalyn Sussman Yalow (Physiology, 1977), Barbara McClintock (Physiology, 1983), Rita Levi-Montalcini (Physiology, 1986), Gertrude Elion (Medicine, 1988), Christiane Nüsslein-Volhard (Physiology, 1995), Linda Buck (Physiology, 2004), Françoise Barré-Sinoussi (Medicine, 2008), Elizabeth Blackburn (Physiology, 2009), Carol W. Greder (Physiology, 2009), Ada E. Yonath (Chemistry, 2009), May Britt Moser (Physiology, 2014), Tu Youyou (Medicine, 2015), Donna Strickland (Physics, 2018), Frances Arnold (Chemistry, 2018), Esther Duflo (Economics, 2019), Andrea M. Ghez (Physics, 2020), Emmanuelle Charpentier (Chemistry, 2020) and Jennifer Doudna (Chemistry, 2020).

7. It is worth noting that, despite this long overdue recognition for women in science, there has been a long-simmering debate that surrounds the CRISPR discovery, with the male-led team of Virginijus Šikšnys also being given credit in some circles, as well as an ongoing battle for patent control (Begley).

8. These awards were established in 1982 by the Mininstry of Education and Science to recognize Spanish researchers performing world-class research. For the Premio San-

tiago Ramón y Cajal in Biology, the female awardees were Ángela Nieto Toledano (2019), María Antonia Blasco Marhuenda (2010), and Margarita Salas Falgueras (1999); for the Premio Leonardo Torres Quevedo en Ingeniería, the awardees were Susana Marcos Celestino (2019) and María Vallet-Regí (2008); and for the Premio Nacional don Juan Carlos I a la Investigación Científico-Técnica, the female awardee was Fàtima Bosch i Tubert (1995).

9. Interestingly, four women were awarded prizes, in 2017 and 2018, following Ansede's publication (Premios Fronteras).
10. Sadly, following the list of successful initiatives, the report also laments that "the organization has been less successful at increasing recognition of women scientists through award nominations" (171), a finding that seems consistent with the concerns of AMIT mentioned above.
11. García-Mainar and Montuenga have noted the increase in female employment, particularly in the services and clerical sectors (807). According to the World Bank compilation of data on the labor force in Spain, females comprised 46.26 percent of the total labor force in 2019.
12. Fox Keller's work on the topic of women and science has been foundational for the field, as evidenced by other ground-breaking publications such as *Reflections on Gender and Science* (1985) and her co-edited work with Jacobus and Shuttleworth, *Body/Politics: Women and the Discourses of Science* (1990). Other key early feminist studies include Lederman and Bartsch's *The Gender and Science Reader* (2001) and Harding's *Whose Science? Whose Knowledge?: Thinking from Women's Lives* (1991).
13. This study, while focused on the disruptions of women and STEM in Spanish culture, also provides a transnational approach with global implications, such that various chapters connect across nation-states, traversing European and non-European theatres (France and Spain, Ch. 3; Mexico and Spain, Ch. 7; Spain and Germany, Ch. 8; Spain and Equatorial Guinea, Ch. 9; Poland and Spain, Ch. 10; Spain, Britain, and Equatorial Guinea, Ch. 13).
14. Banu Subramaniam's *Ghost Stories for Darwin: The Science of Variation and the Politics of Diversity* offers intriguing examples of how women scientists might challenge the norms of established scientific research. The editors discuss Subramaniam's own personal story and its echoes with "la mirada tuerta" in the Acknowledgments section of this volume.

WORKS CITED

Ansede, Manuel. "Las mujeres no existen para los premios científicos." *El País*, 5 July 2016, elpais.com/elpais/2016/06/30/ciencia/1467310162_801306.html.

Balsamo, Anne. "Feminism and Cultural Studies." *Journal of Midwest Modern Language Association*, vol. 24, no.1, 1991, pp. 50–73.

Bazely, Dawn. "Why Nobel winner Donna Strickland didn't have a Wikipedia page." *Washington Post*, 8 Oct. 2018, www.washingtonpost.com/outlook/2018/10/08/why-nobel-winner-donna-strickland-didnt-have-wikipedia-page.

Begley, Sharon. "Three CRISPR Scientists Win Prestigious Award, Fanning Controversy over Credit." *Scientific American* May 31, 2018. www.scientificamerican.com/article/three-crispr-scientists-win-prestigious-award-fanning-controversy-over-credit.

Bermúdez, Silvia, and Roberta Johnson, editors. *A New History of Iberian Feminisms*. U of Toronto P, 2018.

Bernárdez Rodal, Asunción. "Prólogo." *Mujeres, ciencia e información*, edited by María Teresa García Nieto. Fundamentos, 2015, pp. 9–13.

Bian, Lin, Sarah-Jane Leslie, and Andrei Cimpian. "Gender Stereotypes about Intellectual Ability Emerge Early and Influence Children's Interests." *Science*, 27 Jan. 2017, pp. 389–91, science.sciencemag.org/content/355/6323/389.

Bieder, Maryellen, and Roberta Johnson, editors. *Spanish Women Writers and Spain's Civil War*. Routledge, 2017.

Biosca, Patricia, and J. M. Sánchez. "Las mujeres lideran la tecnología en España pero aún queda mucho trabajo para lograr la igualdad." *ABC*, 8 Mar. 2019, www.abc.es/tecnologia/redes/abci-mujeres-lideran-tecnologia-espana-201803072229_noticia.html.

Branch, Enobong Hannah, editor. *Pathways, Potholes, and the Persistence of Women in Science: Reconsidering the Pipeline*. Lexington, 2017.

Creager, Angela N. H., Elizabeth Lunbeck, and Londa Schiebinger, editors. *Feminism in Twentieth-Century Science, Technology, and Medicine*. U Chicago P, 2001.

Eisenhart, Margaret A., and Dorothy C. Holland. "Gender Constructs and Career Commitment: The Influence of Peer Culture on Women in College." *Women, Science, and Technology: A Reader in Feminist Science Studies*, edited by Mary Wyer, et al. Routledge, 2001, pp. 26–35.

Encuesta de Población Activa. Spanish Institute of Statistics (INE), www.ine.es/up/onjRchEq.

Fox Keller, Evelyn. "Making a Difference: Feminist Movement and Feminist Critiques of Science." *Feminism in Twentieth-Century Science, Technology, and Medicine*, edited by Angela N. H. Creager, Elizabeth Lunbeck, and Londa Schiebinger. U Chicago P, 2001, pp. 98–109.

———. *Reflections on Gender and Science*. Yale UP, 1985.

García-Mainar, Inmaculada, and Victor Montuenga. "Over-Education and Gender Occupational Differences in Spain." *Social Indicators Research*, vol. 124, no. 3, Dec. 2015, pp. 807–33.

García Nieto, María Teresa, editor. *Mujeres, ciencia e información*. Fundamentos, 2015.

González Ramos, Ana M. *Mujeres en la ciencia contemporánea: La aguja y el camello*. Icaria, 2018.

Haraway, Donna. "A Manifesto for Cyborgs: Science, Technology and Socialist Feminism in the 1980s." *Simians, Cyborgs and Women: The Reinvention of Nature*, Routledge, 1991, pp. 149–81.

Harding, Sandra G. *Whose Science? Whose Knowledge?: Thinking from Women's Lives.* Cornell UP, 1991.

Jacobus, Mary, Evelyn Fox Keller, and Sally Shuttleworth, editors. *Body/Politics: Women and the Discourses of Science.* Routledge, 1990.

Johnson, Roberta. *Gender and Nation in the Spanish Modernist Novel.* Vanderbilt UP, 2003.

Johnson, Roberta, and Maite Zubiaurre, editors. *Antología de pensamiento feminista: 1726–2011.* Cátedra, 2012.

Kumashiro, Kevin K. "Toward a Theory of Anti-Oppressive Education." *Review of Educational Research*, vol. 70, no. 1, 2000, pp. 25–53.

Lavy, Victor, and Edith Sand. "On the Origins of Gender Human Capital Gaps: Short and Long Term Consequences of Teachers' Stereotypical Biases." *The National Bureau of Economic Research*, Jan. 2015, www.nber.org/papers/w20909.

Lechado García, José Manuel. *Científicas: Una historia, muchas injusticias.* Sílex, 2018.

Lederman, Muriel, and Ingrid Bartsch. *The Gender and Science Reader.* Routledge, 2001.

López, Nuria. "Las mujeres 'asaltan' la ciencia en España." *La actualidad*, 11 Feb. 2019, actualidad.rt.com/actualidad/305218-mujeres-asaltan-ciencia-espana.

López-Sancho, M. Pilar, et al. "Status of Women in Physics in Spain." *Women in Physics, The 3rd IUPAP International Conference on Women in Physics*, 2009, pp. 171–72.

Mayock, Ellen. "Lab-Lit: Ciencia y tecnología feminista-humanista en la novela española actual." *Hispanófila*, vol. 174, June 2015, pp. 187–97.

Moreno Iraola, Laura. "María Wonenburger, la matemática que saltó a la fama con 80 años." *El País*, 27 Aug. 2019, elpais.com/elpais/2019/08/21/ciencia/1566404155_299955.html.

Ortiz, Teresa, and María Jesús Santesmases. *Gendered Drugs and Medicine: Historical and Socio-Cultural Perspectives.* Routledge, 2016.

Premios Fronteras del Conocimiento. "Galardonados." www.premiosfronterasdelconocimiento.es/version/edicion_2017. Accessed 7 Sept. 2019.

"Premios Nacionales de Investigación." Ministerio de Ciencia e Innovación, Gobierno de España, www.ciencia.gob.es/portal/site/MICINN/menuitem.7eeac5cd345b4f34f09dfd1001432ea0/?vgnextoid=82957edcc0186610VgnVCM1000001d04140aRCRD. Accessed 29 August 2020.

Puy Rodríguez, Ana, editor. *Científicas en cifras: Estadísticas e indicadores de la (des)igualdad de género en la formación y profesión científica.* Ministerio de Economía, Industria y Competitividad, 2016.

Rodrigo, Antonina. *Una mujer silenciada: María Teresa Toral, ciencia, compromiso y exilio.* Ariel, 2012.

Roig, Montserrat. *Dime que me quieres aunque sea mentira.* Península, 1993.

Sáinz, Milagros, editor. *¿Por qué no hay más mujeres STEM? Se buscan ingenieras, físicas y tecnólogas.* Telefónica, 2018.

Sáinz González, Jorge. "Prólogo." *Mujeres en la ciencia contemporánea: La aguja y el camello*, edited by Ana M. González Ramos, Icaria, 2018, pp. 9–11.

Sánchez de Madariaga, Inés, Sara de la Rica, and Juan José Dolado. *Libro blanco: Situación de las mujeres en la ciencia española*. Ministerio de Ciencia e Innovación, 2011.

Sánchez Vadillo, Naira, Octavio Ortega Esteban, and Montse Vall-llovera. "Romper la brecha digital de género. Factores implicados en la opción por una carrera tecnológica." Athenea Digital. *Revista de pensamiento e investigación social*, vol. 12, no. 3, 2012, pp. 115–28.

Santesmases, María Jesús. *Circulation of Penicillin in Spain: Health, Wealth and Authority*. Palgrave MacMillan, 2019.

———. *Mujeres científicas en España 1940–1970: Profesionalización y modernización social*. Instituto de la Mujer, 2000.

———. "Towards Denaturalization: Women Scientists and Academics in Twentieth-Century Spain" *Feministische Studien*, vol. 1, no. 11, 2011, pp. 52–63.

Serrano, Elena. "Chemistry in the City: The Scientific Role of Female Societies in late Eighteenth-Century Madrid." *Ambix*, vol. 60, no. 2, May 2013, pp. 139–59.

Subramaniam, Banu. *Ghost Stories for Darwin: The Science of Variation and the Politics of Diversity*. U of Illinois P, 2014.

United Nations Resolution 70/212. "International Day of Women and Girls in Science: A/RES/70/212." United Nations, 22 Dec. 2015, pp. 1–3, www.un.org/en/events/women-and-girls-in-science-day.

Valdés, Isabel. "Especial: Mujeres de la ciencia." 8 Feb. 2018, elpais.com/especiales/2018/mujeres-de-la-ciencia.

Versteeg, Margot, and Susan Walter, editors. *Approaches to Teaching the Writings of Emilia Pardo Bazán*. MLA, 2017.

Viguera, Enrique. "In Memoriam. Obituary: Margarita Salas (1938–2019)." FEBS Societies Network, 18 Feb. 2020, network.febs.org/posts/59791-obituary-margarita-salas-1938-2019.

Woolf, Virginia. *A Room of One's Own*, annotated and introduced by Susan Gubar. Harcourt, 2005.

World Bank. data.worldbank.org/indicator/SL.TLF.TOTL.FE.ZS?locations=ES

Wyer, Mary, et al., editors. *Women, Science, and Technology: A Reader in Feminist Science Studies*. Routledge, 2001.

PART I
ON ROLE MODELS
Female Scientists and Spanish Letters

CHAPTER I

Las chicas raras de STEM

Recuperating #WomensPlace in Spanish Literary and Scientific Histories

DAWN SMITH-SHERWOOD

Shortly before the 2018 UN International Day of Women and Girls in Science, *El País* published a web feature, "Especial: Mujeres de la ciencia" (Special: Women of science), which insisted that women scientists, "Han estado siempre, en todas las ciencias y a todos los niveles. . . . Es el momento de que ocupen su lugar en esa historia" (Valdés; [H]ave always been in all sciences and at all levels. . . . It is time that they occupy their place in that history). Consisting of an interactive chessboard-like graphic display of photographs which, when selected, reveal short biographic summaries highlighting the educational backgrounds and contributions of twenty-four women scientists from around the world, the special report "muestra a algunas pioneras que se abrieron camino en este campo" (Valdés; shows some pioneers that opened the way in this field).

Of the twenty-four women recognized, four, "las españolas de las que nadie suele acordarse" (the Spanish women no one tends to remember), have direct ties to Spain: 1) Sara Borrell (1917–1999), a biochemist and pharmaceutical expert, 2) Josefina Castellví (1935–), an oceanographer, 3) Gabriela Morreale (1930–2017), an Italian-born but Spanish-educated chemist, and 4) María Wonenburger (1924–2014), a mathematician (Valdés). Each summary includes information regarding the woman's formal education, career path, and major contributions to her field. Each summary identifies challenges as well as evidence of steps taken to recuperate each to her rightful place among more well-known members of the scientific community. The bio-summaries reveal certain patterns, and commonalities emerge. In each portrait, there is an underlying tension between private, family support and public, institutional denial of access to a desired field of study.

For example, Sara Borrell "nació en una familia madrileña, republicana y liberal donde siempre se le permitió seguir su propio camino" (was born to a liberal Republican family from Madrid where she was always permitted to follow her own way), but she was unable to study Agronomic Engineering "porque la academia no la admitió por ser mujer" (Valdés; because the academy did not admit her because she was a woman). Another common feature of these personal/professional profiles includes a period of study abroad in order to complete desired coursework and degrees. Not admitted to their home academic institutions, these Spanish women traveled to study in Scotland, England, France, the Netherlands, and the United States. They earned Fulbright awards to Yale and professorships at Indiana University. Eventually, they returned to Spain and found themselves incorporated into institutions and professional organizations such as the Consejo Superior de Investigaciones Científicas (CSIC), the Spanish National Research Council that forms part of the government's Ministerio de Economía y Competitividad (Ministry of Economy and Competitiveness). Their names now title honors and awards given to new generations of scholars in their respective fields.

The 2018 *El País* feature presents two thematic concerns that inform this chapter. First, the assertion regarding women's place in scientific history suggests that, previously, women scientists were lost, hidden, or even actively erased. Second, it represents visually an underlying tension between individual and collective approaches to the recuperation of women to scientific history. Their individual stories distinguish them, yet they share many common personal and professional experiences. Taken together, they present a significant sample, but each one's exceptionality also warrants laud.

A similar dynamic emerges in the 2016 presentation "El no-lugar de las científicas" (The no-place of women scientists) by Encina Calvo-Iglesias, a professor of physics at the Universidade de Santiago de Compostela. Calvo-Iglesias borrows a theory, attributed to the French anthropologist Marc Augé, to explain what she terms the traditional "no-lugar" (no place) of women scientists in the public's collective imagination. In these Augé-theorized "no places," individuals remain anonymous, their experiences insignificant. However, Calvo-Iglesias argues that the appearance of women scientists in recent cultural production suggests their departure from these transient "no places" to "places" in scientific history. She begins her presentation in the US context, showing a photograph of NASA scientists, all of them male, celebrating their having placed the first rocket on the moon. Next, she contrasts that image with the trailer from the 2016 movie *Hidden Figures*, which details the story of Katherine Johnson, the African American female mathematician whose contributions were critical to the mission's success. As the movie's title suggests, Katherine Johnson was there, engaged in science, but not photographed, and so her work was hidden. Additionally, though Johnson was individually meritorious, she was also only one of a significant number of African American women mathematicians then in NASA's employ.

The effort to celebrate and recuperate the contributions of marginalized women is not exclusive to the scientific community. While Calvo-Iglesias, a working female scientist, dedicates part of her research agenda to the historiography of her profession, Carmen Martín Gaite, a working female writer, also dedicates part of her literary agenda to the historiography of literary exclusions. In the 1986 essay "La chica rara" (The odd girl), published in the collection *Desde la ventana: Enfoque femenino de la literatura española* (From the window: Spanish literature's feminine focus), Carmen Martín Gaite analyzes Carmen Laforet's 1944 novel *Nada* (Nothing) and asserts Laforet's unique contribution to the history of Spanish women's literature, not only as the first winner of the then recently inaugurated Premio Nadal, but more importantly as a figure taking the first step to more openly, even at the height of Francoist censorship, occupy a public, literary place for women.

This chapter considers how Martín Gaite's original Spanish literary "odd girl" finds parallels in the scientific realm, tracing similarities between the post-Franco era efforts of Spanish women humanists to recuperate lost voices from literary history and the current efforts of Spanish women scientists to recognize contributions of their own female forebears. In addition to contributing original work in their respective scientific and literary fields, these women engage in the work of recuperating lost (grand*m*) *others* for Spanish scientific and literary histories. This chapter argues that practicing Spanish women scientists of the early twenty-first century, like Calvo-Iglesias and María Jesús Santesmases of Spain's CSIC, are engaged in a project similar to that of practicing Spanish women writers of the late twentieth century, such as Martín Gaite.[1]

Santesmases, a senior member of the Department of Science, Technology, and Society at the CSIC's Institute of Philosophy, has published numerous academic journal articles within the field of biology, as well as book-length works that blend scientific, political, sociological, and economic disciplines. In her 2000 work *Mujeres científicas en España 1940–1970: Profesionalización y modernización social* (Scientific women in Spain 1940–1970: Professionalization and social modernization), Santesmases presents a qualitative analysis of a representative sample of Spanish women scientists of the Franco era. However, rather than identifying and celebrating any individual "odd girl" scientist, Santesmases elevates the place in science of a women's collective. Her project reveals what had previously become hidden from plain sight: contrary to popular belief, there were a significant number of women scientists in public professional practice during the Franco era in Spain.

Additionally, this chapter explores potential parallels between these scientific and literary movements and the processes underlying the larger, ongoing project of recuperation of historical memory that has been occurring in Spain. As Patricia O'Byrne has asserted in her work *Postwar Spanish Women Novelists and the Recuperation of Historical Memory*, "the recuperation of the historical past was a considerably more complex process for many Spaniards as they had first to erase ingrained false memories" (12). Through study of works by Martín Gaite and Santesmases, questions emerge regarding which "ingrained false memories" of Spanish women's literary and scientific contributions have been erased and which endure. Finally, this chapter concludes that the genre-bending hybridity and hashtag-laced intertextuality of Rosa Montero's 2013 work of creative non-fiction, *La ridícula idea de no volver a verte* (The ridiculous idea of never seeing you again), lies at the intersection of a generative Spanish literary and scientific #WomensPlace.

In the opening paragraph of "La chica rara," Martín Gaite notes that with Laforet's *Nada*, "se inicia un fenómeno relativamente nuevo en las letras españolas: el salto a la palestra de una serie de mujeres novelistas en cuya obra, desarrollada a lo largo de cuarenta años, pueden descubrirse hoy algunas características comunes" (101; a relatively new phenomenon in Spanish letters is initiated: the jumping into the ring of a number of women novelists in whose work, developed over forty years, one can discover today some common characteristics). She underscores the incongruity between Laforet's inconsequential background—"era una muchachita de veintitrés años de la que nadie había oído hablar" (she was a young girl, only twenty-three years of age, about whom no one had heard anything)—and her wholly consequential, initial "jump into the ring" on behalf of all the Spanish women writers who would follow her (101). Specifically, Martín Gaite notes how, with *Nada*, Laforet "se descolgaba con una historia cuyos conflictos contrastaban de forma estridente con los esquemas de la novela rosa habitualmente leída y cultivada por mujeres" (101; suddenly introduced herself with a story whose conflicts contrasted in a strident way with the schemes of the romance novel habitually read and cultivated by women).² A woman, a young woman of and from whom no one had heard, told the unexpected, breaking with the gendered literary norms of readers and writers. In "La chica rara," Martín Gaite challenges the oft-quoted joking slight, that Laforet wrote "*Nada* y nada más" (*Nothing* and nothing else), and establishes Laforet's rightful, foundational #WomensPlace in Spanish literary history of the early post-Civil War period.

For Martín Gaite then, Laforet must be considered a rebel, "off the hook" (101), distinct from other women writers of the time. Popular Spanish *novela rosa* (romance novel) writers of that era, Carmen de Icaza and Concha Linares Becerra, protested the rosy color associated with their works: "Ambas dijeron que su novela no era rosa, sino blanca y moderna" (102; Both said that their novel was not pink, but rather white and modern), which Martín Gaite acknowledges. Yet she asserts that their novels ultimately concluded in traditional ways, confirming and conforming to established cultural norms regarding gender, women's work, and women's place. According to Martín Gaite, the seemingly modern travel, professional life, and adventure enjoyed by novela rosa heroines never ultimately compromised their moral standing. While Martín Gaite recognizes Icaza, whose works often feature female protagonists with mysterious, turbulent, even exotic backstories, as the most interesting of the novela rosa authors, she criticizes the way in which those innovative elements are ultimately deactivated by the underlying conservative ideology of the Falange's Sección Femenina. In Icaza's novela rosa, Martín Gaite finds reification of Sección

Femenina ideals in the ultimate pacification of their female protagonists' ill-advised explorations, as well as in their idealization of their male character counterparts. Martín Gaite celebrates how, in contrast to the norms of the novela rosa, Laforet's *Nada* obliterates well-worn, gender-based stereotypes: "Fueron estos estereotipos heroicos lo primero que vino a hacer añicos la peculiar novela de Carmen Laforet" (103; These heroic stereotypes were the first to be shattered by Carmen Laforet's peculiar novel). In *Nada*, Laforet retires a stereotype and institutes a prototype, Andrea, "la chica rara," a female protagonist who challenges the norm.[3]

As Martín Gaite builds her case for Andrea's singularity among female protagonists of the early post-Civil War period—marking a transition from the stock heroines of the novela rosa to the individualist protagonists of the works of her immediate predecessors and contemporaries—she highlights a tradition-breaking climactic, cathartic moment late in the novel, when Andrea finally recognizes and accepts her distinct position and attitude, as a spectator more than a participant in the real-life drama that surrounds her. Of the tears Andrea sheds in chapter 23, Martín Gaite asserts, "podemos interpretarlas como símbolo de su aceptación del papel de espectadora, al que Carmen Laforet la tenía destinada desde el principio" (108; we can interpret them as a symbol of her acceptance of the role of spectator, to which Carmen Laforet had her destined from the beginning). Here, Martín Gaite underlines both Laforet's authorial agency as well as Andrea's late-life narratorial self-realization; the first-person retrospective narrator enacts a kind of metacognitive recuperation of her own lost voice. While the woman in the spectator role traditionally has a negative, passive connotation, here it can potentially be read as positive, active, perhaps even scientific if in "espectadora" one additionally finds an "observadora." Martín Gaite asserts, "el destino de Andrea era ese: el de verse cercada por los argumentos de los demás, intervenir en ellos, observarlos y recogerlos" (108; Andrea's destiny was this: seeing herself surrounded by the plotlines of others, to intervene in them, observe them, and collect them). Andrea's disposition would seem that of an objective, observer scientist of interpersonal relationships. In fact, Martín Gaite likens Andrea's interactions with family and friends to data collection; she analyzes what she hears, sees, and witnesses, but always from a position of objectivity or even indifference (110).

Of Andrea's seemingly scientific demeanor, Martín Gaite concludes, "En una palabra, Andrea es una chica 'rara,' infrecuente" (111; In a word, Andrea is an "odd" girl, rare) and proceeds to identify and define several characteristics of this so-called rareness: the "odd girl" has few female friends, prefers the friendship of men, rejects parental authority, prefers the public sphere (especially the street) to the private sphere (home, domestic space). In her analysis of *Nada*, Martín Gaite establishes the direct link between Laforet as literary foremother and her own sisterhood of the "odd girl" protagonists: "Este paradigma de mujer, que de una manera o de otra pone en cuestión la "normalidad" de la conducta amorosa o doméstica que la sociedad mandaba acatar, va a verse repetido con algunas variantes en otros textos de mujeres como Ana María Matute, Dolores Medio y yo misma" (111; This woman paradigm, which in one way or another puts into question the "normality" of romantic and domestic conduct to which society demanded obedience, is going to see itself repeated with some variations in other texts by women like Ana María Matute, Dolores Medio, and myself). Of course, in telling the stories of "odd girl" protagonists, writers like Laforet, Matute, Medio, and Martín Gaite identified themselves as "odd girls" in turn. They wrote against novela rosa norms and forever altered Spanish literary history. In her essay, Martín Gaite not only recuperates Laforet to her rightful place in that literary history but also shows how Laforet's example encouraged subsequent Spanish women writers to embrace their inner "odd girls" and participate in literature as a profession.

According to María Jesús Santesmases's study *Mujeres científicas en España 1940–1970: Profesionalización y modernización social*, Spanish scientific history also included real life "odd girls." The Spanish women scientists that form the basis of Santesmases's study also eschewed the societal expectations of the post-Civil War era. They pursued academic credentials and sought to establish themselves in a male-dominated profession. However, Santesmases's approach differs from Martín Gaite's in important ways. Rather than exalting the efforts of an individual in securing a place for the group, Santesmases relies on examination of a representative sample. Rather than a sudden jump into the ring, Santesmases's study reveals a group of women scientists already occupying a field that societal norms had not encouraged them to enter.

A Spanish woman scientist and prolific science writer who currently occupies the highest academic rank, "profesora de investigación," at the aforementioned CSIC, María Jesús Santesmases is still perhaps best known for her 2001 study *Entre Cajal y Ochoa: Ciencias biomédicas en la España de Franco, 1939–1975* (From Cajal to Ochoa: Biomedical sciences in Franco's Spain, 1939–1975), in which she details a period of Spanish scientific history by exploring the impact of political and social forces (on science generally and on neuroscience particularly) surrounding the conclusion of the Spanish Civil War and the subsequent Franco dictatorship.[4] In his 2005 review of Santesmases's work, Leoncio López-Ocón explains the political dynamics at work at the time of the CSIC's founding in 1939:

> When the CSIC was set up, its executive directors scorned the school of Cajal, most of whose members had allied themselves with the losing side of the Spanish Civil War. Nonetheless, the first Spanish Nobel scientist had left such a profound impression that the reconstruction of experimental science in Franco's Spain centered on the exemplary value of his work. [...] Hence, the work of this distinguished scientist had created a scientific tradition that managed to survive through thick and thin, including a bloody civil war. (464)

Through her study of the complex political dynamic of the post-Civil War era and its impact on scientific advancement, Santesmases contributes to a large body of work that attempts to examine the past from a critical distance of decades. She uses the well-known, individual figures of male scientists Cajal and Ochoa to chronicle the state of Spanish scientific development generally. Just one year earlier, Santesmases had considered the question of Spanish women's professional participation in scientific communities over roughly the same historical period in her 2000 *Mujeres científicas en España 1940–1970: Profesionalización y modernización social*.

In this previous study, Santesmases engages in the work of what Evelyn Fox Keller has described as "writing women back into history ... with the clear expectation that knowledge of the past would empower women in the present and, as such, help to transform women's lives" (100). The anonymous prologue, attributed to the work's publisher, Spain's Instituto de la Mujer (Women's Institute), confirms these goals for the text: "Reescribir el pasado y asegurar un futuro mejor, respecto a la presencia de las mujeres en estos ámbitos, son dos objetivos de igual importancia para el Instituto de la Mujer y con la aparición de este libro, creemos, se está contribuyendo a su consecución" (8; Rewriting the past and assuring a better future, with respect to the presence of women in these fields, are two objectives of equal importance for the Women's Institute, and with the appearance of this book, we believe, we are contributing to their attainment). As expressed in the publisher's prologue, the Women's Institute's goals for Santesmases's text are consonant with those of Martín Gaite and Calvo-

Iglesias. In setting the past record straight, in populating past "no places" with recuperated female figures, future generations of women writers and scientists see themselves represented historically in public, professional positions and are encouraged to participate as contributing members of writing and scientific communities.

Santesmases's approach to writing women back into history, however, differs in important ways from that of Martín Gaite, Calvo-Iglesias, or the authors of the piece in *El País*. Rather than identifying by name and recuperating outstanding individuals, Santesmases studies a representative sample, "una muestra" (3), as she first terms it in the book's acknowledgments, of some fifty anonymous Spanish women science professionals, "un *grupo* de científicas españolas doctoradas antes de 1970" (3, emphasis added; a *group* of Spanish women scientists earning doctorates before 1970). Rather than identifying any individual "odd girl," any individual woman pioneer occupying a scientific space previously reserved for men, Santesmases focuses on the group of "odd girls" already present in classrooms, faculty positions, and research labs during the post-Civil War era and Franco dictatorship. In a manner that resonates with the tenets of first-wave feminism, Santesmases highlights similarities between the academic preparation and research interests of Spanish men and women scientists: "Todas tenían excelente cualificación, formación comparable a la de sus colegas hombres y los mismos intereses en la investigación" (3; All of the women had excellent qualifications, training comparable to that of their male colleagues, and the same research interests). While recognizing and detailing the obstacles faced by the Spanish women scientists in her sample, Santesmases challenges the commonly held notion that the post-Civil War era and the Franco dictatorship offered few opportunities to women in scientific fields. She acknowledges that the efforts of Spanish women scientists have remained largely backgrounded and, through her study, urges their rightful recognition in Spanish scientific history, not due to any singularity or difference, but rather to their commonality, to one another and to their male colleagues.

Santesmases's approach to the text is notably scientific in content and structure. Of the work's development, she explains in the acknowledgments that, having received a grant from the Instituto de la Mujer, she was able to "aumentar la muestra, emprender una *investigación* de carácter estadístico, realizar algunas *entrevistas* y *discutir* algunos *resultados*" (3, emphasis added; increase the sample size, undertake statistical *research*, complete *interviews*, and *discuss results*). Her study relies on qualitative methodologies. She reviews statistics, conducts interviews, and analyzes CVs.

In the Introduction, Santesmases hypothesizes that, contrary to popular belief, women have historically studied science and participated in science as professionals, even prior to the period of the so-called "economic miracle" that preceded the end of the Franco era. Of Spanish women students of science and professional scientists generally, Santesmases asserts, "Su presencia, aunque sea minoritaria, demuestra que ha existido intención en las mujeres por formar parte de dominios de la vida social cuyas normas intentaban por medios legales y por presiones sociales relegarlas al ámbito de lo exclusivamente doméstico" (10; Their presence, even though in the minority, shows that women have formed part of social domains whose norms intended, by legal means and by social pressures, to relegate them to the exclusively domestic sphere). She also takes to task those who would hide or ignore the achievements of earlier successful individuals in order to meet the needs of their own political agenda: "La historia política y social española no ha amparado logros anteriores en períodos posteriores, y esto ha sido así en buena parte por razones ideológicas de los regímenes políticos" (11; Spanish political and social history has not supported previous achievements in subsequent time periods, and this has been so in large part due to ideological reasons of political regimes).

Next, she establishes a paragraph-length list of seven research questions regarding her sample of women scientists from the field of biomedicine. To paraphrase, Santesmases seeks to examine the family origins of these women, what encouraged them to pursue scientific study in the first place, what universities accepted them and under what conditions, how the number of women dedicated to studying science at the university level grew, how women were represented among the different scientific disciplines, how it was possible for them to pursue a doctorate, and how, following their education, they ultimately gained access to the male-dominated professional world of scientific research (11–12). Then, she provides a summary of the chapters that follow and continues to explore her hypothesis: "las cifras de alumnas empiezan a crecer antes del desarrollo económico de los sesenta y que las mujeres acceden desde los primeros años de esa década de *milagro económico* a todos los estudios ofertados, sean de ciencias o de letras, lo que contradice el estereotipo de que las mujeres han optado en general por estudios más de letras que de ciencias" (12; the number of female students begins to grow before the economic development of the

1960s, and from the first years of that decade of the so-called *economic miracle*, women access all courses of study offered, be they sciences or humanities, which contradicts the stereotype that women have generally opted more for studies in the humanities than in the sciences). As evidence, she notes that, in her study, "se reconstruyen sus vidas profesionales y se analizan los factores que las hicieron posibles y los que las mantenían en minoría respecto a sus colegas hombres y, en la mayoría de los casos, en niveles intermedios de reconocimiento académico" (13; their professional lives are reconstructed and the factors that made them possible are analyzed, along with those that kept them in the minority with respect to their male colleagues, and in the majority of cases, at intermediate levels of academic recognition). Among the factors that Santesmases explores are familial context, education, study abroad, access to academic positions, marital status, and childrearing (12–13).

Having established the reason for and methods of her study, Santesmases provides a literature review, exploring cultural, economic, historical, political, and social forces on women and science and women in science. She considers the impact of gender on scientific objectivity, the implications of Spain's late embrace of the Industrial Revolution, and its isolation following World War II. She details how the traditional functions of women as caretakers, teachers, and doctors in the domestic space, prior to the twentieth century, when no profession of doctor yet existed, led many women to studies and careers in biomedicine. She explains how women religious were permitted to operate pharmacies, which eventually led to the acceptability of women studying pharmacy in the university context. Interestingly, Santesmases quotes from and cites both of Carmen Martín Gaite's *Usos amorosos* book-length historiographical projects, on eighteenth-century and postwar Spain, respectively. She also cites Carmen Laforet's four-article series "La mujer médico en España" (The woman doctor in Spain), which appeared in the journal *Tribuna Médica* (Medical Tribune) in 1961. In these ways, Santesmases links her scientific work to related works of literary and social criticism, demonstrating the common struggle to establish women's place in these and other academic disciplines.

Drawing on her STEM training, Santesmases delves into statistics regarding Spanish women's participation in higher education, generally and by academic discipline, providing a review of gender-related stereotypes and traditions. She incorporates numerous data tables and charts to communicate her findings and discusses in depth those factors, such as family support (financial, emotional); opportunities to study abroad; undergraduate, graduate, and post-graduate formation; and eventual incorporation into public research or higher education institutions mentioned in the Introduction. Her discussion includes interesting observations, recurring themes gleaned from transcripts of interviews. For example, regarding the academic and professional aspirations of the daughters of the most objectively successful women scientists in her sample, Santesmases notes, "sus hijas no quisieron seguir sus pasos en <<la ciencia>>" (145; their daughters refused to follow in their "scientific" footsteps). Occupied outside the home, but still responsible for the domestic sphere, "Sus madres eran distintas y estaban en minoría en la sociedad en la que se criaron" (146; Their mothers were different and in the minority in the society in which they were raised). Different, "odd girl" mothers, the women scientists in Santesmases's sample set a professional example for their offspring but also provided a cautionary tale of sacrifice, in time and energy available for personal pursuits, as well as in relative social isolation.

From Carmen Laforet to Pilar Primo de Rivera, Santesmases draws on many of the same Franco-era political and cultural touchstones as Martín Gaite, but her conclusion is especially challenging to second-wave feminists: "Las estrategias de las mujeres trataban de la inclusión sin reivindicación feminista, como personas que se consideraban con las mismas capacidades que sus colegas" (158; The strategies of the women were inclusion without feminist revindication, as persons considered to have the same capacities as their colleagues). Santesmases does not deny that women were underrepresented in the sciences, that they had less access to educational opportunities and diminished avenues to professional advancement. Despite these challenges, the women scientists in her sample achieved professional success, not because they were different but because they were equal to their male colleagues in ability. In its conclusion, Santesmases's study reveals the complexity of the collective Spanish historical recuperation project and perhaps provides an example of an "ingrained false memor[y]" in need of erasure or at least modification (O'Byrne 12). Santesmases demonstrates that, contrary to popular belief, women scientists, while still clearly in the minority, did inhabit public, professional places. The feminist revindication theme that marks the texts of the immediate post-Franco period, like Martín Gaite's essay, has yielded to increasingly nuanced, complex consideration in more recent studies, like this one by Santesmases. While Santesmases's approach to the question of recuperating women's place in scientific communities of the Franco era reveals many of the same societal norms that Martín Gaite notes governed the literary world, it differs dramatically in its revelation that, despite the numerous challenges, a sizeable number of "odd girl" women scientists pursued their "unwomanly" interests and persisted.

In her study's closing pages, Santesmases reflects on the findings and limitations of her study, "Esta aproximación cualitativa para explicar los umbrales de acceso y la constitución de la masa crítica inicial permite comprender ese acceso creciente, posterior, y que se ha recogido en las variadas tablas que se presentan en esta memoria" (161; This qualitative approach to explain the thresholds of access and the make-up of the initial critical mass permits understanding of that later growing access, which has been gathered in the various tables that are presented in this memoir). The use of the term "memoria" here is surprising, especially in the context of this otherwise data-driven, evidence-based text. It is both genre- and discipline-bending. Santesmases's work, presented in the form of a scientific study, with its sample, methods, research questions, and findings, in the end constitutes and contributes to the larger Spanish recuperation of historical memory project, even as it problematizes the narrative of feminist revindication common to previous efforts to rewrite women into literary and scientific histories.

Resonances of both Martín Gaite's and Santesmases's approaches to rewriting women into literary and scientific histories also sound in Rosa Montero's 2013 work of creative non-fiction, *La ridícula idea de no volver a verte*.[5] Like Martín Gaite, Montero focuses primarily on a singular figure, Nobel Prize-winning physicist Marie Curie. Like Santesmases, Montero explores the socio-economic and political factors that affected Curie's educational formation and professional success. What distinguishes Montero's approach, however, is its hybridity and intertextuality; her text integrates multiple media and modes of communication. Taking Curie's diary as creative, cathartic catalyst, in *La ridícula idea de no volver a verte*, Montero explores the intersection of literature, science, and gender, enacting an international, transdisciplinary collage-like text that is generated in coincidence and sustained in connection.

In the work's sixteen chapters, Montero not only recounts the life and times of Curie and her multiple familial (daughter, wife, mother) and professional roles but also includes asides, digressions, and musings. These asides concern not just Curie's experiences but also Montero's own, not only of widowhood, the reported impetus for Montero's work, but also of womanhood. The concluding two-page "Agradecimientos: Unas palabras finales" (Acknowledgments: Some final words) provides insight into the biographical and autobiographical Curie resources—traditional print as well as Web—from which Montero drew. There, she recognizes the assistance of several individuals (including two physicists) who either suggested additional resources or read a draft version "para ver si decía alguna barbaridad científica" (210; to see if I was saying anything scientifically disastrous). The work also includes an appendix, "Apéndice: Diario de Marie Curie" (Appendix: Marie Curie's Diary), from which Montero quotes in the body of her own text (215–33). Additionally, Montero incorporates photographs throughout the text, not only of expected individuals—Marie Curie; her husband, Pierre; their daughters, Ève and Irène—but also of figures from history (Alexander Litvinenko, p. 11 and Jeffrey Dahmer, p. 60) and pop culture (Patti Smith, p. 44 and Lady Gaga, p. 45). Montero even includes photographs of herself (as a young adult, p. 44 and of her arm bearing a salamander tattoo, p. 185). In these ways, Montero both individualizes and universalizes the experience of the singular Curie along a time- and space-informed continuum.

As she opens her acknowledgments pages, Montero insists, "Todos los datos que hay en este libro sobre Marie y Pierre Curie están documentados; no hay una sola invención en lo factual" (209; All facts about Marie and Pierre Curie that are in this book are documented; there is not even one invention). She then suggests, however, that she has let herself "volar en las interpretaciones" (209; fly in the interpretations). Montero contends that Curie's example serves as a reference point from which she "poder reflexionar sobre los temas que últimamente me rondan insistentemente por la cabeza" (209; can reflect on the themes that insistently circle in my head of late). Themes of coincidence, debt, honor, ambition, guilt, weakness —in personal, professional, familial, and (inter)national contexts—are highlighted in Montero's inclusion of some twenty-plus related and recurring hashtags throughout the work. Curiously, while there is no index of images, the work includes an alphabetically ordered "Índice de hashtags" (Hashtag index) sandwiched between the acknowledgments pages and the appendix containing Curie's diary that form the book's back matter (211).

Montero's use of digital textual markers throughout an otherwise analog work communicates hyper-textuality and follows multiple, thematic reading paths. In *Digital Pedagogy in the Humanities*, Marisa Parham details the way in which hashtags function to form new communities, creating intentional, narrowed indices within wider, serendipitous streams. Parham explains, "When a hashtag is clicked, temporally and spatially dispersed posts are brought into immediate content relation and literally onto the same page." In the deployment of textual hashtags, Montero enacts the intersection of women's literary and scientific communities across time and space.

Montero quickly outs Curie as an "odd girl" in the work's opening chapter with one of the recurring hashtags: "Fue una mujer nueva. Una guerrera. Una #Mutante" (20–21; She was a new woman. A warrior. A #Mutant). The next in-text appearance of #Mutante, in chapter 3, also resonates with the "odd girl" and additionally with the women's place themes explored above in the context of Martín Gaite's essay and Santesmases's study. Indirectly voicing the oft-communicated warnings of mothers to daughters, Montero exhorts, "Sé otro tipo de mujer. Sé una #Mutante. Esa hembra sin lugar, o en busca de otro #Lugar" (40; Be another type of woman. Be a #Mutant. That female without a place or in search of another #Place). The "odd girl" and women's place themes appear again in chapter 4 and accompany several intertextual nods toward Carmen Laforet's *Nada*. In a meditation on Laforet, another indirect voicing, or perhaps an interior dialogue that Montero imagines for both Laforet and Curie, links the "odd girl" and women's place, noting for readers "que eres una intrusa, que no tienes el derecho de estar ahí, junto a los varones. Que eres una #Mutante, fracasada como mujer y un engendro como hombre" (52; that you are an intruder, that you have no right to be there, next to the males. That you are a #Mutant, failed as a woman and a freak as a man).

In #Mutante, one sees parallels to both protagonist "odd girl" Andrea and her creator, Laforet. Relatedly, in chapter 6, "Elogio a los raros," Montero bids the reader permit her a digression "para cantar las alabanzas de los #Raros, los diferentes, los monstruos" (82; to sing the praises of the #OddOnes, the different, the monsters). Then, in the following chapter, she praises both Marie and her husband, Pierre, "dos mentes superdotadas, dos personas #Raras" (92; two super-gifted minds, two #OddPersons). With #Raros and #Raras, Montero allows that an "odd girl" can be of any gender. However, societal norms and structures favor men's recognition and place in the public sphere.

One of the most frequently occurring hashtags in Montero's text, #LugarDeLaMujer (#WomensPlace) appears for the first time in chapter 1. As Montero explains her reason for writing the work, she comments, "no estoy hablando de teorías feministas, sino de intentar desentrañar cuál es el #LugarDeLaMujer en esta sociedad en la que los lugares tradicionales se han borrado (también anda perdido el hombre, desde luego, pero que ese pantano lo explore un varón)" (18; I'm not talking about feminist theories, but attempting to disentangle what is #WomensPlace in this society in which the traditional places have been erased [men are also lost, of course, but let a male explore that swamp]). Interestingly, Montero distances herself from feminist revindication in her sincere querying of the tension between traditional and non-traditional women's places. In the text's third chapter, Montero returns to the theme of women's place as she examines the differences between Curie's two daughters, the older, Irène, who followed in her mother's scientific footsteps, and the younger, Ève, who occupied a more traditional women's place. Montero even includes two photographs of the women to underscore the differences between them: Irène appears severe in countenance, while Ève appears more traditionally coifed and made up. Of the photographs, Montero comments, "comparar los retratos de las dos hermanas . . . equivale a un tratado de varias páginas sobre lo que es o no es lo femenino y sobre el #Lugar o el no #LugarDeLaMujer" (42; comparing the portraits of the two sisters . . . equates with a treatise of several pages about what is and is not feminine and about what is or is not #WomensPlace). Like the daughters of Santesmases's female biomedical scientists, Ève saw her mother's "odd girl" status and chose a different, more traditional personal and professional path, while Irène chose to #HonrarALaMadre (#HonorThyMother), another frequently occuring hashtag in the text. Through the singular scientific example of Curie, Montero meditates on the complexity of multiple socio-economic and political factors informing the personal and professional decisions of women, including herself, across time and space.

Coincidentally, Montero's 2013 textual deployment of hashtags thematically anticipated a real-life story regarding the place of women scientists that took on global dimensions. In 2015, National Public Radio (NPR) reporter Bill Chappell described "disappointing" comments by the Nobel Prize-winning biochemist Tim Hunt (aka Sir Richard Timothy Hunt) of Great Britain, who infamously said of the presence of female scientists in the predominantly male space of the laboratory, "You fall in love with them, they fall in love with you, and when you criticize them, they cry." Hunt's comments were followed by an immediate, direct response launched by women scientists from around the world via the #DistractinglySexy Twitter campaign. Still considered out of place, twenty-first-century "odd girl" scientists continue to disrupt social and professional norms. The delicious irony of the #DistractinglySexy event, of course, is revealed in the clever, expert use of technology by those same women scientists whom Hunt disparages. The use of the hashtag unites women scientists as individuals through time and space. Together they form a collective, resonant of Santesmases's sample, but their grouping is actively generated rather than passively gathered by another.

It is notable that, rather than focusing on a Spanish woman scientist, Montero must necessarily recur to the international figure of Marie Curie, perhaps because the work of scholars like Calvo-Iglesias and Santesmases is still ongoing, perhaps because Montero, as she acknowledges in *La ridícula idea de no volver a verte*, is a doctor of #Coincidencias (#Coincidences). Montero, who, like her work's subject Marie Curie, had lost her husband prematurely, was subsequently introduced to Curie's diary, begun in the immediate aftermath of her new widowhood. Through the efforts of practitioner-historiographers from all disciplines, however, the traditional "no-lugar" of women scientists and writers becomes a new, potentiating #LugarDeLaMujer. If, as Patricia O'Byrne has asserted, "Through their literature, women novelists found a forum for reflection, self-analysis and assertion, availing themselves of the cathartic and also the empowering force of literature to question the social and ideological formations they were expected to absorb as the norm" (25), in women scientists like Calvo-Iglesias and Santesmases, a related scientifically based possibility emerges.

On June 2, 2019, in her weekly *El País Semanal* column, "Maneras de vivir," Montero returned to the #LugarDeLaMujer theme in the singular figure of Mileva Marić, Albert Einstein's little-known first wife. The short work's simple title, "Ella también" (She also), echoes Montero's thesis: "No estoy diciendo que Einstein no fuera un gran científico: digo que ella también lo era" (I am not saying that Einstein was not a great scientist: I am saying that she was as well). While Marie Curie's contributions were recognized in her own time, the works of Marić might have been lost forever. They were only discovered following the death of her and Einstein's son, Hans Albert, through a box containing correspondence between his parents. Montero reports on the conscious effort to erase Marić's significant contributions to Einstein's work: "Los agentes de Einstein intentaron borrar todo rastro de Marić; se apropiaron sin permiso de cartas de la familia y las hicieron desaparecer. También desapareció la tesis doctoral que Mileva presentó en 1901 en la Politécnica y que, según testimonios, consistía en el desarrollo de la teoría de la relatividad" ("Ella también"; Einstein's agents attempted to erase all trace of Marić; they appropriated the family's letters without permission and made them disappear. The doctoral thesis that Mileva presented in 1901 at the Polytechnic and that, according to testimonies, consisted of the development of the theory of relativity also disappeared). Montero regrets that, although there is evidence to support Marić's rightful place in scientific history, an active campaign to cover up or even erase her significant contributions to Einstein's success means that she remains a mostly hidden figure: "Pese a ello, Mileva sigue aplastada bajo el rutilante mito de Einstein" ("Ella también"; Despite this, Mileva remains crushed under the shiny myth of Einstein). Montero, however, rewrites Marić into history, not only in scientific accounts, but also literary and gendered history more generally. If Spanish women scientists like Santesmases and Calvo-Iglesias persist in their efforts to inspire future women to study and pursue science as a profession, perhaps a future Spanish woman writer, inspired by Montero, may pen a creative memoir based on the first Spanish Nobel Prize-winning female scientist. What has already occurred, however, through the works of female Spanish literary and scientific historians, offers great hope. May Spanish women (and all women) pursuing careers as writers, scientists, or in occupations yet to exist, no longer be regarded as singular "odd girls" but freely and commonly inhabit whatever professional places they choose.

NOTES

1. A chapter-length interview with María Jesús Santesmases concludes the first section of this volume.
2. For a socio-political analysis of *la novela rosa* (the romance novel) in the context of the dominant gender discourse in post-Civil War Spain, see the contribution by Miguel Soler in this volume.
3. See Ellen Mayock's 2004 *The "Strange Girl" in Twentieth-Century Spanish Novels Written by Women* for a detailed study of this phenomenon in the Spanish literary context.
4. Santiago Ramón y Cajal, the first Spaniard to receive the Nobel Prize for Physiology or Medicine in 1906, and Severo Ochoa, the second, in 1959, serve as touchstones of twentieth-century Spanish scientific histories. Ochoa spent much of his career in exile in the United States. Following the receipt of the Nobel, however, he returned to Spain.
5. See the contributions of Ellen Mayock and Maryanne Leone in this volume for deep dives into this and other of Montero's scientifically informed works.

WORKS CITED

Calvo-Iglesias, Encina. "El no-lugar de las científicas." *Mujeres con ciencia*, 4 Dec. 2016, mujeresconciencia.com/2016/12/04/el-no-lugar-de-las-cientificas.

Chappell, Bill. "#Distractinglysexy Tweets Are Female Scientists' Retort to 'Disappointing' Comments." *National Public Radio*, 12 June 2015, www.npr.org/sections/thetwo-way/2015/06/12/413986529/-distractinglysexy-tweets-are-female-scientists-retort-to-disappointing-comments.

Fox Keller, Evelyn. "Making a Difference: Feminist Movement and Feminist Critiques of Science." *Feminism in Twentieth-Century Science, Technology, and Medicine*, edited by Angela N. H. Creager, Elizabeth Lunbeck, and Londa Schiebinger, U Chicago P, 2001, pp. 98–109.

Hidden Figures. Directed by Theodore Melfi, Twentieth Century Fox, 2016.

Laforet, Carmen. *Nada*. 9th ed., Destino, 1987.

López-Ocón, Leoncio. "Book Review: *Entre Cajal y Ochoa: Ciencias biomédicas en la España de Franco, 1939–1975*." *Isis*, 2005, pp. 463–64.

Martín Gaite, Carmen. "La chica rara." *Desde la ventana: Enfoque femenino de la literatura española*, Espasa Calpe, 1987, pp. 101–22.

———. *Usos amorosos del dieciocho en España*. Anagrama, 1972.

———. *Usos amorosos de la posguerra española*. Anagrama, 1987.

Mayock, Ellen C. *The "Strange Girl" in Twentieth-Century Spanish Novels Written by Women*. UP of the South, 2004.

Montero, Rosa. "Ella también." *El País Semanal*, 2 June 2019, elpais.com/elpais/2019/05/27/eps/1558955111_252877.html.

———. *La ridícula idea de no volver a verte*. Seix Barral, 2013.

O'Byrne, Patricia. *Postwar Spanish Women Novelists and the Recuperation of Historical Memory*. Tamesis, 2014.

Parham, Marisa. "Hashtag." *Digital Pedagogy in the Humanities: Concepts, Models, and Experiments*. MLA Commons, digitalpedagogy.mla.hcommons.org.

Santesmases, María Jesús. *Entre Cajal y Ochoa: Ciencias biomédicas en la España de Franco, 1939–1975*. CSIC, 2001.

———. *Mujeres científicas en España 1940–1970: Profesionalización y modernización social*. Instituto de la Mujer, 2000.

Valdés, Isabel. "Especial: Mujeres de la ciencia." *El País*, 8 Feb. 2018, elpais.com/especiales/2018/mujeres-de-la-ciencia.

CHAPTER 2

"The Doctor Is In"

Elena Arnedo Soriano (1941–2015), Women's Health, and the Cultural History of Gender and Medicine in Spain[1]

SILVIA BERMÚDEZ

In 1976, historical studies on science and technology in Spain became formalized via the creation of the Sociedad Española de Historia de las Ciencias (SEHC; Spanish Society of Science History). The objectives of the organization were to promote the study of the history of science and technology in Spain, and in 1986 the organization changed its name to Sociedad Española de Historia de las Ciencias y de las Técnicas (SEHCYT; Spanish Society of Science and Technical History). While the impact of the SEHCYT in promoting and safeguarding scientific history cannot be underestimated, the gender gap in the narration of scientific accounts was and continues to be a persistent problem.[2] To begin to remedy that disparity, this essay explores gender and medicine in Spain in the last three decades of the twentieth century by focusing on two pivotal contributions by feminist gynecologist and breast cancer expert Dr. Elena Arnedo Soriano (Madrid 1941–2015). Arnedo was a leading figure in the defense of sexual and reproductive rights in Spain, as well as a committed health advocate, seeking ways to improve the lives of women by bringing attention to gender, health care, and policy through her accessible publications and writings. First, this study underscores her engagement in support of women's sexual and reproductive rights through her involvement in the opening of the first (underground) Family Planning Center in Madrid in 1976, when contraceptives were still banned and considered taboo.[3] Second, these reflections focus on Arnedo's dedication to engaging the general population, particularly women, by publishing important reference volumes in the 1990s. She first published *Cuestiones de mujeres*, the Spanish 1991 and 1995 editions of *Questions de femmes* (1989; Women's questions), written by famed French gynecologist Anne de Kervasdoué, and subsequently, in 1997, Arnedo published the massive *El gran libro de la Mujer: Salud, psicología, sexualidad, nutrición y derechos de la mujer* (The great book of women: Health, psychology, sexuality, nutrition, and women's rights).[4]

El gran libro de la Mujer warrants particular attention in the context of Arnedo's legacy. The 575-page collection of previously published essays includes texts by twenty-eight experts in all matters of interest to women in Spain at the turn of the twenty-first century. The work is the result of Arnedo's feminist coordination and editing, a putting-together-of-pieces, or what she terms "stitching together," as described in detail below.[5] The volume offers a comprehensive look at women's sexuality and reproductive health from menarche and the first gynecological exam through perimenopause, menopause, and postmenopausal care. The volume also addresses mental health issues and provides valuable information on legal and policy matters, offering guidance on how best to achieve political and cultural changes that improve women's lives. The impact of this publication should be equated to that of the revolutionary *Our Bodies, Ourselves* (1971) in the United States, the ground-breaking feminist book on women's health and sexuality first produced by the nonprofit organization the Boston Women's Health Book Collective.[6] Indeed, through *El gran libro de la Mujer*, Dr. Elena Arnedo Soriano appears to have spearheaded almost singlehandedly the Spanish publication of women's health volumes that heed the contributions of epoch-changing global feminist texts. In Spain, after the transition to democracy (1975–1982) and the many processes that transformed the nation into a postmodern society, Arnedo's publications and activism clearly identify her with the global feminist movement.

Arnedo's feminist stance is identified explicitly from the very beginning of *El gran libro de la Mujer*, and this study will pay particular attention to Arnedo's Introduction (23–31), where she labels the lengthy tome as a feminist enterprise: "Este es un libro complejo: actual y útil, ecológico y *feminista*, clásico y disparatado" (23, emphasis added; This is a complex book: current and useful, ecological and *feminist*, classic and offbeat). Additionally, this chapter will analyze two works selected by Arnedo to conclude the volume: "Afrontar los retos" (545–555; Facing the challenges) and "¿La liberación era esto?" (557–559; Was women's liberation this?). Both works are reflections by medical doctor Carmen Sáez Buenaventura (1938–)—psychiatrist, feminist activist, and author—extracted from her *¿La liberación era esto? Mujeres, vidas y crisis* (1993; Was this liberation? Women, lives, and crisis). Dr. Arnedo's selections highlight intriguing insights stemming from Sáez Buenaventura's scientific and medical work. The selected texts were written as a result of discussions that took place during psychotherapy groups for women that Sáez Buenaventura organized and supervised in the late Francoist years when working at the hospital then known as Ciudad Sanitaria Provincial Francisco Franco (Provincial Health City Francisco Franco).[7] All three essays (Arnedo's introduction and her two concluding selections by Sáez Buenaventura) bring to the fore the commitment of Arnedo and other female Spanish medical doctors to provide women in Spain with the tools necessary to empower themselves, achieve real equality, and live full, healthy lives.

Before analyzing *El gran libro de la Mujer* in depth, it will be helpful to understand more about the multifaceted physician and feminist activist who organized the volume. Dr. Elena Arnedo Soriano was the daughter of feminist writer Elena Soriano (1917–1996) and political activist and successful businessman Juan José Arnedo (1918–2015). She was a gynecologist, a breast cancer specialist, a writer, a women's rights activist—militant of the Frente de Liberación de la Mujer (FLM, Women's Liberation Front), and a proponent of equality feminism.[8] She also actively participated in local politics, serving as a member of the city council of Madrid from 2003 to 2007 representing the Partido Socialista Obrero Español (PSOE, Spanish Socialist Worker's Party), which she joined during her teenage years. The first president of the Asociación de Centros de Planificación Familiar de España (Spain's Association of Family Planning Centers), Dr. Arnedo also laid the foundation for the prevention of breast cancer through early diagnosis as one of the founding members of the Sociedad Española de Senología y Patología Mamaria (SESPM; Spanish Society of Senology and Breast Pathology).[9] Most importantly for many, she played a leadership role in addressing women's health and well-being via engaging essays—non-academic medical publications—such as *Desbordadas: La agitada vida de la elastic woman* (2000; Overwhelmed: The hectic life of the elastic woman) and *La picadura del tábano: la mujer frente a los cambios de la edad* (2003; The horsefly's bite: Women facing the changes that come with age). In the former, Arnedo confronts the prejudices and blind spots to which older women are subjected, particularly the prevailing and competing representations of aging women in social and literary contexts. In the latter, Dr. Arnedo calls attention to the gender gap not only in the work force but also in how it applies to housework and the division of labor at home.

Indeed, Arnedo's own home life and the complex legacy of her mother's literary work affected her own literary endeavors. A talented writer, Arnedo's gifts in this arena are also evident in the compiling and editing of her mother's posthumous study *El donjuanismo femenino* (2000), which included her eloquent "Advertencia Preliminar" (11–16; Preliminary Note). Undeniably, a feminist genealogy between mother and daughter needs to be addressed since, much like her mother as a novelist and important cultural figure under Francoism, Dr. Arnedo has yet to receive the critical attention she deserves, particularly within what María Jesús Santesmases, Montserrat Cabré i Pairet, and Teresa Ortiz Gómez define as an "epistemología histórica feminista de las ciencias" (382; historical feminist epistemology of science).

Arnedo's feminist perspectives were deeply informed by her mother's writings and also by the political dynamics of Francoist Spain. Her mother's struggle for literary recognition stemmed in part from being the victim of the arbitrariness of Francoist censorship in Spain's post-Civil War literary landscape. Elena Soriano, born in 1917, was active just prior to the 1950s generation of writers such as Carmen Laforet and Ana María Matute, also known as the "niños de la guerra" (children of war).[10] Soriano is the author of, among other works, the novels *Caza menor* (1951; Small game) and the 1955 trilogy *Mujer y hombre* (Woman and man), which includes *La playa de los locos* (The crazies' beach), *Espejismos* (Mirages), and *Medea 55*. She also founded the prestigious literary journal *El Urogallo* in 1969 and directed its publication until 1975, demonstrating an independent streak and a valiant progressive stance for the repressive times in which she lived. Indeed, repression marked her literary work. *La playa de los locos* was censored by the Francoist regime in a bizarre case by which, while permission was given for the printing of the book at the author's expense and for circulating it among critics, the novel was not to be sold to the public (Winecoff 315). Until 1984, when the editorial house Arcos Vergara published the second edition of the novel, it was subjected to a "zombie-like-existence"—it went into print with an informal authorization which was later revoked, actually preventing the novel from circulation after its publication (Kebadze 75).

In 1986, the editorial house Plaza y Janés published the single-volume trilogy *Mujer y hombre* (Cepedello Moreno). However, as the author explains in the prologue, neither she nor her trilogy ever recovered from the damage done by the Francoist censorship. Soriano's trauma from this irrevocable harm is very much evident in her preface to the 1984 edition, when finally, now that Spain was fully functioning as a democratic constitutional monarchy, she could express her guilt, shame, and utter humiliation for what transpired in 1955:

> Pues bien, mi novela *La playa de los locos* jamás consiguió la tarjeta de autorización para imprimirse legalmente. Fue rechazada en su totalidad, de principio a fin, a pesar de que recurrí a todos los medios a mi alcance por salvarla, en un absurdo forcejeo solitario con invisibles enemigos en todos los escalones jerárquicos del Ministerio de Información y Turismo, a lo largo de casi un año; mientras, en mí se formaba y crecía un 'complejo' de culpabilidad, de humillación, de persecución y de impotencia, realmente kafkianos; yo no comprendía ni nadie me explicaba mi delito ni escuchaba mis protestas ante mi evidente discriminación con otros escritores que por aquellos mismos días publicaban y difundían en España novelas mucho más 'fuertes'—uso el eufemismo típico de entonces—, como también lo eran, a mi entender, las otras de mi trilogía, que pasaron la censura sin dificultades. (8)[11]

> Well, my novel *La playa de los locos* never got the authorization card to be legally printed. It was rejected in its entirety, from beginning to end, despite the fact that I used all means at my disposal to save it, in an absurd and lonely struggle with invisible enemies throughout all the hierarchical steps of the Ministry of Information and Tourism, spanning almost a year; while in me a "complex" of guilt, humiliation, persecution, and impotence was forming and growing, really Kafkaesque; I did not understand my crime, nor did anyone explain to me or listen to my protests against the obvious discrimination with respect to other writers in Spain

who published and disseminated much "stronger" novels—I use the typical euphemism of that time—as were, in my opinion, the other two novels of my trilogy, which passed the censors without difficulties.

More than a decade and a half later, daughter Elena Arnedo, in her "Advertencia preliminar" to Soriano's *El donjuanismo femenino* (2000), refers to the same events by echoing her mother's assessment and signaling her feminist perspective: "*La playa de los locos* sufrió una odisea, kafkiana, inconcebible hoy" (13; *La playa de los locos* suffered an odyssey, Kafkaesque, inconceivable today). Arnedo grew up witnessing her mother's hope and struggle for a more egalitarian, caring society, one hopefully finally free of political censorship and *donjuanismo* at the cusp of the new millennium (16). Arnedo's worldview on women and feminism—principally focused on gender equality, community building, and the feminist goals of social justice—is both a legacy inherited from her mother and a personal choice to which she committed her life as a doctor, socialist feminist activist, and author.[12] Her feminist stance led to important contributions in several fields, not only in the realm of publication but most importantly in the development of women's medical services.

Launching the Clandestine Family Planning Center Federico Rubio

Dr. Arnedo's significant feminist legacy stems in part from her early efforts to open a family planning center and from her success—along with that of the team providing free services—in offering women's health services soon after Franco's death and at the beginning of the transition to democracy. As is well known, contraceptives in Spain were prohibited and penalized during the Francoist dictatorship and continued to be banned until October 7, 1978 (Ferreira 788). Prior to this, and as per article 416 of the penal code, anyone who prescribed, offered, sold, distributed, published, or publicly exposed objects, instruments, devices, means, or procedures intended to facilitate abortion or prevent procreation could be sentenced to substantial jail terms and fines—from 5,000 to 100,000 pesetas (Ferreira 788). The October 1978 amendment of the penal code finally authorized the sale, distribution, and use of contraceptives. It is against this legal background and amid the tumultuous years of the Spanish transition to democracy—the so-called *Transición*—that one must understand the historical significance of the clandestine opening, in the early months of 1976 in Madrid, of the first Centro de Planificación Familiar in all of Spain. The Federico Rubio Center was founded thanks to feminist activist Pilar Jaime (also of the Frente de Liberación de la Mujer), other committed PSOE feminists such as Delia Blanco, and feminist health-care providers such as Elena Arnedo. The underground center, which functioned by word of mouth, was named after the street, Federico Rubio, where it was located, and the name carried symbolic meaning: Federico Rubio y Galí (1827–1902) was also a medical doctor, a famed surgeon known for his dedication to those in need. Thus, the center's location and name both highlighted the important legacy of Spanish medical professionals dedicated to serving the public.

Dr. Arnedo was instrumental in carrying on this important medical advocacy via her work at the Federico Rubio Center, and she also wrote a brief history of the center and its impact, providing important insights into her feminist motivations and aims. In "Centro de planificación familiar Federico Rubio," Dr. Arnedo explains that from the early seventies, the multiple and diverse organizations that constituted Spain's varied feminist collective began focusing on claiming the rights of women to decide what to do with their own bodies. Diverse groups began addressing sexualities, divorce, and other pressing issues while diverging, sometimes contentiously, on how to go about enacting progressive changes made possible after Franco's death in November 1975. It is within this particular historical moment that Dr. Arnedo describes the founding of the center:

> Un grupo de mujeres de Madrid asumió una tarea de acción directa y externa de información y divulgación dirigida a otras mujeres que no tenían nada que ver con los debates internos del movimiento feminista. Ese grupo se llamó Grupo de Planificación Familiar de Madrid. Se buscaba informar a mujeres con escasos recursos, amas de casa, trabajadoras de baja cualificación, mujeres ajenas a movimientos reivindicativos sociales, sindicales, o políticos. Se daban charlas en las que había un inevitable mensaje de teoría feminista, sobre la libertad y los derechos de las mujeres, sobre la posibilidad de disociar sexualidad y maternidad, pero el contenido era sobre todo práctico. ("Centro" 64)

A group of women from Madrid took on the task of direct and external action to inform and educate other women who had nothing to do with the internal debates of the feminist movement. That group was called the Family Planning Group of Madrid. The aim was to inform women with scarce resources, housewives, low-skilled workers, women who were not involved in any social movements, trade unions, or political parties. Talks were given in which there was an inevitable message of feminist theory, about the freedom and rights of women, about the possibility of dissociating sexuality and motherhood, but the content was above all practical.

Arnedo's brief description of the organization provides readers with several insights. First, it brings to light how, for many feminists during the Transition, the commitment to the cause for change and equality required both concrete social actions as well as knowledge and dissemination of feminist theories.[13] Arnedo's history emphasizes that the group took on the task of "acción directa" (direct action) and that the content of the conversations with diverse female groups was "sobre todo práctico" (above all practical) in order to challenge inequality and oppression through practice. The goal was to reach disenfranchised women and to propagate feminism beyond academic/professional worlds. This focus on concrete social action combined with theoretical reflection has been an important part of Spanish feminisms, and Arnedo's historical account confirms the focus on both practical as well as intellectual advocacy.[14]

Part of the social insight offered in Arnedo's history involves the way the founders of the Federico Rubio Center understood the class issues that have long divided women within feminist movements. Arnedo notes that the founders looked in particular to connect with low-skilled workers, "trabajadoras de baja cualificación," and to spread the message of feminism to women "con escasos recursos," or of limited economic means. Last but not least, the above history attests to Dr. Arnedo Soriano's thorough dedication to the feminist struggle by engaging in underground activities to provide much-needed free medical health care to women, including free contraceptives and sought-after advice on all kinds of personal issues. As a clandestine location, engaged in illegal activities penalized by articles 416 and 416 bis of the Spanish penal code, one can understand why the center, as Dr. Arnedo specifies in her piece, became identified only by its address as the Centro de Federico Rubio ("Centro" 64; Federico Rubio Center).

The entire enterprise was supported by private donations as well as from the coffers of the feminist movement. All medical personnel donated their time, as did all the other women from diverse professional backgrounds who, along with taking turns in giving the feminist talks, multitasked as social workers, psychologists, lawyers, receptionists, and secretaries. According to Arnedo, two other medical doctors (Javier Martínez Salmeán and Renée Suárez) helped with exams, while known socialist feminists such as Pilar Jaime, Delia Blanco, and Charo Ema offered their time and expertise ("Centro" 64–65). As a safe haven for disenfranchised women, the Federico Rubio Center provided much more than just free contraceptives; it offered psychological support, medical and legal information, and, most importantly, preventive medicine by which women were carefully examined and offered proper cytology exams and other needed analysis ("Centro" 65). Dr. Arnedo argued that, given the excellent medical care provided at the facility, it served as a model for the series of family planning sites that were to be created soon after in Vicálvaro and Vallecas, in Madrid, as well as in Andalusia, the Basque Country, and Catalonia, ultimately leading to the opening of such locations throughout the Spanish State ("Centro" 65).[15] Dr. Arnedo's groundbreaking work, first in Madrid, enabled many other subsequent centers to begin offering medical and personal advice, free gynecological access, and contraceptives to disenfranchised women. Arnedo's pivotal role, and her writings about that work, must not be forgotten in the history of women in science in Spain.

Writing and Publishing for Equality: Rethinking Women's Health and Reshaping Public Discourses on Gender and Health Care

Arnedo's steadfast presence as a driver for change in issues relevant to gender and health care is evident not only in her medical work at the Federico Rubio Center but also through her publications and writings. As early as 1979 she began contributing to important volumes on the history of science in Spain, such as the Spanish translation of C. U. M. Smith's *Molecular Biology: A Structural Approach*, published by Alianza Editorial.[16] Her involvement with Alianza continued with the 1991 publication of the best seller *Cuestiones de mujeres* (Women's issues), which subsequently led to her participation as a coordinator for the 1995 edition, translated into Spanish by Carmen Santos Fontenla. The volume, originally written by French gynecologist Anne de Kervasdoué, was aimed at females of all ages and offered an important educational tool guiding women through the stages of their reproductive lives from pre-pubescence, menarche, perimenopause, menopause, and postmenopause. De Kervasdoué's volume sought to help women better understand their bodies by communicating in clear and accessible language, a preoccupation that defines Dr. Arnedo's own writings. Moreover, the format of answering specific questions appears to have so inspired Elena Arnedo that in 1997 she committed to the publication of a 575-page tome, *El gran libro de la mujer: Salud, psicología, sexualidad, nutrición y derechos de la mujer*, related to women's rights, health, nutrition, psychology, and sexuality. Arnedo's varied publications demonstrate her commitment to make readily available in Spain a wealth of information on women's issues that had yet to be transmitted in a systematic manner.

The encyclopedic volume *El gran libro de la mujer* aimed to provide, under one rubric, all the information needed for women in Spain to become fully actualized twenty-first century feminist citizens. As Arnedo states in the introductory remarks, the goal was to allow women to "conquista[r] conocimiento, seguridad y autonomía para sentirse más libre, más integrada en la sociedad y ser más feliz" (25; conquer knowledge, security, and autonomy in order to feel more free, more integrated into society, and to be happier). The book is comprised of previously published essays by twenty-eight experts in all manner of interests for twenty-first century Spanish women, and it is divided into seven parts: Part 1: "¿Qué es ser mujer?" (33–56; What is it to be a woman?); Part 2: "Cuerpo de mujer" (57–200; Women's bodies); Part 3: "Alma de mujer" (201–300; Women's souls); Part 4: "De amor y sexo" (301–400; Of love and sex); Part 5: "El gran reto: Salud y belleza" (401–460; The great challenge: Health and beauty); Part 6: "La mujer en un mundo de hombres" (461–511; Women in a man's world); and Part 7: "La difícil tarea de ser mujer" (512–559; The difficult task of being a woman).

Following a feminist strategy akin to sewing, Dr. Arnedo describes her work in *El gran libro de la Mujer* as "stitching together" diverse publications in order to provide everyday women with access to a wealth of critical information on women's health (21). The ultimate objective was to engage women and the community, not solely via interactions with experts and doctors, but also via the practice of self-care. Each of the more than four hundred entries had been originally published as part of either single-authored texts by experts in their fields or, in a few instances, as co-edited volumes by several authors, as documented in the section labeled "Procedencia de los textos" (561; Origin of the texts). Elena Arnedo's work consisted of feminist curation, or the practice of arranging, ordering, and "putting together" diverse materials into one single piece. This creative and practical task echoes the process of sewing—a cultural practice that, while often associated with conventional femininity, can in fact be seen to engage feminist points of view, as argued by critics like Jessica Bain in her analysis using the metaphor of dressmaking ("'Darn right I'm a feminist . . . Sew what?'" 59).

Arnedo explicitly addresses the language of sewing and its relationship to feminist practice in the volume's acknowledgments section. First, she thanks Irene Echevarría for her helping in "stitching" the volume together (21). Subsequently, Arnedo chooses to describe her work using the diction of sewing, noting that her job involved piecing together previously published works in order to "*confeccionar* esta especie de enciclopedia-guía de la mujer" (24, emphasis added; *make/sew* this sort of women's encyclopedia-guide). The use of the term "confeccionar" (to make, to put together, to sew) is deliberate, as Arnedo openly identifies the gendered practice of "corte y confección" (dressmaking) as one of the strategies she has used for fabricating the encyclopedia-guide: "con mi labor de corte y confección" (25; with my work in dressmaking).

Expanding on the metaphors for her feminist tactic of producing a collection by "stitching together" other pieces, Dr. Arnedo also alludes to reusing and salvaging "materiales procedentes de otros libros" (materials from other books), emphasizing her work as a turn-of-the-century recycling activity that attests to the volume's "máxima actualidad" (current relevance). This emphasis on the present moment and on keeping up with contemporary concerns is put to good use in the marketing of the tome when insisting that this is "un libro que hay que tener y leer en este confuso y ecléctico fin de milenio" (23; a book that one must have and read in this confused and eclectic end of the millennium). Arnedo offers two reasons for the volume's contemporary relevance: one has to do with this being a book about "*la mujer*" (23, italics in the original; women)—explaining that "en este siglo, la mujer ha pasado de la invisibilidad al protagonismo" (23; in this century, women have gone from invisibility to prominence). The other reason has to do with precisely the way the volume has been put together: "[é]ste es un libro reciclado, el no va más de moderno y ecológico" (24; This is a recycled book, the most modern and ecological kind possible).

Included in the materials that Arnedo "recycles" are two important selections from Dr. Carmen Sáez Buenaventura, a psychotherapist whose work ¿*La liberación era esto?* (1993) provides the pieces that Arnedo stitches together to create the concluding remarks for her massive volume. In the first concluding piece, titled "Afrontar los retos" (Confronting challenges), Arnedo chooses a segment by Sáez Buenaventura that calls attention to how women have been socialized toward "una ética o unos valores del cuidado y de la ocupación, preocupación o reflexión respecto de los demás, antes que respecto a sí mismas" (545; an ethics or value of care and activity, concern, or reflection about others, rather than about themselves). Arnedo emphasizes this basic feminist notion of self care in the conclusion, and she offers as context the observations undertaken by Dr. Sáez Buenaventura of women suffering from mental health issues and expressing their concerns at a psychotherapy session organized and supervised at the Provincial Hospital, a treatment center named for Spain's long-time dictator and enforcer of oppressive patriarchal systems: Ciudad Sanitaria Francisco Franco (City Health Francisco Franco). Indeed, the women gathered for the study at this hospital made constant references to "guilt" and/or "punishment," and the inclusion of references to the psychotherapy study recalls Dr. Sáez Buenaventura's other publications, such as *Mujer, locura y feminismo* (1979; Women, madness, and feminism), where, along with her own work, she compiled that of, among other feminist psychiatrists and psychologists, Jeannete F. Tudor on adult sexual roles and mental illness and Phyllis Chesler via her groundbreaking *Women and Madness* (1972). In the 1979 publication, coming out in the wake of Spain's official transition to democracy, Dr. Sáez Buenaventura evaluated data while analyzing the influence of gender roles in the psychological identity of women, the patterns of mental illness derived from said roles, as well as the decision-making of the doctor and the choice of treatment. Arnedo's decision to conclude the volume with this emphasis on mental health and self care recalls the social, political, and historical pressures that had long oppressed Spanish women.

References abound to the ongoing challenges facing Spanish women in Arnedo's concluding selections for the volume, and after addressing the issue of gender roles and mental health in "Afrontar los retos," Arnedo continues to use Sáez Buenaventura's texts to invite women to address the complexities of socialization and the expectations of the feminist movement. For example, her selections call on Spanish women to "comenzar a hablarse y escucharse" (549; begin to talk and listen to each other) and to develop alliances—diverse sisterhoods—as a way of finding valid alternative solutions. According to *El gran libro de la mujer*, such communal practices are essential for overcoming mental health issues and becoming independent—goals evident not only in the text but also at the heart of women's liberation: "to build strength to change the situation that causes the conflict" (Berkun 512).

It is within the notion of feminist sisterhood that Arnedo's selected texts tackle what is now considered the outdated category of "women's liberation." Arnedo concludes her collection with a segment by Dr. Sáez Buenaventura in which the pychotherapist responds to the question she was most often asked when giving lectures to the general public in the 1980s and early 1990's: "¿Y yo cómo puedo liberarme?" (558; And how can I liberate myself?). In this final selection, titled "¿La liberación era esto?" (557–559; Was women's liberation this?), Arnedo brings to the fore issues related to the socialization of women and offers Buenaventura's analogy of the "fairy godmother" as an example of the need for systemic and communal change. Buenaventura's essay describes how women attending feminist conferences during the 1980s expected the presenter to function as "hada madrina, o la madre sabia, buena y milagrera que con una frase (la varita mágica de marras) la saque de su confusión, su angustia, su revolución interna en un abrir y cerrar de ojos" (558; a fairy godmother, or the wise, good, and miraculous mother who, with a phrase (magic wand of yore), does away with her confusion, her anguish, her inner turmoil in the blink of an eye). Arnedo highlights Buenaventura's ironic comments about "the magic wand" and "the fairy godmother" in order to underscore how traditional processes of socialization were at the core of women's expectations to seek and gain equality and independence, hoping that such things could be achieved quickly and easily, as through incantation.

Via selections by Buenaventura, Arnedo underscores that liberation is not something that can be easily obtained: "[e]videntemente *la liberación* no es una *cosa* u objeto de consumo que pueda adquirirse en el supermercado, sino que es un proceso, una transición, un cambio" (557, italics in the original; [o]bviously, *women's liberation* is not a *thing* or object of consumption that can be acquired in the supermarket; it is a process, a transition, a change). This selection, and particularly the use of italics, signals that Dr. Elena Arnedo Soriano fully agrees with Sáez Buenaventura's understanding that the women's movement must be done by each woman in solidarity with other women, but also, eventually, with men if one is to live in an equal, fair, democratic, postmodern Spain at the turn of the twenty-first century. Dr. Arnedo Soriano's deliberate choice to conclude *El gran libro de la Mujer: Salud, psicología, sexualidad, nutrición y derechos de la mujer* with her peer's reflections are to be understood as an invitation to adhere to the sentence that begins Sáez Buenaventura's concluding paragraph, which ends up being *El gran libro de la Mujer*'s concluding one, too: "*La liberación se hace, no nace*" (559, italics in the original; Women's liberation is made, not born).

Ultimately, at the heart of Dr. Elena Arnedo Soriano's involvement with both *Cuestiones de mujeres* and *El gran libro de la Mujer* is her belief that women need as many sources of information as possible, from healthcare providers, legal experts, founders of grassroots organizations, and health educators, as well as feminist authors and activists, to allow them to take charge of their well-being while also informing policy and healthcare practices. Indeed, one might argue that for both Arnedo and Spanish women scientists more generally, the ultimate goal is "the conversion of gender inequality from a 'women's problem' to a 'societal concern'" (Annandale and Kuhlmann 460).[17] To that end, Arnedo worked tirelessly to bring information and access to Spanish society not only via her work as a medical doctor but also via her important publications. Dr. Elena Arnedo Soriano consistently and compellingly argued for a feminist agenda that brought to the forefront the need for careful analysis of public health issues in order to effect changes in policy that would close the gaps in medical services for women. Through a medical trajectory committed to public health activities—as in the Federico Rubio Center—and by offering accessible medical, legal, and psychological information to women—as in *El gran libro de la Mujer*—Elena Arnedo Soriano dedicated her life to the common good. Today, she must be remembered in the cultural history of women in science as a feminist pioneer, a publishing political advocate, and a most committed medical doctor.

NOTES

1. This article benefited from the support of a research award by the University of California-Santa Barbara Academic Senate Grant. It is part of a book-length project on Doctor Elena Arnedo Soriano and the cultural history of gender and medicine in Spain.
2. According to Elena Hernández Sandoica, one can find a gender component in recent historiographies, generally presented as a "*historia de las mujeres*" (women's history) without any methodological chronology or periodization (264, italics in the original). For other efforts to narrate the history of women and science see, among others,

Cabré; Santesmases et al. See also the *Libro Blanco. Situación de las Mujeres en la Ciencia Española* (White book: The situation of women in spanish science), published in 2011 by the then Ministerio de Ciencia e Innovación (Ministry of Science and Innovation), which no longer exists.

3. The Spanish Penal Code, as per articles 416 and 416 bis, prohibited the use of contraceptive methods, except for abstinence, until 1978 (Arnedo Soriano, "Centro" 64).
4. The Spanish editions were published by Alianza Editorial. Anne de Kervasdoué's *Questions de femmes* was published in Paris with subsequent editions in 1996, 1999, and 2004.
5. For sewing/stitching as feminist practice see, among others, the articles by Stella Minahan and Julie Wolfram Cox (2007); and Jessica Bain (2016).
6. Originally printed in 1970 as a booklet titled "Women and Their Bodies: A Course," both texts are actually the same. Hence, the 1970 and 1971 prints are considered the first edition.
7. Like Arnedo Soriano, Dr. Sáez Buenaventura committed early on to feminism, combining her dedication to medicine with grassroots activism. In 1978, in the midst of the Spanish Transition to democracy, and when there was still no right of assembly, Carmen Sáez Buenaventura was one of the driving forces (along with other leading feminists such as Jimena Alonso, Celia Amorós, and Carlota Bustelo) in a cooperative format that included over two hundred women and involved the creation of the pivotal Librería de Mujeres (Women's Bookstore) as a meeting point. Through many upheavals and changes, the bookstore remains in its original location of Calle de San Cristobal 17, having celebrated its forty-year anniversary in 2018. Buenaventura also served as Directora General de la Mujer de la Comunidad de Madrid (General Director of Women of the Community of Madrid) from 1989 to 1991. See also Valiente Fernández for a study of El Instituto de la Mujer and feminist activism.
8. The Women's Liberation Front was created as a direct result of the Primeras Jornadas Nacionales por la Liberación de la Mujer (First National Conference for the Liberation of Women) held in Madrid in early December 1975, a few weeks after Franco's death (Ferreira, Bermúdez, and Bernárdez Rodal 302).
9. On July 31, 1980, thirty founding partners signed the document instituting the society. Only two women, Dr. Arnedo Soriano and Dr. Montserrat Herranz Marti, were among them.
10. Among the more recognized authors from this generation are Carmen Laforet (1921–2004), Ignacio Aldecoa (1925–1969), Carmen Martín Gaite (1925–2000), Ana María Matute (1925–2014), Josefina Aldecoa (1926–2011), Rafael Sánchez Ferlosio (1927–2019), and Juan Goytisolo (1931–2017), among others. For Soriano's status vis-à-vis her female and male counterparts, see Winecoff (309–10); for how Soriano works inside and outside conventions of genre and gender in *La playa de los locos*, see Arkinstall (102); for information on her legacy and how she is viewed by the Spanish public, see Fernández de la Vega Sanz and Pérez Oliva.
11. The prologue, cited here from the 1986 edition, was originally published in the 1984 second edition of *La playa de los locos* and was included in the 1986 trilogy by Plaza y Janés.
12. Arnedo consistently identified herself as a socialist (from the PSOE) and as a feminist: "soy socialista y feminista, ambas cosas coexisten irremediablemente desde siempre"

("Mujer y socialismo" 221; I am a socialist and a feminist, both aspects have always coexisted irremediably).

13. The history of the first ten years of the feminist movement during the transition to democracy, following the chronology established by María Ángeles Durán and María Teresa Gallego, has been divided into three phases: phase one, from 1975 to 1979, which consisted of the creation, organization, and expansion of the movement; phase two, from 1979 to 1982, defined by fierce divisions due to infighting regarding objectives and positions; and phase three, from 1982 to 1985, when organized feminism fragmented into hundreds of organizations and minicollectives (Ferreira, Bermúdez, and Bernárdez Rodal 302; see also Johnson and Zubiaurre 399).

14. For more details about concrete social actions as an integral part of Spanish feminisms, see "Historical Overview" by Bermúdez, Bernárdez Rodal, and Ferreira, 345–58.

15. See Ferreira for more information on the creation of centers in Catalonia between 1976 and 1982 and Esteban for a personal account from a medical doctor who worked in family planning centers between 1984 and 1996.

16. The English version was originally published in 1968, and the author, Christopher Upham Murray (C.U.M.) Smith (1930–2013) was a visionary, much like Arnedo herself. He published, among other pivotal works, *The Problem of Life: An Essay in the Origins of Biological Thought* (1976) and *Biology of Sensory Systems* (2005). All of his publications underwent multiple editions and were translated into multiple languages.

17. See also Isabel Rubio for a more recent (2018) analysis of societal concerns related to women and science in Spain.

WORKS CITED

Annandale, Ellen, and Ellen Kuhlmann. "Gender and Healthcare: The Future." *The Palgrave Handbook of Gender and Healthcare*, edited by Ellen Kuhlmann and Ellen Annandale, Palgrave Macmillan, 2010, pp. 454–69.

Arkinstall, Christine. "Rewriting Genre, Re-Reading Gender in Elena Soriano's *La playa de los locos*." *Monographic Review/Revista Monográfica*, vol. 8, no. 1, 1992, pp. 99–113.

Arnedo Soriano, Elena, coord. *El gran libro de la mujer: Salud, psicología, sexualidad, nutrición y derechos de la mujer*. Temas de Hoy, 1997.

———. *Desbordadas: La agitada vida de la elastic woman*. Temas de Hoy, 2000.

———. "Advertencia Preliminar." *El donjuanismo femenino*, edited by Elena Soriano, Península, 2000, pp. 11–16.

———. "Centro de planificación familiar Federico Rubio." *Enciclopedia Madrid, Siglo XX*, edited by Carlos Sambricio, Ayuntamiento de Madrid, 2002, pp. 64–65.

———. *La picadura del tábano: La mujer frente a los cambios de la edad*. Aguilar, 2003.

———. "Mujer y socialismo." *El movimiento feminista en España en los años 70*, edited by Carmen Martínez Ten, Purificación Gutiérrez López, and Pilar González Ruiz, Fundación Pablo Iglesias/Ediciones Cátedra, 2009, pp. 219–45.

Bain, Jessica. "'Darn Right I'm a Feminist . . . Sew What?' The Politics of Contemporary Home Dressmaking: Sewing, Slow Fashion and Feminism." *Women's Studies International Forum*, vol. 54, 2016, pp. 57–66.

Berkun, Eileen. "Meaning of Women's Liberation." *Social Work*, vol. 19, no. 4, 1974, p. 512.

Bermúdez, Silvia, Asunción Bernárdez Rodal, and Ana Paula Ferreira. "Historical Overview." *A New History of Iberian Feminisms*, edited by Silvia Bermúdez and Roberta Johnson, U Toronto P, 2018, pp. 345–58.

Boston Women's Health Book Collective. *Our Bodies, Our Selves*. Boston Women's Health Book Collective and New England Free Press, 1971.

Cabré, Montserrat: "Mujeres científicas e historias 'científicas': Una aproximación al pasado desde la experiencia femenina." *Mujeres de ciencias: Mujeres, feminismo y ciencias naturales, experimentales y tecnológicas*, edited by Teresa Ortiz Gómez and Gloria Becerra, Universidad de Granada-Instituto de Estudios de la Mujer, 1996, pp. 13–32.

Cepedello Moreno, María Paz. *El mundo narrativo de Elena Soriano*. Publicaciones Universidad de Córdoba, 2007.

Chesler, Phyllis. *Women and Madness*. Doubleday, 1972.

Esteban, Mari Luz. "La atención específica a las mujeres: 15 años de Centros de Planificación Familiar." *Cuadernos de Salud Pública*, vol. 15, 1994, pp. 1–15.

Ferreira, Ana Paula, Silvia Bermúdez, and Asunción Bernárdez Rodal. "Historical Overview." *A New History of Iberian Feminisms*, edited by Silvia Bermúdez and Roberta Johnson, U Toronto P, 2018, pp. 301–11.

Ferreira, Sílvia Lúcia. "El movimiento feminista y la salud de las mujeres: la experiencia de los Centros de Planificación Familiar (CPF) en Catalunya (1976–1982)." *Estudos Feministas*, vol. 16, no. 3, Sept.–Dec. 2008, pp. 785–807.

Fernández de la Vega Sanz, María Teresa. "In Memoriam: Elena Arnedo, el compromiso de cuidar a las mujeres." *El País*, 8 September 2015, elpais.com/politica/2015/09/08/actualidad/1441664216_697623.html.

Hernández Sandoica, Elena. "¿Hacia una historia cultural de la ciencia española?" *Ayer*, vol. 38, 2000, pp. 263–74.

Johnson, Roberta, and Maite Zubiaurre, eds. *Antología del pensamiento feminista español (1726–2011)*. Cátedra, 2012.

Kebadze, Nino. "Censorship and Sense-Making in Elena Soriano's Trilogy *Mujer y hombre*." *Bulletin of Spanish Studies*, vol. 92, no.1, 2015, pp. 65–89.

Kervasdoué, Anne de; Elena Arnedo Soriano. *Cuestiones de mujeres*. Translated by Carmen Santos Fontenla, Alianza Editorial, 1995.

Libro Blanco. Situación de las Mujeres en la Ciencia Española. 2011. www.idi.mineco.gob.es/stfls/MICINN/Ministerio/FICHEROS/UMYC/LibroBlanco.

Minahan, Stella, and Julie Wolfram Cox. "Stitch'nBitch: Cyberfeminism, a Third Place and the New Materiality." *Journal of Material Culture*, vol. 12, no. 1, 2007, pp. 5–21.

Pérez Oliva, Milagros. "Elena Arnedo, impulsora de los derechos sexuales." *El País*, 8 September 2015, elpais.com/politica/2015/09/07/actualidad/1441628465_429835.html.

Rubio, Isabel. "Las políticas de igualdad en España también fracasan en la ciencia." *El País*, 8 February 2018, elpais.com/elpais/2018/02/06/ciencia/1517911722_504136.html.

Sáez Buenaventura, Carmen. *Mujer, locura y feminismo*. Dédalo Ediciones, 1979.

Santesmases, María Jesús, Montserrat Cabré i Pairet, and Teresa Ortiz Gómez. "Feminismos biográficos: aportaciones desde la historia de la ciencia." *Arenal; Revista de historia de las mujeres*, vol. 24, no. 2, 2017, pp. 379–404.

Soriano, Elena. *La caza*. La Nave, 1951.

_____. *Espejismos*. Calleja, 1955.

_____. *Medea*. Calleja, 1955.

_____. *La playa de los locos*. Editorial Argos Vergara, 1984.

_____.*Mujer y hombre: La playa de los locos-Espejismos-Medea*. Plaza and Janés, 1986.

_____.*El donjuanismo femenino*. Península, 2000.

Smith, C. U. M. *Biología molecular: Enfoque estructural*. Alianza, 1979.

_____. *Biology of Sensory Systems*. Wiley, 2005.

_____. *Molecular Biology: A Structural Approach*. Faber, 1968.

_____. *The Problem of Life: An Essay in the Origins of Biological Thought*. The MacMillan Press LTD, 1976.

Valiente Fernández, Celia. *El feminismo de Estado: El Instituto de la Mujer (1983–2003)*. Institut Universitari d'Estudis de la Dona, 2006.

Winecoff, Janet. "Existentialism in the Novels of Elena Soriano." *Hispania*, vol. 47, no. 2, 1964, pp. 309–315.

CHAPTER 3

Gender and the Critique of "Ascientific Traditions"

Science as Text and Intertext in Rosa Montero's
La ridícula idea de no volver a verte

ELLEN MAYOCK

Rosa Montero, a much-decorated Spanish journalist and author, declares in a 2016 interview that "hay más magia en la ciencia que en los cuentos de hadas.... Es la magia pura de la vida porque la creatividad de la ciencia es monumental" (Martín; there is more magic in science than in fairy tales. ... It's the pure magic of life because the creativity of science is monumental).[1] In this same interview, Montero praises the efforts of the Instituto de Astrofísica de Canarias (IAC) for "intentar divulgar y unir la ciencia a la vida intelectual y literaria" (Martín; trying to publicize the sciences and uniting them with intellectual and literary life). Montero mentions two of her literary works, *Instrucciones para salvar el mundo* (2008; Instructions for saving the world) and *La ridícula idea de no volver a verte* (2013; The ridiculous idea of never seeing you again), as particularly engaged in this mission of science awareness and creativity.[2] Her 2011 novel *Lágrimas en la lluvia* (Tears in the rain), featuring the return of protagonist Bruna Husky, also underscores current literary-scientific themes, such as the cyborg, research into other galaxies, and technological advances in artificial intelligence. Bradley Nelson states, "One of the more potent critical gestures in *Lágrimas* is the substantiation of the knowledge and self-consciousness of the manufactured *other*, who holds a privileged view of the unrestrained barbarity of the technocratic imperialism whose vanguard she was created to buttress and extend" (129).[3] Montero's body of work demonstrates a sustained interest in identity development and the interactions between self and other. More recent works, such as those mentioned above, reveal Montero's fascination not only with the realm of scientific inquiry, but also with the major innovators in scientific discovery.

La ridícula idea de no volver a verte is a captivating hybrid memoir that braids Rosa Montero's story of the loss of her husband and recuperation of the love of life with Marie Curie's short diary about love, loss, and science. The book includes photographs of Rosa Montero and of Marie Curie, e-mail and Facebook communications between the author and her literary friends, and memories of childhood illness, schooling, and friendships. Readers immediately become attached both to Rosa Montero and to her astute reading of Marie Curie's roles as scientist, two-time Nobel Prize winner, wife, mother, and lover. Despite being a hybrid work, somewhere between memoir and novel, *La ridícula idea de no volver a verte* won the 2014 Premio de la Crítica de Madrid for best novel and to date has been translated into Dutch, French, and Portuguese.

This chapter analyzes Rosa Montero's literary courtship with scientists and scientific concepts and evaluates her incursions into hybrid genres. The first half examines gender and science in Spain in terms of history, laws, and data, and it addresses questions of mentorship, invisibility in the media, perceptions of women in science, and the "second shift" (a phrase popularized by Arlie Hochschild and Anne Machung in their eponymous work on women who work long hours both in and outside the home). The second half then analyzes how Montero develops a textual relationship with Marie Curie, how she vindicates and celebrates women scientists' participation in physics, chemistry, and biology, and how she maps biological, anatomical, and physiological concerns through both medicalization and memory. The focus of this analysis highlights Spanish attitudes toward women in science, as well as Montero's preoccupation with and uses of the body in concrete, physical, and visceral ways to humanize the discussion of science, illness, dying, and death.

The "Ascientific" Tradition

In the aforementioned interview with Martín of *Diario de Avisos*, Rosa Montero responds to a question about Spain "turning its back on research" by retracing Spain's history:

> Llevo años diciendo que España es un país de una tradición acientífica brutal y eso viene de unos polvos muy antiguos: en el siglo XVIII y XIX, mientras toda Europa y el mundo entraba en el futuro, nosotros teníamos las universidades controladas por la Iglesia católica. En la Universidad de Salamanca no había cátedra de Matemáticas porque se consideraba que la ciencia era cosa del diablo, un pecado . . . de ahí tenemos que un intelectual como Unamuno dijera—y repitió varias veces—su famosa frase "que inventen ellos," alabando nuestra pasividad ante la ciencia que promovían los extranjeros. (Martín)

> For years, I've been saying that Spain is a country with a fierce ascientific tradition and that this tradition comes from ancient sources: in the eighteenth and nineteenth centuries, while all of Europe and the world were welcoming the future, our universities were controlled by the Catholic Church. At the University of Salamanca there was no endowed chair in mathematics because science was considered a thing of the devil, a sin . . . and that's how we end up with an intellectual of the stature of Unamuno saying—and he repeated it several times—his famous line of "let other countries be the inventors," thus praising our passivity in response to science promoted by foreigners.

Similarly, twentieth-century psychologist Pedro Laín Entralgo expresses the following about the great Spanish scientist Ramón y Cajal: "Cajal, hombre-mito bajo forma de lavaconciencias, chivo expiatorio a la inversa del difuso sentimiento de culpa que en relación con el cultivo de la ciencia existe y opera en los senos morales de la sociedad española" (qtd. in Aladro Vico et al. 183; Cajal, a myth-man in the form of a social placebo, an inverse scapegoat for the diffuse feeling of guilt that exists in relation to the cultivation of science and operates under the moral wings of Spanish society). Aladro Vico et al. describe Spaniards' approach to science a century ago as stemming from a "falta de apremio moral en la sociedad española, al desarrollo de la ciencia" (183; lack of moral urgency in Spanish society toward the development of science). This sociocultural dichotomy between morality (based in conservative Catholicism) and scientific inquiry (deemed to be amoral) speaks both to a science/religion divide still existent in some areas in the twenty-first century and to a marked gender divide.[4] In fact, literary critics know this tension between religion and science to play itself out through the debates on literary naturalism that ricocheted from France to Spain and back through the end of the nineteenth and the beginning of the twentieth centuries. Spanish culture, so influenced by a hierarchical Catholic morality, already prescribed strict gender roles for women and men. Men were taught to live and work in the public sphere. Women were the angels of the hearth, and their role was to live privately, carrying out the duties of the home and church. Scientific laboratories existed in the public sphere, and thus were a realm assigned primarily to men.

Gender and Science in Spain

If scientific inquiry, even initiated by men, was thought in Spain to be bordering on sin, then one can understand even more clearly the social strictures on women who aspired to be scientists. Aladro Vico et al. state, "La ciencia entre las mujeres ha sido, durante siglos, una cuestión semioculta" (183; Science among women has been, for centuries, a semi-covert issue), and they note that "el caso español es, probablemente, de los más obvios que existen" (Spain offers one of the most obvious cases). Suzanne Le-May Sheffield notes that Spanish women needed special permission to attend university in the late 1800s. Some who favored women's access to university education did so because they believed it would liberate women from the influence of the Catholic Church and make them better "mother-educators" (109). Le-May Sheffield goes on to note that "Spanish women were not allowed full access to university until 1910, but once they were given access, their numbers in science increased faster than in other disciplines, suggesting that institutional barriers and not a lack of interest or intellectual ability had been keeping women out of science in Spain" (109). More recent trends in Spain reveal that institutional barriers, when recognized by legislative bodies, can be removed.

In 2007, the Spanish government passed a law "para la igualdad efectiva de hombres y mujeres" (Ley orgánica, *BOE*; for effective equality of men and women). The official website of the Consejo Superior de Investigaciones Científicas (CSIC), sponsored by the Spanish government, cites this 2007 law and then states that "El CSIC no ha permanecido al margen de las demandas que la sociedad española ha venido reclamando con fuerza, desde hace años" (The CSIC has not been on the sidelines of the demands Spanish society has been making vociferously for years). The CSIC website makes clear that they have been working on the issue of women's participation in science since 2002. Most of the research completed on women and science in Spain has been in the wake of the 2002 work by the CSIC and/or the 2007 equality law. This includes 2006 empirical research on women and men in science sponsored by the Instituto de la Mujer (government-sponsored), a major white paper from 2011 (*Libro Blanco*), and a 2013 study on the culture of science in Modern Spain (Bauer and Howard; sponsored by BBVA), accompanied by increased media coverage of the issue and statements by professional academies and associations about the status of women in the profession.

This greater visibility in the Spanish context parallels highly promoted works on women in science published in the United States in 2016—Dava Sobel's *The Glass Universe* and Margot Lee Shetterly's *Hidden Figures*. In addition, Harvard physicist Lisa Randall published an opinion piece in *The New York Times* with the convincing headline "Why Vera Rubin Deserved a Nobel."[5] Rubin proved the existence of dark matter, and Randall says that, "of all the great advances in physics during the twentieth century, surely this one should rank near the top, making it well deserving of the world's pre-eminent award in the field, the Nobel Prize." Indeed, the need to bring to light the achievements of high-level women scientists is acute in more than one national context.

Status of Women in Science in Spain

The White Book published by Unidad de Mujeres y Ciencia in 2011, titled *Libro Blanco. Situación de las Mujeres en la Ciencia Española*, reinforces the European Commission's 2010 stated goal for 2030: "señala como objetivo para el año 2030 que la mitad de todo el personal científico, en todas las disciplinas y en todos los niveles del sistema científico, sean mujeres" (4; states the objective of having one half of all science personnel, in all disciplines and at all levels of the scientific system, be women). The authors add, "Se trata de romper la segregación horizontal y vertical que hoy existe en la ciencia europea y, también, en la española" (4; This means breaking down the horizontal and vertical segregation that exists today in the sciences in Europe and in Spain). The White Book takes as a point of departure information from the United States' National Science Foundation and the National Academies of Sciences and Engineering in order to propose possible measures for change in Spain (8–9). In addition, the White Book cites statistics from Spain's Instituto Nacional de Estadística (INE) to demonstrate that, controlling for other factors, male scientists are 2.5 times more likely to be promoted to tenured and endowed positions than women scientists (12–15). The picture is even more acute when one factors in parental status. Male scientists with children are four times more likely to be tenured and promoted than female scientists with children (12–15).

The dense 2006 study titled *Mujeres y hombres en la ciencia española. Una investigación empírica* by García de Cortázar y Nebreda, Arranz Lozano, Val Cid, Agudo Arroyo, Viedma Rojas, Justo Suárez, and Pardo Rubio begins with an epigraph by renowned scholar of science and technology studies Donna Haraway in order to destabilize the notion of science as universal, neutral, or objective: "Parte de nuestra reconstrucción como seres humanos socialistas y feministas consiste en rehacer las ciencias que construyen la categoría 'naturaleza'" (22; Part of our reconstruction as socialist and feminist human beings consists in remaking the sciences, which construct the category of "nature"). The study's authors base their work on the notion that many individuals believe science to be neutral, objective, and unable to discriminate and, therefore, that they must help make the discipline's invisible prejudices visible.[6] They cite physicist and feminist scholar Evelyn Fox Keller: "'la ideología de género puede ser considerada como mediadora crítica entre los orígenes sociales, políticos e intelectuales de la ciencia moderna'" (García de Cortázar y Nebreda et al. 24–25; gender ideology can be considered a critical mediator between and among the social, political, and intellectual origins of modern science). The authors also criticize sociologist Pierre Bourdieu's hypocrisy in establishing his theory of habitus and the lack of recognition of its androcentric focus (25–29): "'¿Es Bourdieu un representante más de la tradición intelectual masculina occidental que refunfuña al teorizar sobre los privilegios que conllevan el estatus de dominante, pero que no pone en peligro su propio *statu quo*?' (Hurtado, 1996, cit. por Apfelbaum, 2002: 78)" (27; Is Bourdieu just another representative of the Western male intellectual tradition who grumbles as he theorizes about the privileges of dominant status but neglects to examine his own status quo?). These questions become more important in the Spanish context, given the historical trajectory of the lack of respect for or faith in science treated above.

Bauer and Howard's 2013 *The Culture of Science in Modern Spain* traces several major sociocultural phenomena in order to understand the support for the sciences in Spain. These include (1) an understanding of medicine, biology, and physics as "the prototype sciences" (13–14); (2) increased numbers of Spaniards reporting interest in the sciences and recognizing their role in the protection of the environment (14); (3) the tension between secularism and progressivism and comparisons between Europe and the United States (14); and (4) "engagement with and attention to science" across age cohorts (15). One interesting gender distinction made in this study is the following: "While highly knowledgeable Spanish women are more secular in their outlook, this is not so clear for Spanish men. In effect, less scientifically literate men are more secular than women of similar literacy, while highly literate women are more secular than men of similar standing, everything else being equal" (31). These data reinforce the assertions about the tension between science and religion made by Rosa Montero in her interview with Verónica Martín. If Spain's centuries-old legacy of religion over science prevails, then women scientists, found in this study to be more "secular," might be confronted with even greater doubts about their legitimacy.

The status and success of women in science in Spain are also influenced by the lack of mentorship and sponsorship, the invisibility of women scientists in Spanish media, and women's "double shift." An example of the lack of sponsorship comes from the Royal Academy of Exact, Physical, and Natural Sciences of Spain, a group that declares: "There are no proactive measures to increase the number of female members. At the moment only two out of our fifty-four National members are women. This ratio is much smaller than that within the whole scientific community" (IAP). Given that the Academy's main functions include the dissemination of scientific research through conferences, journals, consultations with the Spanish government, and education, its lack of women implies an absence of women scientists in all of these academy-sponsored public activities. To this point, the organization Spanish Women in Physics declares in its 2009 report that it "has been less successful at increasing recognition of women scientists through award nominations" (López-Sancho et al. 171), again emphasizing the invisibility of women scientists' participation at the national level. This inequity is noted by Aladro Vico et al.:

> El pequeño grupo de artículos que se centra en la mujer científica (14,3%) y el que se dedica a ambos sexos (15%) exponen las investigaciones que desarrollan estas mujeres sin darles protagonismo, a diferencia de lo que ocurre con los hombres científicos. Además, se representa a la mujer de forma aséptica, prescindiendo de adjetivos calificativos que la identifiquen en un sentido positivo como líder, gurú, o brillante, que sí se aplican a sus homólogos masculinos. (190)

> The small group of articles focused on female scientists (14.3 percent) and on both genders (15 percent) describe the research work of female scientists without giving them relevance, unlike what happens with the description of male scientists. In addition, female scientists are described aseptically, without the use of positive evaluative adjectives (e.g., leader, guru, or bright), unlike what happens in the description of their male counterparts.

While professional academies and organizations in Spain struggle to nominate and recognize women scientists in fair numbers, Spanish media reinforce the invisibility of women scientists and the positive visibility of men scientists. For example, a September 23, 2015, headline from *El País*—"El 63% de los españoles cree que las mujeres no valen para científicas de alto nivel" (63 percent of Spaniards believe that women aren't capable of being high-level scientists)—reinforces sociocultural stereotypes about women's intellectual capabilities. Add to these limitations the fact that 55 percent of Spanish women at the beginning of the new millennium reported doing "the entire workload of the household" (FECYT), and it becomes evident that Spanish society (like many others) has a long road ahead to recognize women scientists and promote their work.

Marie Curie as Text and Intertext

Absent homegrown science heroines in their own countries, many people have turned to Polish-born Marie Curie as the universal model for women scientists. In fact, Curie could serve as an icon for all scientists, given her astounding work as a physicist and chemist, her groundbreaking research in radioactivity, and her not one but two Nobel Prizes (the 1903 Prize in Physics, shared with Pierre Curie, and the 1911 Prize in Chemistry, won by Marie alone). In her important book *Women and Science: Social Impact and Interaction*, Suzanne Le-May Sheffield features Marie Curie in the book's introduction, providing a detailed account of Curie's life as a scientist and meshing it with the gender-based struggles Curie experienced. She concludes the introduction by saying about Curie: "It is in this complete story that we find a truly useful icon for women scientists, attesting to what women can do, reminding us of the social barriers she had to face and overcome, and showing us that despite her icon status, she was a woman of deep feeling and varied experience" (xxxii). Like Montero, Le-May Sheffield enriches the brief biographical sketch of Marie Curie with photos of her in the professional and personal realms. Frequent mention of Marie Curie in numerous studies on women and science underscores the singularity of her accomplishments (for men and for women) and of her recognition. Furthermore, one informal Google search reveals the ubiquity of Marie Curie in fictional and documentary films and television shows, on postage stamps, and in the names of cancer centers. Several sites also recognize her as a "great" feminist. Montero and Le-May Sheffield insist on examining what I will call "the full life" of Curie.

While Rosa Montero had written about and incorporated the sciences into several of her novels before *La ridícula idea*, it was Marie Curie's diary that attracted Montero, rather than (or maybe in addition to) Marie Curie's renown as a celebrated scientist. Montero expresses wonder at Curie's amazing humanity—her professional successes, move from Poland to France, deep curiosity and intellectual acumen, roles as wife, mother, lover, and widow, and her short, beautifully wrought diary that tells so much about both the obstacles for women in science and the process of mourning. The first chapter of *La ridícula idea* begins with a rumination about the pain of loss, yet Montero readily admits on the second page: "Pero éste no es un libro sobre la muerte. En realidad no sé bien qué es, o qué será" (10; But this is not a book about death. I don't really know what it is, or what it will be).[7] One paragraph later, just one and a half pages into the narration, Montero exposes that "La santa de este libro es Marie Curie" (10; This book's saint is Marie Curie). Montero here winks at the beatification of this science powerhouse and, in so doing, signals a new embrace of science in Spanish letters.

At this very early moment in the narration, Marie Curie becomes both text—a real, live person whose biography is both commonly known and briefly recounted by Montero (10–11)—and intertext—a figure whose autobiography weaves through Montero's work, which includes some of Montero's own autobiographical details but confounds genre categories. Lucy Lorena Libreros summarizes the work in this way: "relato a medio camino entre el recuerdo personal y la biografía de una de las mujeres más extraordinarias de todos los tiempos" (a story halfway between personal memory and the biography of one of the most extraordinary women of all time). Montero's playful approach to genre allows the author to study the facts about Marie Curie, imagine more details of her life by analyzing Curie's diary and photographs, and explore her own feelings of loss and regeneration.

In 2015 (two years after the publication of *La ridícula idea*), Montero gave a talk at the Universidad Internacional Meléndez Pelayo focused on women and science. In the talk, Montero lists Curie's struggles, which include general discrimination, an unwillingness on the part of the Nobel Committee to award the science prize to a woman, and the incessant linking of Marie Curie's work to that of her husband ("Rosa Montero reivindica"). Montero then makes clear that the challenges Marie Curie faced in the early twentieth century have not disappeared, that they still plague women scientists in Spain: "el número de mujeres licenciadas es de un 59 por ciento, mientras que el de mujeres investigadoras tan solo es del 33 por ciento, que se queda en un minúsculo 20 por ciento si de lo que hablamos es de directoras de proyectos científicos que reciban financiación de fondos públicos" ("Rosa Montero reivindica"; the number of women with advanced degrees is 59 percent, while that of women researchers is only 33 percent, which is a miniscule 20 percent if we look only at women directors of publicly funded scientific projects).

It is relevant that Rosa Montero herself has been described often in glowing terms, but this is sometimes limited to the context of an outstanding woman among women (with men as the universal, women as the exception). In the introduction to her interview with Rosa Montero, Estrella Cibreiro describes Montero as "una de las novelistas y periodistas españolas más leídas y estudiadas tanto en el ámbito nacional como internacional" (49; one of the most widely read and studied women novelists and journalists in the national and international realms). At this point, one can say all of this about Montero in a universal (not only among women) context. Therefore, one could say that she is "una de los novelistas y periodistas" or "una de l@s / lxs novelistas y periodistas." In this same interview, Montero insists, "La mujer tiende a vivir en el deseo de los otros, y no considera su propio deseo como la prioridad que debe ser" (52; Women tend to live in others' desires, and they don't consider their own desires as the priority they should be). Montero here shows us that internationally acclaimed women novelists, like women scientists, still are not immune to externally imposed gender scripts.

Marie Curie is not the only highly accomplished scientist Montero's *La ridícula idea* foregrounds; rather, the work mentions many other women scientists who never gained the fame they deserved, including Rosalind Franklin (12–14), "eminente científica británica que descubrió los fundamentos de la estructura molecular del ADN" (12; eminent British scientist who discovered the bases of DNA's molecular structure); Henrietta Swan Leavitt, "una brillante pero por lo general ignorada astrónoma" (14; a brilliant and generally unknown astronomer), who "logró dar un salto gigante en la medición de la distancia a las estrellas" (Walter Lewin qtd in Montero 14; made a giant step in measuring the distance to the stars); Lise Meitner, who "ayudó a descubrir la fisión nuclear" (14; helped to discover nuclear physics), and Jocelyn Bell, who "descubrió los púlsares" (14; discovered pulsars).

Montero subtly incorporates the mention and brief stories of these scientists, without turning the work into a bio-bibliography of scientists. She uses the hashtag #LugarDeLasMujeres (#WomensPlace) as a way to talk about how women were subordinated to men, firmly encouraged to fulfill gender-based roles. For example, Montero recounts Marie Curie's teen years, when she grieves the loss of her mother and sister, finishes up formal schooling, and is funneled into what was considered a proper profession for women who needed to work—teaching or serving as governesses (54). Montero writes: "O sea que estaban [las mujeres que necesitaban trabajar] en una especie de limbo social, supuestamente respetadas como iguales por los señores pero ocupando una posición tan falsa que la realidad cotidiana se encargaba de *ponerlas en su sitio*, como se decía cruelmente" (54; In other words, they [women who needed to work] were in a type of social limbo, supposedly respected by the men as equals, but occupying such a false position that everyday realities took over and *put them in their place*, as was cruelly said back then). As Montero does throughout this work, here she looks at the case of Marie Curie, extrapolates to make general statements about women and work (55–57), and then circles back to the specifics of Marie Curie's life and circumstances (by page 58).

Throughout *La ridícula idea*, Montero marvels at Curie's many accomplishments and her brilliance, tenacity, and humanity. She also expresses admiration for Curie's appreciation for and skills in the arts: "Al parecer a Marie le encantaba la literatura y escribir (escribía sorprendentemente bien) y sopesó durante cierto tiempo dedicarse a ello. Pero al fin se decidió por la física y la química" (38; It seems Marie loved literature and writing (she wrote surprisingly well), and for some time she considered devoting herself to the literary enterprise. But, in the end, she chose physics and chemistry). In this sense, Montero is advocating for what some United States academics have called a "STEAM" (versus "STEM") approach—the union of science, technology, engineering, the arts, and mathematics. Curie's decision to pursue the sciences, of course, led to extreme challenges and even more outstanding accomplishments. Montero admires how, even at a young age, Curie embraced Auguste Comte's positivist philosophy (50–51), and then Montero relates Curie's struggle to break into the sciences with Spanish novelist Carmen Laforet's need to understand her own literary ambitions and early successes on a mostly male stage (52–53). Montero insists that great scientific brains are often creative geniuses, prone to the expression of their passions and curiosities. She writes of Marie Curie:

> Volviendo a nuestra Marie, pienso que, por debajo de su rígida contención, y justamente por eso, era un verdadero torrente pasional. Rebosaba sentimientos volcánicos en las cartas que escribía en su juventud; en el diario que hizo tras la muerte de Pierre; en las pocas líneas que mandó a su amante, Langevin, y que casi originaron una tragedia. (63)

> Returning to our Marie, I think that, under all that rigid, contained posture, and precisely because of it, existed a real torrent of passion. She spilled over with volcanic feelings in the letters she wrote in her youth; in the diary she created after Pierre's death; in the few lines she sent to her lover, Langevin, and which almost started a tragedy.

The "volviendo a" ("returning to") underscores the meandering mode of the work's narration (from Curie as example or model, to extrapolation to general points about women and women in the workplace, and back to

Curie), while the use of "nuestra Marie" ("our Marie") reinforces the intimate relationship that Montero establishes between and among herself, Marie Curie, and her readers. Montero's appreciation for Curie's depth of feeling and penchant to express it establishes again the author's insistence that the creative arts are an essential element of the hard and natural sciences.

Science and the Body in *La ridícula idea de no volver a verte*

Physics and chemistry tell us about our physical world—how things move, how compounds form, how colors are made and transform in chemical processes—generally stated, why the world is as it is. Montero's fascination with Marie Curie's scientific work and diary is not surprising, especially in the wake of the loss of Montero's husband Pablo. She seems to need to hold onto proven truths and logic—staples of the hard and natural sciences—while at the same time coming to understand her own emotions. Montero's attachment to Curie and the diary allows Montero to remember a physical touch that is no longer present and to think about material items and their hold on us, especially as they lead us to further memories of loved ones. *La ridícula idea* becomes almost a memory metonymy of the physical world—a concatenation of objects and their place in our brains—to guide readers through notions of love and loss, gender and ambition, and science and the arts. Material objects and the physical body undergird Montero's narration and anchor readers to her and Curie's compelling accounts. In fact, Montero's inclusion at the end of *La ridícula idea* of a Spanish translation of Curie's brief diary serves to give authentic voice to Curie's material loss and feelings of devastation. This strategy also reveals again the bridge between the creative arts and the sciences: Montero writes copiously about scientists and their work, while Curie through her diary enters the realm of the affective, usually considered to "belong" to the creative arts.

Montero's interest in the physical sciences shines through in the opening pages of *La ridícula idea*, in which she gives radioactivity a literary debut through this artful description: "Además Marie descubrió y midió la radiactividad, esa propiedad aterradora de la Naturaleza, fulgurantes rayos sobrehumanos que curan y que matan, que achicharran tumores cancerosos en la radioterapia o calcinan cuerpos tras una deflagración atómica" (11;

Furthermore, Marie discovered and measured radioactivity, that terrifying property of Nature, stunning, superhuman rays that cure and kill, that shrink cancerous tumors through radiation or scorch bodies after an atomic combustion). This poetic rendering of radium and its properties reminds us of the chemical element's terrific beauty ("terrifying," "stunning," "superhuman"; and the searing verbs used in this passage) and of the risks Marie and Pierre Curie took in handling it. Montero cites Marie Curie's observations of the luminous effects of their "products" and then observes of Marie and Pierre: "Estaban encantados, ésa es la palabra; embrujados, atrapados por el hechizo del fulgor verdiazul. A veces, después de cenar, corrían al laboratorio para disfrutar con la visión de sus fantasmitas luminosos. Y en la cabecera de la cama tenían una muestra de radio, supongo que para adormecerse con su fosforescencia" (127; They were enchanted, that's the word; haunted, trapped by the spell of the greenish blue glint. Sometimes, after having dinner, they would run to the laboratory to enjoy the vision of these luminous little ghosts. And at the head of their bed they had a radium sample, I guess so that they could fall asleep to its phosphorescence). This passage underscores the novelty of it all, the giddy excitement of discovering something so significant, the passion for scientific research, and the seeming other-worldliness of the substance itself. Of course, given that Marie Curie and her scientist daughter Irène died from constant exposure to radium, this joy of scientific discovery and beauty is tempered by the fatal effects of the compound.

The physical and corporeal dimension of Curie's work is further emphasized in several passages from Montero's homage, and Montero links these corporeal elements to a connection she has with another great science fiction writer and thinker: Ursula K. Le Guin. Outstanding author and Montero's close friend, Le Guin was also enraptured by Curie's diary and recommended to Montero a particular passage from it that highlights the physical reality of death, the literal pieces of a person's body that decay and disappear. Intertextuality dominates Montero's text at every level, and here we see an example not just of intertext with the great scientific mind of Marie Curie, but the double intertext of analyzing Marie Curie's journal through the lens of another important science-fiction giant, Le Guin. This move models the incorporation of the arts into the world of STEM or, in this case, STEAM. Montero cites the recommended passage and comments on its connection to Le Guin: "Con mi hermana quemamos tu ropa del día de la desgracia. En un fuego enorme arrojo los jirones de tela recortados con grumos de sangre y los restos de sesos. Horror y desdicha, beso lo que queda de ti a pesar de todo" (27; My sister and I burn your clothes from that tragic day. I throw on the enormous fire the shreds of cloth stained with bits of blood and brains. Horror and misfortune, I kiss what is left of you in spite of everything). She then recounts how Le Guin's comments made her realize that "beso lo que queda de ti" is not a metaphor, but the actual physical touch of Marie's lips on the remains of brains and blood on the cloth two months after the day of Pierre's accident and death. This yearning for any physical piece of the lost loved one is so fundamental, so visceral, so what Marie needed to do at that moment, that it affects Montero (who, one page earlier, cites Le Guin's poem about her own loss: "I see broad shoulders, / a silver head, / and I think: John! / And I think: dead") (26). The passage, as memory metonymy, leads Montero to tell of her own interaction with the armchair that belonged to Pablo, her deceased husband:

> Yo nunca llegué a eso, desde luego; al contrario, quise 'portarme bien' en mi duelo y agarré el hacha: me deshice inmediatamente de toda su ropa, guardé bajo llave sus pertenencias, mandé tapizar su sillón preferido, aquel en el que siempre se sentaba. Me pasé de tajante. Cuando llegó el tapicero para llevarse su sillón, me senté en él desesperada. Quería disfrutar del sudor adherido a la tela, de la antigua huella de su cuerpo. Me arrepentí de haber llamado al operario, pero no tuve el coraje o la convicción suficiente para decirle que ya no quería hacerlo. Se llevó el sillón. Aquí lo tengo ahora, recubierto de un alegre y banal tejido de rayas. Jamás he vuelto a usarlo. (28)

> I never went that far, of course; quite to the contrary, I tried to "behave myself" in my mourning, and I grabbed hold of the ax: I got rid of all of his clothing, I locked up his belongings, I arranged to have his favorite armchair, the one he always sat in, upholstered. I was focused on cutting ties. When the upholsterer arrived to carry away the armchair, I sat on it, desperate. I wanted to sense his sweat in the cloth of the chair, feel the trace of his body. I regretted having called the man, but I had neither the courage nor the conviction to tell him that I no longer wanted to carry through with this. He took the armchair. I have it here now, newly covered with a happy, banal striped cloth. I've never used that chair again.

Montero compares her own reaction to Pablo's death with Marie's to Pierre's death. The cloth with blood and brains that Marie kisses goodbye becomes in Montero's account the favorite armchair. She emphasizes how the chair retains a sense of Pablo's sweat and the trace of his body. She laments her too-quick decision to upholster the chair because all she has left at the end is a chair that used to be Pablo's but no longer emits the physical recollection of the deceased. Montero later cites another section of Curie's diary and responds to it, saying, "Cuánta piel, cuánto roce, cuánto deleite en el cuerpo del otro hay en estas líneas. Y cuánta desesperación por haberlo perdido" (87; So much skin, so much touch, so much pleasure in the body of another are in these lines. And so much desperation for having lost him). Again, the metonymy of physical object in one story of loss to physical object in another story of loss continually weaves the stories together and calls to the reader metatextually, while also signaling anew the importance of our understanding of the physical world and its effects on us.

Death (la Muerte, la Parca, as Montero occasionally refers to it [e.g., 109]), and the physical finality it represents are omnipresent in *La ridícula idea*. Montero cites a long passage from Curie's diary in which Marie recounts details of Pierre's funeral service. Montero responds, "Sí, hay que hacer algo con la muerte. Hay que hacer algo con los muertos. Hay que ponerles flores. Y hablarles. Decir que les amas y siempre les has amado. Mejor decírselo en vivo; pero, si no, también puedes decírselo después. Puedes gritarlo al mundo. Puedes escribirlo en un libro como éste. Pablo, qué pena que olvidé que podías morirte, que podía perderte" (174; Yes, you have to do something with death. You have to do something with the dead. You have to put flowers on their graves. And talk to them. To tell them that you love them and that you have always loved them. It's better to tell them when they're alive, but, if you haven't, you can tell them afterwards. You can shout it out to the world. You can write it down in a book like this. Pablo, such sorrow that I forgot you could die, that I could lose you). The anaphoric repetition of "hay que" ("one has to") in this quote communicates the generalized lessons learned through mourning, lessons applicable to any "one," while the movement to the "tú" (informal "you") forms of the verbs ("amas," "has," "puedes") connects Montero to her readers through an intimate apostrophe. The "tú" form finally gives way to "Pablo," thus moving us from the most universal ("hay que") to the most personal (the author's deceased husband being addressed directly).

Marie Curie's deep pain in mourning and her frank, aggrieved expression of it in her diary lead Montero to this moment in *La ridícula idea*, a moment that responds to the meandering question in the opening pages about what this book will be.

Rosa Montero artfully shows Marie Curie to be both bigger than life (through her scientific inquiry, discovery, and accomplishments) and just a regular figure of life (a member of the human race who experiences relationships, emotions, and ups and downs). The format of the hybrid memoir allows Montero to reach across a century to bring Marie Curie to life and enter into dialogue with her, which foregrounds both Curie's greatness and her great humanity. The author's use of Curie as intertext of science, gender-role struggles, and expression of loss combines with a collage-like display of hashtags, diary entries, quotes from other fiction and non-fiction authors, and photographs. The overall effect is an extremely engaging braided tale that models the incorporation of the arts in Science/Technology/Engineering/Mathematics (STEM) issues and undoes or reimagines medicalized prescriptions for the mourning process.

NOTES

1. The translations from Spanish to English are mine. See Rosa Montero's author webpage to see the many awards she has won over her forty-five-year journalistic and literary career.
2. My article "Lab-Lit: Ciencia y tecnología feminista-humanistas en la novela española actual" examines through a feminist lens Montero's use of scientific tropes in *Instrucciones para salvar el mundo*.
3. See also Maryanne Leone's chapter in this volume on the trajectory of science in Rosa Montero's literary works.
4. This is especially true in the context of Darwin's theories from *The Origin of Species*, which was translated to Spanish and published in Spain during the nineteenth century and was met with great consternation for its theory of evolution and the inherent conflict it presented with Catholic precepts.
5. Randall writes, "Dr. Rubin's insight was revolutionary, and she received other awards in her career; in 1993, President Bill Clinton gave her the National Medal of Science. The elephant in the room is gender. Dr. Rubin was not alone in having been overlooked for the Nobel. Every major discovery in the Standard Model of particle physics, perhaps the crowning achievement of 20th-century physics, was awarded a Nobel, except one. Chien-Shiung Wu, who showed that physical laws distinguish between left and right, was overlooked, even though two of her male colleagues won for developing the theory behind her work and an even more subtle follow-up symmetry violation later won the prize."

6. García de Cortázar y Nebreda et.al. say it this way: "La ciencia, y aquí se demuestra, es un campo social más atravesado por la dominación masculina, que se impone y soporta, como consecuencia de la violencia simbólica, violencia amortiguada, insensible e invisible" (37; Science, and you can see it here, is yet another social field influenced by masculine dominance, which is imposed upon us and tolerated as a consequence of symbolic violence, a cushioned, insensitive, and invisible violence).
7. In an interview with Carmen de Eusebio, Rosa Montero also says the following: "Siempre digo que una no escoge las historias que cuenta sino que las historias te escogen a ti. La formulación y el 'sonido' del libro surgido de golpe en el mismo momento en que leí el pequeñísimo diario de duelo de Marie Curie (28 páginas que la científica escribió durante el año siguiente a la muerte de su marido). Supe en ese mismo instante que quería hablar de la muerte de los seres queridos, pero sobre todo de la vida, de la bella vida, de la vida plena que puede vivirse cuando aprendes a sobrellevar la idea de la muerte. Y también supe que quería que el libro tuviera esa estructura libre, fluctuante, sinuosa, que fuera una especie de murmullo sosegado, una conversación íntima con un amigo" (108; I always say that one doesn't choose the stories she tells but rather is chosen by them. The formulation and "sound" of the book came together at the same time that I read the very short diary of mourning of Marie Curie [twenty-eight pages the scientist wrote over the year following the death of her husband]. I knew in that very instant that I wanted to talk about the death of loved ones, but especially about life, this beautiful life, of a full life that you can live when you learn to endure the idea of death. And I also realized that I wanted the book to demonstrate a free, flowing, sinuous structure, that it be a type of quiet murmur, an intimate conversation with a friend).

WORKS CITED

Aladro Vico, E., et al. "La presencia y representación de la mujer científica en la prensa española." *Revista Latina de Comunicación Social*, vol. 69, 2014, pp. 176–94, www.revistalatinacs.org/069/paper/1007_UCM2/10g.html.

Bauer, Martin W., and Susan Howard. *The Culture of Science in Modern Spain: An Analysis of Public Attitudes across Time, Age Cohorts and Regions*. Fundación BBVA, 2013.

Cibreiro, Estrella. "Entrevistas a María Reimóndez, Rosa Montero y Julia Otxoa: El arte de la escritura y el activismo." *Romance Studies*, vol. 34, no.1, Jan. 2016, pp. 43–63.

Consejo Superior de Investigaciones Científicas (CSIC). Mujeres y ciencia, www.csic.es/mujeres-y-ciencia. Accessed 4 Oct. 2016.

Eusebio, Carmen de. "Rosa Montero: 'La literatura consiste en dar vueltas en torno a un centro de silencio.'" *Cuadernos hispanoamericanos*, vol. 762, Dec. 2013, pp. 107–13.

FECYT (Spanish Foundation for Science and Technology). "Women in Spain still work double shifts." *Phys.org*, 26 Oct. 2010, phys.org/news/2010-10-women-spainshifts.html. Accessed 4 Oct. 2016.

García de Cortázar y Nebreda, María Luisa, et al. *Mujeres y hombres en la ciencia española. Una investigación empírica*. Ministerio de Trabajo y Asuntos Culturales/Instituto de la Mujer, 2006.

Hochschild, Arlie, and Anne Machung. *The Second Shift*. Penguin, 2003.

IAP (The Global Network of Science Academies). "Royal Academy of Exact, Physical and Natural Sciences of Spain." Instituto de la Mujer y para la Igualdad de Oportunidades, www.inmujer.gob.es. Accessed 4 Oct. 2016.

Le-May Sheffield, Suzanne. *Women and Science: Social Impact and Interaction*. Rutgers UP, 2006.

Ley Orgánica 3/2007 de 22 de marzo. *Boletín Oficial de Estado (BOE)*, www.boe.es/buscar/pdf/2007/BOE-A-2007-6115-consolidado.pdf. Accessed 9 Dec. 2016.

Libreros, Lucy Lorena. "'La escritura nos salva la vida,' dice la escritora Rosa Montero." *El País*, 26 Jan. 2014, www.elpais.com.co/elpais/cultura/noticias/escritura-nos-salvavida-dice-escritora-rosa-montero. Accessed 4 Oct. 2016.

Libro Blanco. Situación de las Mujeres en la Ciencia Española. Unidad de Mujeres y Ciencia, 2011.

López-Sancho, M. Pilar, et al. "Status of Women in Physics in Spain." *The 3rd IUPAP International Conference on Women in Physics,* edited by B. K. Hartline, K. R. Horton, and C. M. Kaicher, American Institute of Physics, 2009, pp. 171–72.

Martín, Verónica. "Rosa Montero: Hay más magia en la ciencia que en los cuentos de hadas." *Diario de Avisos*, 25 June 2016, diariodeavisos.elespanol.com/2016/06/mas-magia-la-ciencia-loscuentos-hadas. Accessed 4 Oct. 2016.

Mayock, Ellen. "Lab-Lit: Ciencia y tecnología feminista-humanistas en la novela española actual." *Hispanófila*, vol. 174, June 2015, pp. 187–97.

Montero, Rosa. *Instrucciones para salvar el mundo*. Santillana/Punto de lectura, 2009.

———. *Lágrimas en la lluvia*. Seix-Barral, 2011.

———. *La ridícula idea de no volver a verte*. Seix-Barral, 2013.

Nelson, Bradley. "Knowledge (scientia), Fiction, and the Other in Cervantes's *La Gitanilla*." *Romance Quarterly*, vol. 61, no. 2, 2014, pp. 125–37.

Randall, Lisa. "Why Vera Rubin Deserved a Nobel." *New York Times*, 4 Jan. 2017, www.nytimes.com/2017/01/04/opinion/why-vera-rubin-deserved-a-nobel.html.

Rosa Montero Official Website. rosamontero.es. Accessed 8 Dec.16.

"Rosa Montero reivindica el papel de las mujeres en la ciencia." *El Ideal Gallego*, 23 June 2015, www.elidealgallego.com/articulo/coruna/rosa-montero-reivindica-papel-mujeresciencia/20150622233502246649.html. Accessed 4 Oct. 2016.

Shetterly, Margot Lee. *Hidden Figures: The American Dream and the Untold Story of the Black Women Mathematicians Who Helped Win the Space Race*. William Morrow, 2016.

Sobel, Dava. *The Glass Universe: How the Ladies of the Harvard Observatory Took the Measure of the Stars*. Viking, 2016.

CHAPTER 4

From *la santidad de la escoba* to *la trinidad higiénica*[1]

Rosario de Acuña (1851–1923) and a More Inclusive Vision of Spain's Public Health

ERIKA M. SUTHERLAND

In 1887, Guillermina Pacheco set out from the upper-class bastions of Madrid into the alleys and patios of the city's humbler neighborhoods, preaching "la santidad de la escoba, del agua y el jabón" (Pérez Galdós *Fortunata y Jacinta* 1: 592; the holiness of the broom, soap, and water"). The fictional champion of public health, a creation of Benito Pérez Galdós (1843–1920), draws nearly everyone in the novel *Fortunata y Jacinta* to her campaign to purify the city's streets and souls. The community treats Guillermina "con respeto, casi con veneración" (1: 263; with respect, almost veneration). Even Mauricia la Dura, as hard as her nickname suggests, stands in awe, confiding that, "esa doña Guillermina . . . la habrás oído nombrar . . . me cogió por su cuenta y me trajo a este *establecimiento*. La doña Guillermina es una que se ha echado mismamente a pobre, ¿sabes?, y pide limosna y está haciendo un palacio ahí abajo para *los huérfanos*" (1: 609, emphasis in original; that doña Guillermina . . . surely you've heard of her . . . she just went and grabbed me on her own and brought me to this *establishment*. Doña Guillermina is someone who chose to be poor, you know? She begs alms and is building a huge palace down there for *the orphans*). In the character of Guillermina Pacheco, the conflation of cleanliness and godliness is personified, extending women's role as *ángel del hogar* (domestic angel) into the public (and public health) sphere. Benito Pérez Galdós drew his inspiration for Guillermina from a real-life modern saint: Ernestina Manuel de Villena (1830–1886), a noblewoman who abandoned the comforts of her class to create Madrid's Corazón de Jesús orphanage and who, "en concepto público, murió en olor de santidad" ("En honor" 362; according to popular opinion, died in odor of sanctity).

While the discourse of public health in turn-of-the-century Spain presented women as *ángeles del hogar* and focused on the saintliness of hygiene and the benevolence of the upper classes, other voices, such as that of Rosario de Acuña (1851–1923), challenged the norms of literature and the press. For example, the same year that Galdós introduced Spanish readers to *la santa fundadora* (the saintly founder) Guillermina Pacheco, Rosario de Acuña took to the pages of *Las Dominicales del Libre Pensamiento* to lay out what she saw as essential elements for health: "Aire puro, luz directa, espacio anchuroso" ("La ramera" 1; Clean air, direct light, wide open spaces). Acuña expands this simple formula to include elements that make for a life that is not just healthy but well lived: "[s]obriedad, sencillez de alimentos, ejercicio general, de todos los músculos, estudio, meditación, trabajo, aspiraciones a lograr estimación sólida, aprecio inacabable, bienestar continuo, esperanza en lo inmortal por nuestras obras, nuestros hijos, o nuestras creencias" (1; sobriety, simple foods, general exercise of every muscle, study, meditation, work, aspirations of earning a solid reputation, unending appreciation, continuous wellbeing, and the hope of immortality through our works, our children, or our beliefs). Like the fictional Guillermina, whose brooms and mops were destined to Madrid's poorest neighborhoods, the real-life Acuña criticized the poor living conditions that led to disease and degradation. However, unlike Guillermina, Acuña offered the masses practical assistance in order to address scientific realities related to hygiene and public health. Acuña's work, particularly her essays "La ramera" (The whore) and "La higiene en la familia obrera" (Hygiene in the working class family), demonstrate the complex discourse of women's health circulating in nineteenth-century Spain and the ways women like Acuña challenged cultural and class norms in order to provide a more inclusive vision of public health.

While it might appear from the literary example cited from Galdós that the representation and reality of public health were in congruence, the work of Acuña demonstrates a growing awareness in Spain of the complex and problematic aspects of cultural norms related to women, hygiene, and science. In both its domestic (*higiene privada*) and public (*higiene pública*) applications, public health in nineteenth-century Spain was conflated with bourgeois morality, a morality personified and upheld by the domestic angel. As medical treatises, pedagogical manuals, public lectures, press reports, and literature attest, the health of the nation was deemed women's work. Yet while Rosario de Acuña embraced women's role in the promotion of public health, her vision of public health had nothing to do with either charity or sainthood. In life and in her writing, she rejected both official discourses and the notion of women as celestial beings, insisting instead on the social need for, and scientific basis of, public health.

One of Acuña's early essays, focused on prostitution and published in 1887, offers important insights into her unique contributions for challenging accepted discourse related to public health in nineteenth-century Spain. Featured on the full expanse of the newspaper *Las Dominicales*'s first page and part of the second, Acuña's boldly titled essay "La ramera" moves beyond basic notions of cleanliness to challenge gendered and socialized norms. Acuña's subject is prostitution, "un semillero inagotable de males" (an inexhaustible breeding ground of evil) and "gangrene," and the object of her vitriol includes the public health and other authorities who have declared prostitution a "vicio preciso" (necessary vice), a "necesidad de la naturaleza" (natural necessity), or a "mal que evita mayores males" (1; evil that prevents greater evils). Acuña directs her plea not only to public health authorities, but also more broadly to women, arguing that they too need to stand up against prostitutes and their male clients to preserve their families from disease: "¿de esta o de aquella enfermedad? de una o de otra, ¡de todas!" (1; from this disease or that one? from one or the other, from all of them!).

As Christine Arkinstall has noted, while Acuña considers the prostitute an inferior being, she also "rejects the sexual double standard to extend this analogy to the prostitute's male clients" ("Challenging" 25). In so doing, Acuña casts aside moral arguments against the prostitute, expressing her outrage without ambiguity: "¡fuera respetos! El hombre se prostituye tanto, exactamente igual que la ramera" ("La ramera" 1; forget about respect! Man prostitutes himself so much, in the exact same way as the whore). Even worse, as the prostitute spreads disease, so does her client, carrying within him "la ponzoña, con su nidal de Dolores" (poison, with its nest of pain) that proceeds "a roer su sangre y sus huesos, y a arrancarle, una por una, sus prerrogativas todas de ser inteligente" (to eat away at his blood and bones and to pluck, one by one, each and every prerogative of an intelligent being) and leaving behind "su descendencia, raquítica retoñadura dispuesta a ser carne de presidio o de lupanar" (1; his descendants, rachitic offspring destined for prison or the brothel). "La ramera" was not Acuña's first foray into medical matters, nor would it be her last.[2] Yet it is here, in this essay and in later speeches like "La higiene en la familia obrera" that Acuña rejects contemporary expectations and introduces a new approach to the discourse of public health.

Acuña was not the first woman writer to address prostitution as an issue of public health. As early as 1865, Concepción Arenal (1820–1893) had noted that the man who frequents prostitutes "[a]rruina sus fuerzas con los excesos y su salud contrayendo enfermedades repugnantes y dolorosas, que si no le matan, anticipan su vejez" ("Carta XXIV" 307; ruins his strength with excesses and his health by contracting hideous and painful diseases that, if they don't kill him, age him prematurely). According to Arenal, the prostitute puts herself at grave risk:

> arroja su cuerpo al muladar del vicio que le envenena; vende por algunos reales a un hombre repugnante el derecho de trasmitirle una enfermedad asquerosa; y pasa continuamente de los brazos de la lujuria a la cama del hospital, donde a nadie inspira compasión, donde a todos inspira desprecio y asco, donde se le cura para que vuelva a servir como un animal que enferma, y curado, puede ser útil. (*La mujer del porvenir* 284)

> she casts her body to the dunghill of the vice which poisons her; for a handful of coins she will sell a repulsive man the right to transmit a disgusting disease to her; she passes constantly from the arms of lust to the hospital bed, where she inspires compassion in no one, where she inspires disdain and disgust in all, where she is cured so that she can serve again, like an animal that gets sick, and once cured is useful again.

Arenal's bitter description of the hospital here points to her advocacy for the abolition of prostitution, a position she shared with Acuña but few others in Spain.[3] For example, in 1876, Dr. Ángel Pulido (1852–1932) argued that only doctors were qualified to discuss the social and medical impact of prostitution, "pues conocemos algo la fisiología del organismo humano; sus afectos, instintos y pasiones; lo que arrastran sus necesidades y lo que permiten sus virtudes" (115; because we know a bit about the physiology of the human organism; its sympathies, instincts, and passions; what its needs drag in and what its virtues allow). Speaking for the medical establishment, Pulido declares that "desde ahora aseguramos que la prostitución existirá mientras lata con vida la humanidad, del mismo modo que ha existido en todos los tiempos pasados" (115–16; as of now we can assert that prostitution will exist as long as humanity pulses with life, just as it has always existed throughout history). For Pulido, prostitution was "una llaga depuratoria del organismo social, rebelde a toda cicatrización" (116; a cleansing wound of the social organism, resistant to scarring). Even as he acknowledges the existence of yet-incurable diseases spread by venal sex, he rests his argument on religious as well as political authority.

Both Arenal and Acuña challenged the dominant discourse expressed by Pulido, though as we shall see, Acuña advances her challenge in more progressive ways. To understand the radical newness of both women's approach to public health in Spain, it is useful to review the way in which class, morality, health, and women came to be intentionally linked within Spanish discourse. The 1857 Moyano Act, which established obligatory instruction for all children—girls as well as boys—in cities larger than five hundred residents, stipulated that girls' education include the fundamentals of domestic hygiene; not until 1901 would this subject be required for boys (Perdiguero 228). There were several books designed to teach, as part of the full curriculum, the requisite "ligeras nociones de Higiene doméstica" (Ley de Instrucción Pública, Título 1, Artículo 5; basic notions of domestic hygiene).

As one early schoolbook, the *Tratado de economía y labores para uso de las niñas* (Treatise on home economics for girls), explains, "una mujer instruida debe contar, en el número de sus primeros deberes, el poseer unos regulares conocimientos de Higiene" (134; an educated woman should have among her first duties, that of having a basic understanding of hygiene). The book divides its lessons into five chapters—moralidad, orden, inteligencia, economía y aseo (morality, order, intelligence, economy, and cleanliness)—and offers the assurance that "vamos a exponerle prácticamente, sirviéndonos de un estilo fácil, al alcance de las niñas" (v; we will explain this in practical terms, using an easy style, understandable to girls). Laid out in terms that any child would understand, the basic requirements for health were described as straightforward: "El aire, la luz y el calor son indispensables para la conservación de la salud; y así en estas tres cosas deben ponerse nuestros continuos cuidados" (134; Air, light, and heat are indispensable for the conservation of health; and so we should focus our attention always on these three things).

The chapter on cleanliness, the shortest in the 175-page manual, consists of two and a half pages on controlling household pests and three and a half pages on personal hygiene. The advice is basic and includes the instruction that the well-trained girl "procurará que la muda de ropa interior se haga con frecuencia, sin dar lugar a que pasen más de ocho días sin verificarlo" (166; should try to change underwear frequently, not allowing more than eight days to pass without doing so). The brevity of this chapter reflects the secondary status granted to cleanliness, a status that would continue even after 1901, when hygiene was incorporated into the required elementary curriculum for boys as well as girls. In 1906, Rogelio Francés y Gutiérrez introduced his elementary *Fisiología e Higiene* (Physiology and hygiene) textbook with the lament that because these topics were considered to be "las más áridas e indigestas para las tiernas inteligencia infantiles" (the driest and least digestible for children's tender brains), as a school subject "se continúa relegándola, si no al olvido, a lo menos a un lugar muy secundario entre el resto de las enseñanzas" (cited in Perdiguero 241; it was still relegated, if not to oblivion, then at least to a very secondary place among the other courses).

The most widely used textbook was Pilar Pascual de Sanjuán's *Flora o La educación de una niña* (Flora, or A girl's education), a complete primer for girls that was first published in 1881 and reedited for more than seventy years after that (Sánchez García and Martínez Rus 41). *Flora* imparts its lessons using a story format, with two lessons out of the fifty-three addressing issues of health and hygiene. This textbook lays out the components that would define much of the hygienist discourse in the years to follow. In lesson 2–33, "Higiene doméstica" (Domestic hygiene), young Flora visits the working-class home of Paquita with her mother and grandmother. The house has low ceilings and little light or ventilation and is located down "una callejuela estrecha, húmeda y lóbrega" (272; a narrow, damp, and gloomy little alley). These conditions spur Flora's grandmother to share her wisdom on healthy living, underscoring as she does so the moral value of her suggestions:

¡Cuántas familias arrastran una vida desgraciada, privadas del precioso don de la salud, comunican a sus inocentes hijos una vida raquítica y enfermiza, y todo por no atender a los consejos de la higiene!

—Yo no sé lo que es higiene, dijo Paquita con desenfado.

—Poco importaría que no supiese V. el nombre, con tal que conociese las reglas de esta interesante parte de la educación de la mujer, que tiene por objeto el atender a la conservación de la salud.

Las principales necesidades de la humanidad, y las condiciones indispensables para no enfermar son:

Aire puro, alimentos sanos y vestidos adecuados a la estación, al clima y a las circunstancias especiales del individuo, amén de una proporción o equilibrio entre el trabajo y el reposo. (273–74)

How many families lead wretched lives, deprived of the precious gift of health, and transmit a rachitic and sickly life to their innocent children, and all because they do not heed the advice of hygiene!

—I don't know what hygiene is, said Paquita nonchalantly.

—It doesn't matter much if you don't know the name, as long as you know the rules of this key part of a woman's education, with the object of tending to the preservation of health.

Humanity's principal needs, the conditions that are indispensable for healthy living, are:

Clean air, healthy foods, and clothing that is appropriate for the season, the climate, and each individual's special circumstances, in addition to proportionality or a balance between work and rest.

A later lesson sees Flora left in charge of the household while her mother and grandmother are away. The young woman has assimilated her education well, as she shows herself to be an excellent housewife and a budding hygienist: she washes the linens once a week, "porque ni es bueno en las habitaciones de las ciudades (regularmente no muy grandes ni ventiladas) retener un foco de infección cual es la ropa sucia" (326; because in bedrooms in the city [which are generally neither very large nor well ventilated], it is not good to retain a source of infection like dirty clothes). As Josette Borderies-Guereña observes, Flora's education shows how cleanliness acquires the enhanced value of both private and social order (304).

The manuals of domestic hygiene were produced for a bourgeois readership (Sánchez García and Martínez Rus 41). Rosario de Acuña found this limited audience problematic, but even more troubling for Acuña was the manuals' insistence that the different classes comport themselves according to their status. An example of this lesson comes from Flora, when doña Ángela reminds a carpenter's family that "toda persona, y en especial la mujer del pueblo" (everyone, and especially the common woman) must take care "de no salirse de su esfera en cuanto a lo que son los trajes" (276; to not leave their social class in terms of their dress) and exhorts them "que no pierdan nunca de vista la decencia y la sobriedad, tan laudables entre la gente del pueblo" (277; to never lose sight of decency and sobriety, so laudable among the common folk). The *Tratado de economía y labores para uso de las niñas* also concludes with an impassioned plea for social order, urging that, "toda mujer, cualquiera que fuere su clase o posición debe contentarse con su estado, y en él atemperarse a las circunstancias; pues el pretender salir de su esfera sin consultar con su haber, es locura que hunde en el abismo a toda persona imprudente y ambiciosa" (171; every woman, no matter what her class or position may be, should be content with her condition and should adjust to those circumstances; trying to leave her condition without consulting her means is a madness that sinks anyone so imprudent and ambitious).

The importance of class here is not a mere question of social standing. It implies obligations tied to moral codes. Thus, when thanked for the advice she had given, Flora's grandmother replies: "Amigo mío, el enseñar al que no sabe es una obra de misericordia" (276; My friend, teaching those who do not know is an act of charity). For her, promoting public health is a moral obligation conferred to her by her elevated status. Her attitude was considered exemplary, as Arenal's 1860 essay on charity illustrates. There, Arenal praises the wealthy women who devote themselves to helping the less fortunate, who rescue:

> millares de niños abandonados por los autores de sus días, consuelan a los pobres enfermos, reúnen fondos para distribuirlos entre los necesitados, establecen colegios donde alimentan y enseñan a los niños pobres, talleres, escuelas, donde a veces sirven ellas mismas de maestras. La gran señora no desdeña llegar hasta la miserable hija del pueblo para instruirla en los principios de la religión y en las reglas de la instrucción elemental; desciende más, y bajando a esa repugnante cloaca moral que se llama prostitución, procura arrancarle y le arranca numerosas víctimas. (*La beneficencia* [Charity] 29)

thousands of children abandoned by their creators, console the ill, raise funds to distribute among the needy, establish schools where they feed and teach poor children and workshops and schools where they themselves sometimes teach. The grand lady does not scorn reaching out to the miserable daughters of the people to teach her the principles of religion and the rules of basic education; she descends even further, lowering herself to the repulsive moral sewer called prostitution, attempting to pull and pulling out numerous victims.

Arenal's praise of Spain's illustrious ladies here is not surprising. She herself served in a number of official capacities, including inspector of prisons, and cultivated relationships with powerful people and the institutions they founded. For example, she was a supporter of Dr. Manuel Tolosa LaTour's seaside camps for sick children and the Centro Protector de la Mujer in Alcira, a shelter founded by the "aristocracia de la caridad Española" (Lacalzada de Mateo 339; aristocracy of Spanish charity) and run by a priest and a woman director, "auxiliada por varias jóvenes de familias distinguidas" (Arenal, "Centro" 418; with the assistance of young women from distinguished families). At the Centro Protector, at-risk women received "auxilios, apoyo y dirección" (assistance, support, and guidance), slept "sin peligro para su honestidad o para su reputación" (without risk to their virtue or reputation), received practical training or even a formal education, and were treated with dignity (416). Arenal notes with particular satisfaction the cleanliness of the center's facilities, "hay orden grande, esmero exquisito y, aunque pobres, aquella elegancia que proviene de la limpieza y del buen gusto; las paredes no están adornadas con cuadros, pero sí con máximas de los libros santos y de autores que han pensado santamente" (416; there is great order, exquisite care, and, though humble, that elegance that comes from cleanliness and good taste; the walls are decorated not with pictures but rather maxims from holy books and authors of holy thought). Arenal's expressions of concern are filtered through the lens of her own privilege and Catholic morality, clearly distinguishing her advocacy from that of Acuña, who rejected any connection with official organs of charity and instead would affirm her "desprecio completo y profundo del dogma infantil y sanguinario, visible e irracional, cruel y ridículo que sirve de mayor rémora para la racionalización de la especia humana" ("Testamento" 272; complete and profound disdain for the childish and bloody, visible and irrational, cruel and ridiculous dogma that serves as the greatest obstacle to the rationalization of the human race).

The bourgeois bias within Spanish discourse on public health can be seen not only in discussions of hygiene but also within the context of education, particularly regarding access for women and girls. In Spain's late nineteenth century, education was neither guaranteed nor consistent. Despite the significant corpus of curricula and textbooks, and although the Moyano Act promised a primary education for girls, it is unclear how many girls actually received a formal education. Irene Palacios Lis points to the "más que precaria realidad escolar" (schools' more than precarious reality) of the period that was marked by an "excesivo número de alumnas por maestra, falta o escasez de materiales de todo tipo, locales inadecuados cuando no abiertamente atentatorios contra la salud infantil, absentismo o ignorancia, incluso, de las propias maestras" (135; excessive number of female students per teacher; a lack or shortage of every type of supply; locations that were inadequate if not openly detrimental to children's health; absenteeism or even ignorance of the teachers themselves). As late as 1930, only slightly more than half of Spanish girls attended school, with the result that illiteracy rates among Spanish women were quite high: some 90 percent in 1860, dropping to 71.5 percent in 1900 and 47.5 percent in 1930 (Perdiguero 229).

Despite the ongoing challenges of public education, girls and women did seek opportunities to learn. Women began to enter Spanish universities in the 1870s, with a majority choosing medicine as their field of study (Ortiz 528). Hygiene was a required topic for medical education in Spain and figured prominently in the dissertations of the first women to graduate as medical doctors in Spain. In contrast, and although Arenal pointed to the United States as a model for the education of women (*La mujer del porvenir* 36; The woman of the future), the first women doctors in the United States were cautioned against working to promote "proper hygienic influences" among "the denizens of filthy localities" because "this is not the chief end and aim of educating women medically" (cited in Wells 50). Indeed, some female American doctors avoided any association at all with the discourse of hygiene, referring to "prevention, hygiene, and temperance" as "'despised things'" (Wells 64). In her 1882 dissertation, the Spanish physician Dolores Aleu y Riera (1857–1913) took a different approach and focused instead on the need for women's education as a means of improving maternal and infant health. For Aleu, the basic notions of hygiene stipulated in the Moyano Act were insufficient: she called for incorporating physiology, medical botany, and household medicine (39). She advocated for ongoing education, beyond the classroom and throughout adulthood: "Se necesita la instrucción entre las mujeres, porque por ella mejora la higiene: los vestidos, los cosméticos, el aire, los alimentos, las bebidas, y en general, todos los agentes de la higiene, serían empleados con conciencia científica y se evitarían la mayor parte de las enfermedades que hoy día afligen al sexo y a la prole, si las madres fuesen mejor instruidas" (45; What's needed is women's education, because it is the woman who improves hygiene: clothing, cosmetics, air, food, drink, and, in general, all of the components of hygiene; these would be used with a scientific awareness and the greater part of the diseases that today afflict women and children would be avoided, if mothers were better educated).

Martina Castells Ballespí (1852–1884), who also completed her doctorate in 1882, wrote her dissertation on maternal health as well. She pressed even more forcefully for better education, enumerating a long list of subjects that could benefit women and especially mothers and affirming that "creemos firmemente que la mortalidad de los niños, está en razón inversa del grado de ilustración de la madre" (273; we firmly believe that infant mortality is inversely related to the level of the mother's education). Of all the components of a woman's education, she holds hygiene above the others, asking, "¿Será menester insistir más todavía para comprender que la Higiene es uno de los puntos principales que debe abrazar la educación de la mujer? Creo que no" (274; Is it really necessary to keep insisting so that people understand that Hygiene is one of the main points that women's education should include? I don't think so). Dr. Castells Ballespí offers a plea to women: "recordemos todos que la mujer depende y dependerá siempre de otra mujer; por lo tanto, edúquese la primera" (278; let's remember that a woman depends now and will always depend on another woman; for this reason, she should be educated). However, this call for feminine solidarity in addition to education would not be shared by all of the new women doctors.

Manuela Solís Claras (1862–1910), who earned her undergraduate degree in medicine in 1889 and her doctorate in 1905, also centered her work on the health of mothers and children. Her *Higiene del embarazo y de la primera infancia* (1907; Hygiene for pregnancy and early childhood) is a manual designed for home reference, written for women facing the countless questions posed by married life and motherhood. The book was conceived to "vulgarizar aquellos conocimientos elementales que deben servir de norma a la mujer para la conservación de su salud y alivio de sus males durante los períodos más interesantes de su existencia y para la del nuevo ser a que ha de dar origen" (Solís 1; popularize those basics that should serve to guide a woman to preserve her health, alleviate her ills during the most interesting periods of her life, and for the health of the children she will bring forth). The celebrated doctor Santiago Ramón y Cajal (1852–1934) wrote the introduction to the book, lending a note of unquestionable authority and praising Solís as a writer who is "discreta y conocedora de su público" (discreet and knows her public well) and is able to express "sus consejos en un lenguaje llano, correcto, exento de tecnicismos y pedanterías" (ix; her advice in straightforward and correct language, free of technicalities and pedantry). However, despite its claim of mass appeal, the book is clearly directed to an elite female readership. This is evident in the unchallenged assertion that "[e]xiste ya una gran parte de la sociedad entre las clases acomodadas en que se cumplen con todo rigor las prácticas de la higiene relativas a la limpieza de los niños" (in terms of the cleanliness of children, a large part of the privileged classes follows hygienic practices scrupulously) and in the blame Solís places on the struggling and rural classes for their poor health: "La falta de aseo es además origen de no pocas enfermedades y la prueba de ello es que algunas de éstas son patrimonio casi exclusivo de esa clase desgraciada de la sociedad que vive en medio del abandono y de la miseria. [. . .] Esta falta de limpieza alcanza su máximum en las gentes del campo" (Solís 280–81; The lack of cleanliness is also the source of more than a few diseases, and proof of this is that some diseases belong almost exclusively to that misfortunate class of society that lives in neglect and misery. [. . .] This lack of cleanliness reaches its maximum among the country folk). Solís is not unique in pointing to the dirty living conditions of the poor as a basis for their ill health, but while one might expect the author to assess how those conditions came to be, Solís instead simply retains traditional class-bound prejudices.

These few and exceptional women who earned university degrees in nineteenth-century Spain faced challenges unimaginable today. As late as 1911, the university could be an openly hostile place for women, as Rosario de Acuña decried in her notorious "La jarca de la Universidad" (The university rabble). Following an ugly incident in which a group of male students attacked a group of women students at Madrid's Universidad Central, an attack stopped by a passing cart driver, Acuña vented her fury and disdain for the bourgeois society—and, in very specific terms, the bourgeois men—that opposed women's education. The article appeared first in Paris, but soon was widely reprinted across Spain, inspiring so much anger that Acuña was forced into a brief exile in Portugal.

While a university education was still limited to only the most talented, confident, and assertive women students, literate women had broad access to books and magazines. Public lectures offered even more women avenues for both learning and sharing their knowledge. These too could be fraught, as an 1884 review shows. Although Rosario de Acuña, the first woman ever invited to present her work at Madrid's Ateneo, was welcomed by a large and enthusiastic audience, the reviewer from *El Salón de la Moda* concluded that the reading "ha tenido poco éxito" (was not very successful). The explanation for this judgement speaks to the intellectual woman's difficult place: "La señora Acuña es para los hombres una literata y para las mujeres una librepensadora, y no inspira, entre unos y otras, simpatías" (cited in Bolado 111; Men see Acuña as a bluestocking, and for women she is a free thinker; she inspires sympathy in neither one nor the other). As we shall see, Acuña would be ever less concerned about the sympathies she inspired among the elites attending this sort of event; her increasing interest in speaking to a broader public would take on a greater role.

Public lectures, even on issues of medicine and domestic hygiene, tended to attract audiences from the upper classes. For example, in 1869 the noted physician Dr. Santiago Casas spoke on women's health as part of the Universidad de Madrid's Sunday conference series on women's education. The fifteen-part lecture series was directed at a female audience, and that group, as an engraving from *El Panorama* shows, was clearly drawn from the upper ranks of society.[4] Dr. Casas's lecture on domestic hygiene provides additional evidence of his elite audience, as the doctor refers to celebrated dermatologists such as Gíbert, Bazin, Hardy, and Cazenave (23). His bid to connect with his privileged audience continues as he names specific products available to women shopping in urban pharmacies and perfumeries, including *el agua de Botot* (22; Botot's water, a distillation of anise, clove, cinnamon, and mint used as a mouthwash), Pelletier's *elixir odontálgico* (22; odontalgic elixir, a toothpaste), Violette's *jabón de Thridace* (27; Thridace soap, a scented lettuce-based soap), *blanco de Thénard* (28; Thénard's or French whitener, a blend of zinc oxide and white clay), and *el rojo líquido* (28; liquid rouge), patented by Sofía Goubet.

Beyond these recommendations, the prospective reader's elevated socioeconomic status is reflected in advice that simply does not make sense in the context of a working person's life. Dr. Casas devotes three full pages to an explanation of why it is important to get up early—ideally by eight a.m. but no later than nine a.m. in the winter and no later than seven a.m. in the summer (15–17)—and to take a post-breakfast constitutional (17). He recommends washing the face and hands daily with cold water, habits that "no solo conservan la salud y preservan de varias enfermedades, sino que son quizás el mejor, el único medio verdadero de dar a las carnes una firmeza, y a la piel una frescura, haciendo desaparecer las arrugas anticipadas" (18; not only preserve health and prevent various diseases, but they are also the best, perhaps the only real means of imparting firmness to the body and freshness to the skin, making premature wrinkles disappear). While cleanliness would have unquestionably been a good practice for everyone, concerns about wrinkles and youthful skin would have been a luxury limited to the leisure classes. For Dr. Casas, "ni la Higiene, ni la Medicina tienen nada absolutamente de misterioso; sus principios fundamentales son tan sencillos, tan claros, tan accesibles a todas las inteligencias, como los de todas las demás ciencias y artes" (8; neither Hygiene nor Medicine are in the least bit mysterious; their fundamental principles are as simple, as clear, as accessible to any mind as every other science and art). He undercuts this claim by adding an essential caveat, "solo en retener las leyes de detalle, y en saber aplicar los principios generales a cada caso particular, es donde reside la dificultad, cuyo vencimiento exige un estudio y una práctica especiales" (8; the only difficulty lies in retaining the details of these laws and in knowing how to apply the general principles in each specific case; specialized study and practice are required to overcome this difficulty). In effect, medical doctors were the ones with access to this knowledge.

With the important exception of Spain's few women doctors, how was a woman to know when common sense was an insufficient guide to good health? How were these best practices to be learned? Arenal acknowledged this dilemma in her *La mujer del porvenir*, when she asked, "¿cómo puede aprender higiene la mujer sin tener algunas nociones de fisiología, ni tener conocimiento de la fisiología careciendo de todos los auxiliaries" (186; how can a woman learn the principles of hygiene if she has no notion of physiology? How could she have any knowledge of physiology if she lacks all of the underlying subjects?).

While the dominant discourse of hygiene remained firmly associated with the bourgeois establishment, functioning "como si fuera un baluarte doble, primero contra el índole impulsivo e inquietante de la mujer, segundo contra un posible desorden social, tan temido por las clases dominantes" (as if it were a double bulwark, the first against women's impulsive and disquieting nature, the second against any possible social disorder, so feared by the ruling classes), a second model was emerging in which the working classes were taking a more active role in the discourse and practice of public health (Borderies-Guereña 304). The Spanish government formed the Comisión de Reformas Sociales in 1883 to study the current issues facing the working classes and to explore means of improving their health and well-being (Campos 502). Early on, the Partido Socialista Obrero Español (PSOE; Spanish Workers' Socialist Party) embraced the issue of workers' health and assumed an active role in the work of the Comisión, focusing "no sólo en las insalubres condiciones de trabajo y de vida sino en la propia condición de trabajador como causa de la enfermedad (Campos 506; not only on the unhealthy working and living conditions but also in the very condition as worker as a cause of disease). In rejecting the idea that workers themselves were solely responsible for their problems, they called into question the previously unassailable, challenging "la visión unilateral y hegemónica de la medicina en cuestiones de salud e higiene, generando un contradiscurso en el que la salud adquiría un alto contenido político y reivindicativo" (medicine's unilateral and hegemonic view on issues of health and hygiene, generating a counter discourse in which health took on a heightened political and vindictive content) in which the "supuesta neutralidad del higienismo y de la salud como valor supremo daba paso a una impugnación del sistema socioeconómico vigente" (Campos 506–07; supposed neutrality of the hygienic movement and health as a supreme value gave way to a criticism of the current socioeconomic system).

The editors of *El Socialista* (The Socialist) declared in 1893, "[l]os pobres no pueden tener salud; se lo prohíbe la sociedad, aunque se diga lo contrario" (cited in Campos 508; the poor cannot be healthy; society prohibits it, though it may claim otherwise). Without a living wage, workers and their families were unable to purchase sufficient and adequately nutritious food. The workers' living conditions posed an even greater risk to their health:

> El aire, la luz, el calor, el terreno, la localidad, el clima y la habitación, son una serie de cosas que influyen directamente en la salud, pero a cuyos efectos perniciosos no pueden sustraerse los proletarios. Tienen que adoptar, aunque les sea perjudicial, el clima de la localidad donde encuentran limitado sustento y tienen —a la fuerza y pagando elevados alquileres— que aceptar las habitaciones que proporcionan los propietarios, aunque reúnan malas condiciones higiénicas. (508)

> Air, light, heat, land, location, climate, and living spaces are a series of things that have a direct influence on health, but the proletariat cannot escape their pernicious effects. They need to adopt, despite the harm it does them, the climate where they are able to find their limited sustenance, and they must—by force and paying elevated rents—accept the rooms that landlords give them, even though they may have unhealthy conditions.

While Galdós, Arenal, and Acuña had identified dry, well-ventilated, and light-filled spaces as the basis of healthy living, now the systems by which these conditions were made accessible or not were targeted. Landlords, employers, municipal authorities, all of the components of bourgeois hegemony were called to task. Acuña would distinguish herself from other writers in her continued insistence on the central importance of light, air, and water combined with an equally strong conviction that this precept needed to emerge from the working classes themselves, with working-class women leading the way.

Indeed, to combat the abuses of the workers—and as a means of organizing their base—labor activists organized lectures in local Centros Obreros (workers' centers) and Casas del Pueblo (houses of the people), affiliated with PSOE and the Unión General de Trabajadores (UGT; general union of workers) across Spain. Issues surrounding workers' health were a particular focus of concern. One example of this is the 1899 "Curso de higiene vulgar" (Course in popular hygiene) offered by Dr. Felipe Ovilo in the meeting rooms of *El Liberal* in downtown Madrid. While pointedly calling out the areas where public authorities and bosses were unable or unwilling to provide adequate guarantees of public health, Ovilo also showed where and how the workers themselves could take responsibility for individual hygiene. This called for a new, more engaging approach to health education. Thus, where earlier advocates of fresh air addressed the issue in general terms, Ovilo addressed the needs of his working-class audience more directly. He stressed the importance of rest and recuperation between shifts, expressing concern about the workers' poor living conditions in homes that "carecen de las condiciones indispensables para que durante el descanso vuestro organismo compense las pérdidas que habéis experimentado en la jornada" ("Curso [2]" 4; lack the conditions needed so that while resting, your organism compensates for the losses you have suffered during the workday). In very specific terms, Ovilo explains the problem:

> El hombre necesita para respirar en condiciones normales un espacio de diez metros cúbicos, ¿en qué vivienda obrera dispone de ellos cada individuo? En el trabajo se consumen más de los 500 litros de oxígeno.
>
> La falta de oxígeno produce el empobrecimiento de la sangre por falta de glóbulos rojos, y como consecuencia fatal la anemia, la debilidad orgánica que predispone a todo género de enfermedades. (qtd. in "Curso [2]" 4)

> Under normal conditions, man needs ten cubic meters of space to breathe; what working-class home has that much space for each individual? At work more than five hundred liters of oxygen are consumed.
>
> The lack of oxygen weakens the blood, producing a lack of red blood cells and its fatal consequence, anemia, the organic weakness that predisposes the body for all kinds of disease.

Ovilo offers suggestions that workers themselves can implement to improve air quality, inviting workers to create small openings in exterior walls, as well as above and beneath doors. He notes a brand-name ventilation system available for factories, but adds that any simple tube, "cuyo coste en Madrid es de 22 céntimos de peseta" ("Curso [3]" 4; whose cost in Madrid is twenty-two cents), can do much to improve airflow. Above all, he reminds workers that they should use their own common sense to seek solutions, "por muy cerradas y ajustadas que estén las puertas y ventanas de un local, vosotros sabéis que hay muy poco materiales de construcción por completo impermeables" ("Curso [3]" 4; no matter how closed or tight-fitting the doors and windows of a place are, you know that there are very few construction materials that are completely impermeable). It is hard to know how many workers followed Ovilo's suggestion that they open holes in the buildings where they worked, and it is easy to question the seriousness with which they were made.

Rosario de Acuña had little patience for advice that she perceived as exclusionary, elitist, or unworkable. When invited to speak at Santander's Centro Obrero in April 1902, she would draw on the previous quarter century of hygienist discourse and workers' movements to create a dynamic and radical new message that would liberate knowledge from the university, where "el privilegio burgués la encerró" ("La higiene" 740; bourgeois privilege locked it away). In what is perhaps her signature work, "La higiene en la familia obrera," she raises up her working-class audience and eviscerates the entitled attitudes found in many of the lectures, dissertations, and books circulating widely at the time and described at length above.

She opens with a disavowal of the notion that health and hygiene are the exclusive province of any intellectual elite, asserting instead that with or without a formal education, the ideas she is about to express will be accessible to all. She shares that her own education was informal, with early childhood lessons from her father, and later the lessons came from books and her own life experiences. She brushes aside any idea of scholarly credentials, affirming that "mis títulos de sabiduría radican sólo en mi inteligencia, en mi voluntad y en mi ternura" (749; my claims to wisdom are based solely on my intelligence, my will, and my tenderness). She does not claim to be of the working class, a claim that, given her elevated profile, would surely ring hollow, but instead positions herself as a go-between among the classes: "He aquí por qué me encuentro en mi terreno, toda vez que yo que soy una *vulgaridad* para los universitarios, resulto una *sabia* para vosotros, y útil, por lo tanto, en mi papel de intermediaria" (749; This is why I find myself at home here, as while I am a *vulgar creature* for university folk, it turns out I am a *wise woman* for you, and therefore useful in my role as intermediary).

From the start, she identifies her target audience as working-class women, underscoring the intentionality of this choice by repeating it: "[D]edico mi conferencia exclusivamente a las obreras, [. . .] dedico esta conferencia a las obreras, porque la higiene de la familia, lo mismo en los palacios que en las chozas, debe estar a cargo de la mujer" ("La higiene" 750; I dedicate my conference exclusively to working women, [. . .] I dedicate this conference to working women because a family's health, be it in a palace or in a hut, should be the responsibility of the woman). This dedication conveys the respect Acuña has for women laborers; by placing them on an equal footing with all other women, she is explicitly rejecting earlier suggestions, as we saw in *Flora*, that women should not seek to reach beyond the bounds of their social standing. The essential role women play is, for Acuña, tied to their nature as women: "Tengo absoluta fe en el destino superior de la mujer, que, lo mismo entre el pueblo que entre la burguesía, reúne, a mi entender, más condiciones perceptibles de mejoramiento que el hombre, tanto porque los sentimientos de la mujer son más generosos, como por estar menos entregada a los vicios, sobre todo al alcoholismo" (750; I have absolute faith in the superior destiny of woman who, the same in the popular classes as among the bourgeoisie, has, as I see it, greater potential for betterment than man, as much because women's feelings are more generous as because she is less given to vice, especially to alcoholism). This approach, once again, elides the boundaries of social class and creates a sense of shared challenges and possibility. For example, alcohol is not presented as a poor man's vice but rather a problem that could affect any family.

Acuña's focus on women here, as in other texts, also reflects her conviction that women are best able to effect change from within the family structure rather than in the workplace. This places, as she acknowledges, a special burden on the woman of the working class,

> que se siente empujada por la necesidad a salir del hogar para traer a él algunas monedas más que unir a los escasos jornales del hombre. [...] ¿[Q]ué valen algunos míseros céntimos, logrados en jornada de trabajo siempre cruel, con los miles de pesetas que representan las vidas infantiles y púberes, abandonas a la degeneración, a la enfermedad, al vicio y a la muerte que se ceban en ellas al encontrarlas indefensas de su natural protector, que es la madre? (757–58)

> who feels pushed by necessity to leave her home to bring home a few more coins that she can add to the man's meager salary. [...] What is the value of a few miserable cents, earned in an always cruel workday, compared with that of the thousands of pesetas represented by the babies and children abandoned to the ravages of degeneration, disease, vice, and death when left unprotected by their mother, their natural defender?

This is not the only time Acuña has advocated for women to focus their energies on the home, but her insistence on this point is not a traditional celebration of the domestic angel.[5] Instead, for Acuña, this is a matter of both good hygienic practice and good stewardship. In terms of health, the double load of remunerated work and housework subjects the working woman to unhealthy pressures, exposes her children to the dangers described above, and subjects the community to the risk factors that these children then represent. In terms of economics, her emphasis is really on the limitations of the workplace, not the idealization of the home. Her point is a valid one: the work that women could get was neither sufficiently compensated nor safe. Citing the groundbreaking study of Parisian prostitutes by Alexander Parent-DuChâtelet (1790–1836), Pulido had said as much: women who leave the home to work in "los talleres de labor, ganando sueldos reducidos que no bastan a satisfacer las más apremiantes necesidades de un solo individuo, cuanto menos de una familia" (sweatshops, earning salaries so low that they cannot satisfy the most pressing needs of an individual, to say nothing of a family) fall easy prey to hunger, desperation, and seducers (147). Despite the fury she had directed at the prostitute in "La ramera," Acuña also recognized the precarious conditions that led women to prostitution: "tantas almas violentamente arrastradas por impuras atmósferas al funesto extravío de sus destinos" ("La ramera" 2; so many souls violently torn by impure settings to a disastrous straying from their destiny).

Acuña's 1902 lecture on public health is structured around a simple group of three items: light, air, and water. She draws on the deep cultural resonance of the trinity to drive home her lesson: "La higiene es una religión humana, con un dios que se llama limpieza y tres personas distintas, que son: la luz, el aire y el agua" ("La higiene" 751; Hygiene is a human religion, with a god called cleanliness and three distinct persons: light, air, and water). In a country where the Catholic Church played a dominant role in communities and schools, the use of religious imagery served as a familiar pedagogical device; it also undoubtedly softened the message from Acuña, well known for her opposition to organized religion. This narrative strategy is sustained within the lecture, as Acuña describes domestic hygiene as "la génesis" (the genesis) and "el dogma que toda familia debe tener a la cabecera del lecho, para saberlo de memoria y rendirle, en todas las horas, el culto y el amor que se merece" (752; the dogma that every family should keep above the bed, to know it by heart and to render it the reverence and love that it deserves). The concepts contained in this dogma go beyond the simple central image, however: "Muchas cosas representa esta trinidad científica de la higiene, porque como toda ciencia posee multitud de elementos complejos, y para formarla se unen la fisiología, la física, la química, la biología, la botánica, la zoología y hasta las artes y la industria le prestan su concurso" (751; This scientific trinity of hygiene represents many things, because just like any science it is made up of a multitude of complex elements; physiology, physics, chemistry, biology, botany, zoology, and even arts and industry join together to make it up).

Acuña invokes a natural metaphor to help demystify this concept, explaining that as the branches of a tree multiply as they spread from the trunk, so the science underlying hygiene and public health is more complicated. However, at the center of the science she is explaining, just as at the center of a tree there is a solid, tangible trunk, "esta complicada ciencia tiene una sola base, un solo cimiento, uno solo, del cual, como de tronco poderoso, surgen ramas, hojas, flores y frutos: este tronco de la higiene es la limpieza, ¡verdadera piedra angular de todo el edificio higiénico" (751; this complicated science has one single base, one single foundation, just one, from which branches, leaves, flowers, and fruit emerge: the trunk for hygiene is cleanliness, truly the cornerstone for the entire edifice of health!).

Acuña is realistic in her approach, accepting that workers face limits on where and how they live, but firm in her insistence that women can find ways to improve living conditions in even the humblest circumstances: "debemos hacer cuanto podamos para mejorar las condiciones higiénicas de nuestra vida privada. La choza antes que la casa, siempre que la choza tenga luz y aire puros" (761; we must do everything we can to improve the hygienic conditions of our private lives. A hut before a house, as long as the hut has clean light and air). Other public health advocates affiliated with the labor movement would repeat this advice. In his lecture on the hygienic concept of cleanliness, given at Madrid's Centro de Sociedades Obreras in January 1903, Dr. Juan Sánchez Ulibarri recommended that "siendo la luz y el aire elementos esenciales para la vida, los obreros debían preferir las habitaciones elevadas, a las bajas y los sótanos" ("En el Centro Obrero" 3; given that light and air are essential elements for living, workers should prefer their living quarters higher up, not at ground level or in basements). He added that because a daily bath "no está al alcance de los obreros" (is not within the workers' means), a weekly one would do; he assured his audience that workers would immediately notice the health benefits of even a weekly bath (3).

Acuña's message stands in contrast to Dr. Casas's promotion of brand-name commercial products to maintain a clean body and home. Acuña acknowledges the new science of bacteriology but rejects the new antibacterial cleaners. She argues instead that regular, careful cleaning with soap and water goes a long way, at low cost, to combat germs:

> El microbio patógeno y la desinfección, o sea la destrucción del microbio por medio del desinfectante, pertenece casi exclusivamente a la higiene pública y colectiva, y puede pertenecer a la higiene de la familia, cuando

esta familia cuenta con medios económicos para el gasto del desinfectante; mas en la familia obrera, apenas hay dinero para el cotidiano pan, ¿cómo ha de haberle para el cloruro de cal, el ácido fénico, la creolina, el zotal, el sublimado, las fumigaciones sulfurosas y esterilizantes? ("La higiene" 766–67)

The pathogenic microbe and disinfection, that is, the destruction of the microbe by means of a disinfectant, these belong almost exclusively to public and collective health and can belong to family health, when this family has the economic wherewithal to afford the disinfectant; but in the working-class family, where there is barely money for the daily bread, how would there be money for calcium chloride, carbolic acid, creolin and zotal disinfectants, sublimate, sulphurous and sterilizing fumigations?

The advice that follows on how to maximize workers' access to light, air, and water is straightforward, without jargon and centered on the reality facing the rural workers to whom she is speaking.

As she had noted in her earlier article, "La ramera," the environmental challenges of limited light and air are complicated by individual bad behaviors. Acuña had focused on the urban prostitute, in whom she saw the greatest risks to society; for her 1902 talk, she shifts the focus of her concern to alcohol, ascribing nearly identical results to that vice. Like the prostitute who spreads disease that is passed on to future generations, alcohol is a cursed poison:

no le basta un solo ser para aplacar su furia destructora, sino que sigue a través de los hijos, a través de los nietos, a través de una y otra generación, ablandando, y deshaciendo, y licuando; y los hijos idiotas, o los hijos criminales, esos dos tipos de cerebro blando, deformado o desnutrido, que son castigo y baldón de la especie humana, se engendran en las entrañas de los alcoholizados. (769–70)

a single being is not enough to calm its destructive fury; instead, it continues on through the children, the grandchildren, through one generation and another, softening, undoing, and dissolving; engendering in the loins of the alcoholic idiot sons, or criminal sons, those two types with soft or deformed or underdeveloped brains, punishment and stain on the human race.

While in "La ramera" Acuña had raised the specter of diseased children to amplify her outrage at regulated—and therefore government- and Church-sanctioned—prostitution, here her goal is to use the science of hygiene to educate about the effects of a common problem. Even as she called out the public health authorities who were unwilling to address the destructive health and social impacts of prostitution, Acuña embraced the science and principles of hygiene to take on the long-term social impact of alcoholism.

This strategy could be found within the labor movement as well. One example comes from a series of articles on "La higiene del obrero" (Workers' health), appearing in *El Socialista* from July 1902 to January 1903 (Campos 523). In his first article, Dr. Vicente Pérez Cano calls for workers' empowerment in their health, placing his argument within a clearly Socialist framework:

> si el Socialismo lanza a los cerebros un axioma: "haz *conscientemente* el bien de todos, el bien colectivo," también dice la Higiene: "cuida de tu aseo, de tu policía personal, sé metódico, odia el alcohol." El higienismo particular, por otra parte, es colectivo en cuanto el bien resulta para todos, se evita la infección, la *sepsis*, en lenguaje científico, se mata el microbio; el Socialismo, no pudiendo dejar de ser *higienista*, rechaza los cerebros incultos o anhela pulimentarlos, según el caso. (3)

> if Socialism teaches the axiom "do good *conscientiously* for all, for the collective good," Hygiene adds this: "take care of your cleanliness, of your habits, be methodical, abhor alcohol." On the other hand, individual hygiene is collective, in that the positive results are for everyone: infection, or *sepsis* in scientific terms, is avoided, microbes are killed; Socialism, which cannot help but *embrace hygienic principles*, rejects uncultured minds or seeks to improve them, according to each case.

The sense that workers needed to commit to "automejora" (self-improvement), to valuing individual health and wellbeing "como un acto solidario con el resto de compañeros en la lucha colectiva por el socialism" (as an act of solidarity with the rest of the workers in the collective fight for Socialism), represented a new approach for workers and their representatives (Campos 511–12). Workers would hear ever more clear affirmations of their personal responsibility for the health of themselves, their families, and their communities. Given that public health practices were widely seen as a form of discipline of the working classes, the embrace of these basic hygienic principles in the pages of *El Socialista* shows an evolution in popular thinking, a wider move away from assigning blame to empowering individuals and communities.[6]

The intentionality with which Acuña presented her scientifically sound ideas outside the bounds of official, academic, and politically motivated discourse is quite striking. Her embrace of women, regardless of their social condition or education, set her apart from other promotors of public health. It may also help explain why this aspect of her work has been overlooked. Although most of the recent biographies of Rosario de Acuña highlight her patriotic writing, she deserves recognition for both her championing of women, women workers, and public health and for her critical approach to discourse about women and hygiene.[7]

Acuña spent most of her life advocating on behalf of the disempowered and embracing progressive ideals, joining a Masonic lodge, and retreating to rural Cantabria, where she lived and wrote in virtual solitude. Following her death, *El Noroeste* described her as "un ejemplo de vida diáfana y severa consagrada a un ideal de arte y de reforma social" (Díaz Fernández 3; an exemplary model of a life of clarity and severity, consecrated to the ideals of art and social reform). Other obituaries and tributes celebrated her work in public health. For example, one of the earliest obituaries, in *La Voz*, highlights Acuña's efforts to teach "a los rústicos para que no despreciaran la hygiene" (uncultured folk to not disdain the principles of hygiene), adding that the public health authorities "harían obra beneficiosa editando un folleto de poco costo para divulgar los artículos de doña Rosario de Acuña" (Castrovido 1; would do a good deed by editing an inexpensive booklet to disseminate doña Rosario de Acuña's articles). The *Heraldo de Madrid* opened its 1925 appreciation of the writer by observing, "Ahora que tan difusa propaganda se hace de los preceptos higiénicos debe recordarse que ella escribió—cuando sólo los profesionales se ocupaban de ello—una serie de artículos tratando de reglas higiénicas, divulgándolas y procurando hacerlas llegar a los rústicos entendimientos" (Canto 4; Now that the precepts of hygiene are so widely promoted, we should recall that she wrote—when only professionals were dealing with the issue—a series of articles addressing the rules of hygiene, disseminating them widely and trying to make them accessible to uncultured minds). Acuña's work promoting an alternative discourse for public health, one focused on hygiene and women's empowerment, is a legacy that merits recovery.

That alternative discourse is tied closely to Acuña's recognition of both the gender and class dynamics affecting women in turn-of-the-century Spain. As the nineteenth century came to a close, "[t]raditional prejudices, the nation's calamitous lack of scientific education and cultural capital, and the inability of elite social groups to find common ground from which to address satisfactorily the needs of the socially disadvantaged" left Spain in an untenable situation (Arkinstall, *Histories* 83–84). The solutions identified by Spain's many concerned ladies, physicians, and public health workers were simply not functional for the vast majority of the Spanish population. The privilege underpinning their good works made it impossible for their efforts to be fully realized. For example, when Guillermina Pacheco sweeps dust—or people—from the streets of Galdós's Madrid, her tool of choice, the broom, is insufficient for the bigger tasks at hand. What she can do is ultimately limited by what critic Denah Lida calls her "falta de intuición y sensibilidad humanas" (lack of human intuition and sensitivity), with the result that "lo que resuelve son necesidades materiales" (21; what she does resolve are material needs).

As Galdós, Arenal, and Acuña all attest, the material needs of the dispossessed were significant, but while some, like the fictional Guillermina and the historical Concepción Arenal, worked from within the elite power structures to effect change and alleviate suffering, Rosario de Acuña chose to work from outside those systems, with a blend of free thinking, socialism, social commitment, and womanly concerns. Rejecting connections with organized charities, the Church, and universities, she resituated the discourse within a framework of individual initiative and collective responsibility. Acuña advocated for more and better options for women of all backgrounds to learn and become empowered to preserve and improve the health of their families, communities, and nation. In the closing words to her 1902 lecture, Acuña reminds working women that they themselves are at the front lines of public health: "No olvidaros jamás de que en vuestras frentes, por muy sudorosas que las ponga el trabajo, por muy inclinadas que las tenga la pobreza, por muy entenebrecidas que las vuelva la ignorancia, en vuestras frentes de proletarias y desheredadas han tejido los siglos la diadema de la soberanía humana" ("La higiene" 776; Don't ever forget that the centuries have woven the crown of human sovereignty on your brow, on your proletarian and disinherited brow, and no matter how sweaty it might be from work, how bowed it might be by poverty, how dark it might be from ignorance). As she began her talk, so she concluded it, with a reminder of the dignity and value of the working woman, a dignity and value equivalent to that of any other woman. With the sovereign condition of all people as her guiding principle, she reminded her listeners and readers that this previously unrecognized sovereignty implies some important personal responsibilities: "es preciso que esa brillante diadema, ante la cual el mundo comienza a rendir su homenaje, no se arrastre enfangada por el lodazal de las insanias, de los vicios, ni de los odios" (776; it is necessary that that shining crown, before which the world is beginning to bow in homage, not be dragged through the mud of unhealthy living, of vice, or of hatred!).

Acuña rightly noted that times were changing and that women would need to embrace their full potential in a new, more equitable, and healthier society. It would no longer be acceptable for society's elites to introduce public health measures by force, as Galdós's Guillermina did, "sin más razón que su voluntad [. . .], sin dar explicaciones a nadie de aquel atentado contra los derechos individuales" (Pérez Galdós *Fortunata y Jacinta* 1: 628; with no other right than her will [. . .], without explaining that attack on individual rights to anyone). As she had argued before in "La ramera," women play an essential role in creating "la vida ordenadora, higiénica, racional, que el hogar le ofrece cuando en él reside una compañera" (1; the orderly, hygienic, and rational life that home can offer when a female companion lives there). With her simple, direct, and unambiguous message of "la trinidad higiénica" (the trinity of health), consisting of light, air, and water, Acuña made the complex science and practice of public health a familiar concept, neither threatening nor intimidating ("La higiene" 756). Taking to heart Dr. Castells Ballespí's plea that women educate women—taking it beyond the university, beyond the elegant salons of the capital, beyond the circle of family, and even beyond the bounds of social class—Rosario de Acuña brought public health education to the masses.

NOTES

1. From "the holiness of the broom" to the "trinity of health."
2. In 1882, *La Higiene* reported that Acuña and her then husband Rafael de la Iglesia had sponsored an essay contest on the topic of mental illness and the ability to discern madness from bad behavior (Bolado 99). Two years later, Acuña would fund medical students who lost their scholarships during a university strike over changing requirements ("Los estudiantes" 3).
3. Jean-Louis Guereña ascribes Spain's abolitionist movement to outside agitators, calling it "una doctrina importada del extranjero por extranjeros, de ideología protestante

además" (345; a doctrine imported from foreign lands by foreigners, a Protestant ideology as well). From its beginning with Josephine Butler in 1875, the crusade to abolish prostitution, first in England and later throughout Europe, was dominated by female figures; in Spain, however, the movement was never defined by "grupos feministas estructurados ni tampoco con muchas mujeres" (Guereña 339; organized feminist groups nor by many women). Instead, abolitionism was closely associated with the Masons, whose members included both Arenal and Acuña.

4. The Catholic paper *El pensamiento español* described the Sunday lectures with no small amount of irony: "Las sesiones están concurridas de curiosos, y asisten también señoras, a quienes especialmente se dedican las conferencias. Recordamos que un revistero, hablando de estas sesiones, decía: «allí se puede ir a todo, menos a buscar novia.» Y es la verdad" ("Las conferencias" 3; The sessions are filled with curious onlookers, and women, to whom the conferences are specially dedicated, were also in attendance. We recall that one reviewer, speaking of these sessions, would say: "You can go there for anything, except to find a girlfriend."And that is the truth). The engraving that accompanies *El Panorama*'s reporting on the conferences shows the university lecture hall filled with elegant men and women, with a couple in the foreground engaging in what appears to be flirtation ("Conferencias dominicales" 123).

5. This thinking can be seen in Acuña's dramatic works as well as in earlier essays. For example, in 1887, the author addressed her female readers in *Las Dominicales del Libre Pensamiento*: "vosotras, todas las que en el silencioso retiro del hogar, de donde ha de surgir la nueva era, sentís en vuestras almas el latido de este siglo y respiráis esta atmósfera regeneradora que comienza a estremecer las sociedades, anunciando a la mujer que su sitio está al lado de la libertad y del progreso" ("A las mujeres" 1226; you women, who from the silent retreat of your home, from where the new era will spring, all of you who feel in your soul the pulse of this century and who breathe in this regenerative air that is beginning to shake up society, announcing to women that their place is on the side of freedom and progress).

6. Beyond labor's activist core, the notion that workers were learning about public health and personal hygiene was received with some skepticism. Following the lecture by Dr. Sánchez Ulibarri cited above, the mainstream newspaper *El Imparcial* described the talk as, "perfectamente arreglada al nivel intelectual del auditorio" (perfectly matched with the intellectual level of the audience) and concluded that, "dentro del medio económico en que vive, puede el obrero, con un poco de buena voluntad, cumplir sus más rudimentarias prescripciones, atenuando de esta suerte los inconvenientes de las circunstancias desfavorables en que se desenvuelven su actividad y la existencia de los suyos" ("Sección de noticias"; within his economic means, the worker can, with a bit of willingness, fulfill the most basic requirements, thereby lessening the disadvantages of the unfavorable circumstances in which he and his family live and work).

7. José Bolado's extensive biography is an essential resource, as are the studies by Christine Arkinstall, Marta Fernández Morales, Macrino Fernández Riera, and María del Carmen Simón Palmer.

WORKS CITED

Acuña y Villanueva, Rosario de. "A las mujeres del siglo XIX." *Obras reunidas*. Vol. 2, *Artículos (1885–1923)*, edited by José Bolado, KRK, 2007, pp. 1225–40.

———. "La higiene en la familia obrera." *Obras reunidas*. Vol. 3, *Prosa*, edited by José Bolado, KRK, 2008, pp. 745–78.

———. "La jarca de la universidad." *Obras reunidas*. Vol. 2, *Artículos (1885–1923)*, edited by José Bolado, KRK, 2007, pp. 1609–18.

———. "La ramera." *Las Dominicales del Libre Pensamiento*, año 5, no. 234, 28 May 1887, pp. 1–2.

———. "Testamento." *Obras reunidas*. Vol. 5, *Lírica*, edited by José Bolado, KRK, 2009, pp. 635–41.

Aleu y Riera, Dolores. *De la necesidad de encaminar por nueva senda la educación higiénico-moral de la mujer*. La Academia, 1883.

Arenal, Concepción. *La beneficencia, la filantropía y la caridad*. Imprenta del Colegio de Sordo-Mudos y de Ciegos, 1861.

———. "Carta XXIV. Delitos contra la hostidad. Artículos 358 al 362." *Cartas a los delincuentes. Obras completas*. Tomo 3, Victoriano Suárez, 1894, pp. 307–19.

———. "Centro protector de la mujer." *Obras sobre beneficencia y prisiones*. Vol. 5. *Obras completas*. Tomo 22, Victoriano Suárez, 1902, pp. 414–18.

———. *La mujer del porvenir y Artículos sobre las Conferencias dominicales para la educación de la mujer*. Eduardo Perié, 1870.

Arkinstall, Christine. "Challenging Pasts, Exploring Futures: 'Race,' Gender, and Class in the Fin-de-siècle Essays of Rosario de Acuña, Concepción Gimeno de Flaquer, and Belén Sárraga." *Intersections of Race, Class, Gender, and Nation in Fin-de-siècle Spanish Literature and Culture*, edited by Jennifer Smith and Lisa Nalbone, Routledge, 2017, pp. 23–44.

———. "Configuring the Nation in Fin-de-siècle Spain: Rosario de Acuña's *La voz de la patria*." *Hispanic Review*, vol. 74, no. 3, 2006, pp. 301–18.

———. *Histories, Cultures, and National Identities. Women Writing Spain, 1877–1984*. Bucknell UP, 2009.

———. "Writing Nineteenth-Century Spain: Rosario de Acuña and the Liberal Nation." *Modern Language Notes*, vol. 120, 2005, pp. 294–313.

Bolado, José. "Introducción. Rosario de Acuña. Escritora y vida aventurada." *Obras reunidas*. Vol. 1, *Artículos (1881–1884)*, by Rosario de Acuña y Villanueva, KRK Ediciones, 2007, pp. 19–463.

Borderies-Guereña, Josette. "El discurso higiénico como conformador de la mentalidad femenina (1865–1915)." *Mujeres y hombres en la formación del Pensamiento Occidental. Actas de las VII Jornadas de Investigación Interdisciplinaria*, Vol. 2, edited by Virginia Maquieira D'Angelo, Ediciones de la UAM, 1989, pp. 299–309.

Campos, Ricardo. "'El deber de mejorar': Higiene e identidad obrera en el socialismo madrileño, 1884–1904." *Dynamis*, vol. 31, no. 2, 2011, pp. 497–526.

Canto, Rosa. "Recordando a Rosario de Acuña." *Heraldo de Madrid*, año 35, no. 12,260, 19 May 1925, p. 4.

Casas, Santiago. "Sexta conferencia sobre la higiene de la mujer." *Conferencias dominicales sobre la educación de la mujer*, 28 Mar. 1869, M. Rivadeneyra, 1869.

Castells Ballespí, Martina. *Educación de la mujer. Educación física, moral e intelectual que debe darse a la mujer para que esta contribuya en grado máximo a la perfección y la de la Humanidad. Memoria leída por en el acto de recibir la Investidura de Doctor en Medicina. Madrid, octubre 1882.* "La educación de la mujer según las primeras doctoras en medicina de la universidad española, año 1882." Consuelo Flecha García. *Dynamis*, vol. 19, 1999, pp. 241–78.

Castrovido, Roberto. "Rosario de Acuña. También era mucho hombre esta mujer." *La Voz*, año 4, no. 894, 9 May 1923, p. 1.

"Conferencias dominicales sobre la educación de la mujer." *El Panorama*, año 3, no. 16, 30 Aug. 1869, p. 123.

"Curso de higiene vulgar [2]" *El Socialista*, año 14, no. 720, 22 Dec. 1899, pp. 3–4.

"Curso de higiene vulgar [3]" *El Socialista*, año 14, no. 721, 29 Dec. 1899, pp. 3–4.

Díaz Fernández, José. "Una mujer muy mujer." *El Noroeste*, año 28, no. 9904, 4 May 1924, p. 3.

"En el Centro Obrero. Concepto higiénico de la limpieza." *El Socialista*, año 18, no. 881, 23 Jan. 1903, pp. 3.

"En honor de Ernestina." *La Ilustración Católica* Época 4, año 12, tomo 10, no. 31, 5 Nov. 1887, p. 362.

"Los estudiantes." *El Día*, no. 1645, 8 Dec. 1884, p. 3.

Fernández Morales, Marta. *Rosario de Acuña. Literatura y transgresión en el fin de siècle*. Asociación Milenta Muyeres y Ayuntamiento de Gijón, 2006.

Fernández Riera, Macrino. *Rosario de Acuña en Asturias*. Trea, 2005.

Guereña, Jean-Louis. *La prostitución en la España contemporánea*. Marcial Pons, 2003.

Lacalzada de Mateo, María José. *Concepción Arenal: Mentalidad y proyección social*. Prensas de la Universidad de Zaragoza, 2012.

"Las conferencias dominicales de la universidad central." *El pensamiento español*, año 10, no. 2865, 24 May 1869, p. 3.

Ley de Instrucción Pública. *Gaceta de Madrid*, no. 1710, 10 Sept. 1857, pp. 1–3.

Lida, Denah. "Galdós y sus santas modernas." *Anales galdosianos*, año 10, 1975, pp. 19–30.

Ortiz, Teresa. "Profesiones sanitarias." *Historia de las mujeres en España y América Latina*, edited by Isabel Morant Deusa, Vol. 3, *Del siglo XIX a los umbrales del XX*, Cátedra, 2006, pp. 523–46.

Palacios Lis, Irene. "Mujeres aleccionando a mujeres. Discursos sobre la maternidad en el siglo XIX." *Historia de la Educación*, vol. 26, 2007, pp. 111–42.

Pascual de Sanjuán, Pilar. *Flora o La educación de una niña*. Faustino Paluzíe, 1889.

Perdiguero, Enrique. "Popularización de la higiene en los manuales de economía doméstica en el tránsito de los siglos XIX al XX." *Malaltia i cultura. 1ª Trobada Interdisciplinaria sobre Malaltia i Cultura*, edited by J. L. Barona, Seminari d'Estudis sobre la Ciència, 1995, pp. 225–50.

Pérez Cano, Vicente. "La higiene del obrero." *El Socialista*, año 17, no. 855, 25 July 1902, pp. 3.

Pérez Galdós, Benito. *Fortunata y Jacinta*. Edited by Francisco Caudet, Cátedra, 2006. 2 vols.

———. "Santos modernos." *Cronicón. (1886–1890). Obras inéditas*. Vol. 7, edited by Alberto Ghiraldo, Renacimiento, 1924, pp. 7–18.

Pulido, Ángel Fernández. *Bosquejos médico-sociales para la mujer*. Víctor Saiz, 1876.

Sánchez García, Raquel, and Ana Martínez Rus. *La lectura en la España contemporánea*. Arco/Libros, 2010.

"Sección de noticias." *El Imparcial*, año 37, no. 12,855, 18 Jan. 1903, p. 3.

Simón Palmer, María del Carmen. "Introducción." *Rienzi el Tribuno. El padre Juan. Teatro*. By Rosario de Acuña. Castalia, 1990.

———. "Rosario de Acuña y Villanueva." *Escritoras españolas del siglo XIX. Catálogo bio-bibliográfico*, Castalia, 1991, pp. 4–11.

Solís Claras, Manuela. *Higiene del embarazo y de la primera infancia. Para las madres*. Prologue by Santiago Ramón y Cajal, Imprenta F. Vives Mora, 1907.

Tratado de economía y labores para uso de las niñas. Parte componente del Educador publicado por la Casa de José González, 3rd ed., Museo de la Educación, 1861.

Wells, Susan. *Out of the Dead House. Nineteenth-Century Women Physicians and the Writing of Medicine*. U of Wisconsin P, 2001.

CHAPTER 5

Science, History, and Gender

An Interview with María Jesús Santesmases

VICTORIA L. KETZ
AND DEBRA FASZER-MCMAHON

Victoria L. Ketz and Debra Faszer-McMahon interviewed María Jesús Santesmases on Friday, May 31, 2019 in her office in Madrid at the *Consejo Superior de Investigaciones Científicas*. Santesmases is a member of the Department of Science, Technology, and Society at the Institute of Philosophy of the Spanish National Research Council (CSIC). She has published numerous academic journal articles and books focused on the history of science, women, and gender in Spain, as well as contributed to contemporary news sources such as *El País*. Santesmases holds a PhD in Organic Chemistry from the Complutense Univeristy in Madrid. She has employed her training as a scientist to analyze and critique the development of the biomedical sciences and their gender dynamics. Her publications include *Mujeres científicas en España 1940–1970: Profesionalización y modernización social* (2000), *Entre Cajal y Ochoa: Ciencias biomédicas en la España de Franco, 1939–1975* (2001), "Towards Denaturalization: Women Scientists and Academics in Twentieth-Century Spain" (2011), and with Teresa Ortiz-Gómez *Gendered Drugs and Medicine: Historical and Socio-Cultural Perspectives* (2014). Her most recent work, *The Circulation of Penicillin in Spain: Health, Wealth, and Authority*, was published in 2018 by Palgrave.[1] The transcript below, translated from the original Spanish, includes key selections from Santesmases's interview, which was condensed in the interest of space.

VLK: Tell us about your professional trajectory. What attracted you to the field of scientific research?

MJS: I was always attracted to the sciences. When I was young and went to school, it was difficult to decide between the sciences and the humanities because everything interested me. I liked studying and reading a lot. Then mathematics became enthralling, but Latin also. In Spain, we all studied Latin when I was a teenager; it was common and mandatory. I thought that was cool, and it was difficult for me to decide. In fact, I started in humanities but later changed over to sciences when I became intrigued with the formulas from the girls in the class next door at my Catholic school in Barcelona. Later on, I had a good chemistry teacher, and when I came to Madrid during my senior year, I decided to pursue chemistry.

I studied chemistry at the Complutense and received my doctorate in organic chemistry. I did a postdoc in Berlin, though I had to return early for personal reasons. It was very difficult for me, but I left it all, and I began to search for work. I did a variety of things: I worked in administration, at a newspaper, in television, in a government minister's cabinet. I left the academic world for five years, doing a little bit of everything. The jobs were good, but they were things that I had never done before. I learned a lot, and it was very interesting, but the last position I held was very close to political decisions, and I became very disappointed. That is when I decided to return to the academic world. At that moment, I was expecting my first daughter, and everything happened at once. I started at zero again, earning very little money and working a lot. That was during the years when my children were born, which was a period of around eight years.

During those years, I began my project on the history of biochemistry and molecular biology in Spain. Since I had already received my doctorate, at least I had that. Emilio Muñoz, who had been a high-ranking member of the scientific policies team in Felipe Gonzalez's government, was very supportive of me. We applied for funding, and I was able to return to research. It was a topic that I knew practically nothing about. At that moment, few people did. The community that writes about the history of molecular biology in the West and the people that I knew from Europe and the Americas were very few, but interest in the subject would grow from that point on, along with the community of scholars. It was then that I entered into contact with them. They are a group of extraordinary people with whom I maintain contact. They were at Princeton, Paris, Berlin, Boston, and Cambridge. Those are the people that went before me. Lucky for me, of course.

In Spain, during the '60s and the '70s, while there were many productive researchers, they all had the sense of lagging behind what was occurring in other countries. Thus, it was very interesting to study the Spanish scientific community, their trajectory, their publications, where they had studied, and who had influenced them. In the end, that is how I entered this field, that is, the history of biochemistry and molecular biology. There, I studied people and wrote a reconstruction of the field in Spain, as well as two biographies, of Alberto Sols and of Severo Ochoa.

It was always about men at the beginning, until there was a moment when Emilio Muñoz said to me, "Hey, there is this conference in Barcelona about women scientists." That was a long time ago. Now, everybody talks about it, but at that time, not many people did, except for an interesting but small group of feminist scholars. [. . .] At that time there was an interesting journalist from Barcelona who now has retired, Vladimir de Semir, who worked for *La Vanguardia*. He organized a meeting with Cristina Ribas, a journalist as well, I think. I was invited to speak about Spanish women scientists. There I met some of the feminists who were already doing research and writing about women and gender in science and technology: Carme Alemany, Carmen Magallón, Ana Sánchez, Eulalia Pérez Sedeño, and Capi Corrales, among others. At that time, I had some research that I had done on the topic. There were many women, and I had not worked on all of them yet. So I made a list, and I found two things that impressed me: one, the laboratories; even though they were directed by male scientists, everyday practice and work in the lab were directed and organized by women scientists. What I'm saying is that the person who ran the day-to-day in the laboratories was always a woman, and in some cases it was the scientist's own wife, a scientist herself. This led me to consult the work of Pnina Abir-Am, who has been a pioneer in scientific studies. *Creative Couples in the Sciences* was the first book that she published.[2] Pnina extracted this from the first investigations of Margaret

Rossiter. In her first book, Margaret Rossiter notes that there are a group of scientific couples, but she says very little, just enough that it served as an inspiration for us, especially for Pnina at that moment. Since then, Pnina has written two books with other colleagues, which served as a reference for me to explore this topic. Even though I wasn't able to explore them in a thorough manner, I did a sample of about fifty to sixty scientific women who had received their doctorates before 1970 in the areas of biology and biomedicine. I did this study utilizing testimonies in an anonymous way.

It was a very difficult time. I had begun to study sciences in Franco's Spain, and some scientists were mad at me because I used to call the time period Francoist. I was very young, and in Spain, historian colleagues were still not exploring the history of science during the Franco era, but I was not aware of that. Things just happened that way. I also had Emilio Muñoz, who had formed part of that community and served a little bit as a memory of all that for me. I would consult Emilio, who still is an emeritus professor here. I was lucky that when the money ran out, I was able to apply for funds for the research project from the Women's Institute. That is when I did my research and wrote my book that was published in 2000. Many people supported me, since one cannot achieve these things alone. The most fierce critics of the manuscript on Spanish women scientists were two sociologists, Julio Carabaña and Luis Garrido, who had published in the fields of history and sociology related to women's work and education. Their critiques were useful for my discussion. At the Women's Institute, the support of Ana Mañeru, a high-ranking officer there and a poet, was very influential.

Even though I was studying women scientists, I did not belong to any community, because my formation in history was not formalized. I did not take any courses, except one when I worked as a journalist, about the history of twentieth-century physics (taught by José Manuel Sánchez Ron), and I did not complete a master's degree because I already had a doctorate. Therefore, I began from scratch, with the daring that one has when one is very young. I believed that by reading, I could overcome this. Of course, it took a lot of hard work. I had to read until exhaustion, but even though I did not know everything that I should have learned, I had some systematic formation. I was lucky in that I entered into contact with very interesting people, among them Eulalia Pérez Sedeño, who is here [at CSIC]. Later, especially when I was studying the history of medicine, I met Teresa Ortiz of the University of Granada, who has been a pioneer in Spain regarding gender studies, medicine, and women's health.

In the end, little by little, I entered into contact with all of these women. My work was published in English. It was about the history of biochemistry and molecular biology, citing women's research achievements but not making this the principal theme. It was only later that I started to work on the topic of women and gender. My main research, done in a very slow fashion, was in the book that was read by more people than I could have ever expected (*Mujeres científicas en España 1940–1970: Profesionalización y modernización social*, 2000). More than I thought would read it. Now, so much time has passed that I realize what I did not realize at that moment. I thought that what was going to happen was what happens to everyone, that nobody would read it. Then of course, you say, "Who reads what we write? No one, or very few people."

It is true that in the '90s in Spain there were master's and doctoral programs, pioneering programs, focused on feminism and gender studies. The dictatorship did a lot of damage, but it did not occupy every space. There still was intellectual research in Spain, even when Franco lived. The political and economic difficulties that existed are very clear, well-known, and documented. Among them was included the lack of explicit support for research as a priority in the government's politics. It was not until the first government of Felipe González, in the 1980s, that support was offered by growing the budget and reorganizing the structure of the Spanish research system.

At some point during the Franco dictatorship (1939–1975), a growing research community started the reconstruction of what had been lost during the Spanish Civil War and the early decade and a half of the dictatorship. The great areas of investigation that nowadays exist in Spain began their ways of working and managing research at that time, and although there were few groups, they became very influential. Many women and men stood out in many areas, with scientific production of the highest quality with absolutely scarce resources. This was the community that I studied, which is part of the historiographical debate about the conditions imposed by the dictatorship and the conditions imposed by foreign relations. It has always been my thesis that there coexisted two conditions and influences: those imposed by the dictatorship, and those received from foreign relations, including both scientists as well as the political diplomacy of the Franco regime, which involved membership in international organizations and geopolitical practices. The dictatorship took advantage of the growing recognition abroad of some Spanish scientists in order to demonstrate to the international community its best face. These were educated scientists, people who had expert research experience, who had been trained abroad, who spoke different languages, and who published their research in internationally distributed foreign journals. They created new communities, new ways of doing research, and new ways of sharing results, as was occurring in all parts of the world.

DFM: We had read a bit about your biography, but this gives us a much deeper understanding. Related to your experiences, we wanted to ask you about your work here at the CSIC. Tell us about the faculty and programs here. How does your unit work, with its focus on women, philosophy, and social justice related to the sciences?

MJS: Like all of my colleagues at the CSIC, which is a research institution, I have practically no teaching responsibilities. I only participate in an inter-university master's program. I began to give classes in the master's when Eulalia Pérez Sedeño invited me to speak about science and gender. I focused on the classes that had to do with biology, especially related to hormones, the history of sex hormones. [. . .] I started teaching as a half assignment until I was offered a course about the history of biology. When I came here, I was in the Institute of Advanced Social Studies that was created in CSIC. [. . .] Then I moved to the Institute of Philosophy, when the Institute was in a very nice building on the Serrano campus, where the Council is still located. We were expelled from paradise because humanities and social sciences did not belong to the areas of priority for science policy in any European country, and thus we were sent to this center. This location has very good administrators and managers as well as amazing library facilities, but it is far from any academic setting in Madrid, making it difficult to attract students and visitors for our academic activities, even if, through a lot of effort, we sometimes are able to make it happen.

What has become important is the approximation of the political authorities to the particularities of research and its governance. I think that knowing a little bit about history and sociology would help the humanities to receive a little more support. Governments usually like to believe in the direct relationship of science to application and use, but it hasn't always worked like that.

For me, industry was never an option because no opportunity ever presented itself. Besides, I was never interested. When I wrote my dissertation, I was still very interested in research. After graduating, I suffered a lot for some months until I obtained a scholarship. I believed that if I did not receive funding, I would be very unfortunate. [...] Then, I went from one scholarship to another.

I worked in diverse areas—that was the '80s in Spain, a very fun time period. We were always in the streets. We would work, we would study a lot, and we would have a good time. You let yourself be carried away, even though you worked a lot. I worked a lot, and I was a laughing stock to my friends. They would say, "Leaving already?" "No, I have to start an experiment. I'll be back." "Don't go!" "No, no I'm sorry!" I would go and start the experiment. I was very organized, but those were the years when everybody was in the streets. Well, that's what we did then. We slept a little bit, worked a lot, and also went out a lot to concerts, political meetings, art exhibitions, book presentations, and all that.

Then when I left the research laboratory and moved, or tried to shift to the history of science, I continued to work a lot because there was no other way to overcome my formative deficiencies in the area except by doing research and publishing. I adopted the strategy of the people whom I studied, that is, publishing in foreign journals. This allowed me to create a profile that current science policy criteria values in a positive manner. My colleagues have mostly published in Spanish, people much more wise than I was, even though I think they respect my strategy. I think it would have been much more enriching for me if I had published in Spanish, surrounded by people with sufficient knowledge to criticize my work. That is not the way it happened. I think I progressed as I was able to, and now I am at a different moment.

Now I do belong. I have my colleagues and friends from my work area, and all these exchanges are permanent. I collaborate with some colleagues in Spain and also abroad, and we've worked together and written things together. We are always trying to do things together for security and for entertainment because having colleagues at your side is crucial. You're always making mistakes, because not everything is always well formulated in a way you can discuss it. Also, it's much more fun to work together. It's more difficult, but much better to stay in contact with people, to discuss, to talk, to try to convince the other person. Because the people who investigate alone are bored, right? It is easier, but a lot more boring. One person reads some things, and you read others; you exchange your readings, and you say, "Hey, haven't you seen this?" I have colleagues in Mexico at the UNAM, Edna Suárez and Ana Barahona, with whom I've collaborated often, and also with my colleague, Angela Creager. Therefore, I've been lucky to collaborate, including with medical historians like Christoph

Gradmann, with whom I shared membership in the steering committee of a European research project.

Finally, I was able to integrate my research internationally because of new European directives, since for the past few decades European research projects must include people from other countries, not only the English, French, and German. They started to require the participation in every project of researchers from Southern Europe, and then from Eastern Europe, as well, and they started to invite me. I participated in one with colleagues from Paris that was very interesting, a project entitled ITEMS. We had a great time. People are amazingly smart, and that was when I started to think about my project on the history of prenatal diagnosis. I do not want to call it that anymore because it seems to require different terminology. That was its inception and also the origin of the work of my colleague Ilana Löwy, who wrote two books on prenatal diagnosis after we jointly organized a session at a meeting of ITEMS.

Later, another project that was very important was that of the history of pharmaceuticals in Europe. It was titled DRUGS and funded by the European Science Foundation. I was the only woman on the steering committee composed of colleagues in the history of medicine and biomedical sciences (among them Toine Pieters, Jean-Paul Gaudillière, Christoph Gradmann, Christian Bonah, and Volker Hess). We had our first meeting in Strasburg a few years ago, and it was a super interesting project. When we had the first meeting, we spoke about everything, and I suggested, actually blurted out, that this project should perhaps consider something that we had not talked about but that was important, and that was the issue of gender and medicine. I noted that pharmaceuticals do not work the same in men as women, and women should participate in the trials. I emphasized that women had been participants in the history of drugs and medicines, and the history of contraceptives already belonged to the area of gender studies. And everybody was silent. The mistake was mine, because I should have negotiated this with them previously. During the preparation of the project, they had sent me the draft, but I was not paying attention to it at that moment. Nonetheless, they agreed to support my proposal, and I think that was good for the project and also for me. What happened afterwards is that Teresa Ortiz and her PhD student at the time, Agata Ignaciuk, both inspiring colleagues of mine at the University of Granada, began a research project on the history of contraception in Spain, through which they participated in the European project. Then I acquired a better understanding of the problem of gender and the history of pharmaceuticals. Everything turned out okay because the committee supported me. It was good because we organized a few successful academic events, and one of the outcomes of the ESF project was the book that Teresa and I edited: *Gendered Drugs and Medicine*.

VLK: *If you had a chance to reflect on all of your scholarly production, which work do you think was the most important or has made the most significant impact on society?*

MJS: Well, I am not convinced about the impact of my work. I have been doing things as they emerged, since I do not have a planned trajectory. Now I am growing older, and I would say I am going to write a book this year. However, I have just been doing things as they have come up. The thing with women came up, so I applied for funding, and then I wrote my manuscript. Then there have been people who have proposed things to me, and I would say, "Yes." I do not know. I am happy with the book on penicillin. That is only because I have done it recently, as an older and wiser person. Then there are other things that I wrote before that are also okay. Sometimes I read them and say, "Hey, that wasn't so bad!" They were simpler things. I would say that my work on penicillin was conceptualized as something more purposeful and attributes to penicillin a symbolic value as well as a medical value.

DFM: *That relates to another question we had. You have this new publication that came out in 2018. Why did you select the topic of penicillin? Also, what are you thinking about for your future projects?*

MJS: The study of penicillin came about when I was at the Rockefeller Archive Center [RAC]. I wanted to see what they had about Spain. Many of my colleagues who do research on the history of biochemistry and molecular biology have gone to those archives of the Rockefeller Foundation. This has been the great legitimization of that archive. That is to say, the archive has conserved so much documentation, so many details about so many people, even though they have thrown away information on the projects that they refused. They only preserve those that they have funded, and these records are available to researchers who request access. One can consult them, and working with the foundation is very pleasant. So, one day I wrote to them and said I wanted to know what information was available about Spain over a certain time period.

During the period that I was asking about, among the few archival materials at the RAC related to Spain was information about a physiologist at the University of Madrid, Antonio Gallego, who had asked for a little funding to buy scientific journals. The material about Spain was scarce because the Rockefeller Foundation did not want to help Spain during the Franco dictatorship. Well, they never said that. In fact, there was no formal declaration as far as I can say, but there was also never any support, except this and some other minor support. So, after I found the information about Antonio Gallego, I went to see Margarita Salas, an influential and generous molecular biologist who very recently passed away; she has helped me a lot. She gave me information about herself. I have talked to her many times about female scientists in Spain, as well as regarding her work with Severo Ochoa in the 1960s, when I wrote Ochoa's biography. The chapter dedicated to the initiation factors of protein synthesis was based on the information Salas gave me; that led me to propose the importance of her influence for Ochoa's scientific trajectory, and also that of Marianne Grunberg-Manago, a woman scientist who also worked with Ochoa at NYU in the 1950s. [. . .]

But returning to the issue of the archival materials related to Antonio Gallego at the RAC, she [Margarita Salas] said to me, "Well, his [Antonio Gallego's] son is a professor of physiology at the University of Alicante. Call him on my behalf. Write to him." I wrote to this person, Roberto Gallego, and it turns out that his father, Antonio Gallego, in addition to being a professor of physiology at the University of Madrid, was the research director of one of the first Spanish penicillin and antibiotics companies, which meant that he had participated in the organization of one of the two penicillin factories that existed in Spain. He was a fascinating character. Roberto Gallego helped me understand his father and the origins of the industry. Then I had found out about the women working in that factory, and I decided to focus on them. I refocused the chapter in the book about the history of the factory, and it became a chapter about women's history. Antonio Gallego was a very important man. He had his role in the story, even though I did not convert him into a protagonist. If at some point I do a Spanish version, maybe when I retire, who knows, I would like to write more about the topic. I believe that doctors read books, and if it is in Spanish, Spanish doctors will read this book. [. . .] When I collaborated with Alianza Editorial, they taught me that doctors are the professional community that reads the most. So one should write books that are interesting to doctors. They are a big audience, something we don't often have.

VLK: So you consider this Spanish version one of your future projects. Anything else?

MJS: I really want to do a brief history about the origins of medical genetics in Spain. Something about the history of what Barbara Duden has called the public fetus. At what point does the fetus become a social character, a social medical figure, in a culture where women's bodies are practically invisible in anatomy and in medicine? That is something that other colleagues are exploring, and I think I would like to continue with that. I believe I'm going to drop everything else and dedicate myself to writing on that topic for a few years. Well, that is never possible because temptations arise everywhere, but now I have a few things going. I have published some papers on this issue in English, and now I have the project to write a book about the issue in Spanish.

There is another area that I would like to study, which is to continue with the history of penicillin, exploring the history of resistance to antibiotics from a more biological point of view. Many women had their start in microbiology. There is a question of gendering in the description of the manner in which they work with plates and cultures that has to do with qualities assigned to women, which are care, attention to detail, patience, and modesty. I would like to explore the meaning of all of this, and the participation of women in microbiology, a science that has many women in it. Besides, many women have really stood out in this field with the incredible things they have achieved. Right now, I'm studying some of them; they are amazing women.

DFM: We also wanted to talk about your perspectives related to the way women approach different forms of knowledge, particularly in the sciences. You commented on your interest in group research, in collaboration. That is obviously important in the sciences, but we have noticed that in your own work you have a preference for an interdisciplinary approach. Do you think social or environmental factors play a role, or what do you think might explain some of the patterns related to women and their interests in this type of collaboration for interdisciplinary research, or even whether you see this marking your own work?

MJS: I am not sure when I say that women are supposedly more detailed in their work if this is true. What I am saying is that these are stereotypes upon which I want to reflect. I am thinking more about strategies that women have used to forge for themselves an unforeseen space and one that still is not planned for them. In other words, I want to address the lack of recognition that women's work as scientists has received. The fact is that the authorial world shares poorly with women. We still have some problems, and we have to make it permanently explicit so people will be wary. It is very cumbersome, very cumbersome, and it is very tiring. Sometimes we get mad, because what we want to say cannot be said smiling, you know? And there are many times that you don't smile, right? So then, I believe that collaborative works emerge because of those two things. [. . .] I see some of my other male colleagues work collaboratively, so I don't know if that's interdisciplinary. [. . .] I believe that collaboration is a characteristic feature of the community of historians of science who are studying the scientific developments of the twentieth century. Perhaps it is because you need to know a little bit about biology, about chemistry, and about those things so that you can start thinking about, for example, DNA and hydrogen bonds. These things entertain us because we are a little bit freaky about chemistry and biochemistry. [. . .] I am a sociable person, and it also makes me calmer to work with someone. It is security, a little refuge, because in good company your work enormously improves, and you get a lot of inspiration, and alone you can make a lot more mistakes. And I am afraid of missing particular dates and particular concepts and approaches, and I have done it many times because I was young and because I was daring. Well, it is always good to continue to be daring, but one has to protect oneself from oneself, and a good, wise group of colleagues keeps your work in perspective.

DFM: You mentioned challenges that scientific women have experienced, and we wanted to ask you how you see the current environment in Spain for women who are interested in the sciences? How has the cultural dynamic changed for women scientists over the years?

MJS: Oh, I believe what has happened is that we have had a time when everything seemed to be going better. The world was progressing toward justice and equality. We [as a society] have had this hopeful period. But since I know the history of women in science, I already knew that social and professional respect for women had experienced highs and lows throughout history. Women have lost the authority they had in medicine. For example, the authority they had in medieval times was lost when the great temples of knowledge that were the medical schools were built. I know that these things happened. So, as they say in anthropology, there is no full progress. There are always debates, risks, and inequalities.

We've had moments when we believed, when everyone believed, that women's numbers were increasing, but the data on this is very clear in that respect. The betterment is not observed on a statistical level. The personnel at CSIC remain at almost the same proportion of women over the last several decades. Now, we do talk more about the situation. We are more careful when we select committees, and there are commissions that are in charge of being vigilant about these things. It forms part of the political agenda, but the situation is that we mostly talk more about women. Journalists now feel ashamed to only ask for men's opinions. Finally, they are starting to realize, when they ask for experts and things, that they have the obligation to seek women's opinions and advice. Then it seems that things are going better for women.

The research community in Spain has had many difficulties in having their grants renewed because there has always been very little money for research personnel during the last decade especially, and there have been very few calls for new scientific personnel at the CSIC during the last decades. This means that the research community is growing older, and many researchers are retiring without there being any replacement policy. In times of few resources, the position of women worsens. In other words, even though we talk about the problem, and it has entered the political agenda and occupies every space of the scientific and academic agenda, the result is that the situation isn't better. It's better because we talk more, we have more duties, we have more voice, people listen to women more, and people pay attention to us more. Nevertheless, it has been decades since women have been in the majority among medical school students, to exceed fifty percent of the whole university student body, and I don't see tenured female faculty in medicine, anywhere. The problem persists, doesn't it? This is a question that will take many years to resolve, but like now, the fight for recognition is very strong. Voices clamor in order to reclaim things, not only workplaces and research grants but also social respect, rights, and visibility. All of this produces reactions.

DFM: Do you see the same thing happening at an international and global level, when you speak with colleagues from, for example, Mexico, or other parts of Europe, or the United States, or do you think that the situation in Spain is unique?

MJS: I think these are problems that we have in general. What we have here [in Spain] happens in different countries. For example, there was a time in Spain when the research profession did not have enough prestige, and then there were more women in it than perhaps in other countries. This has been shown in women's history, that as soon as a workplace occupied by women obtains prestige and social recognition, then it begins to be occupied by men as well, and sometimes women are expelled from these workplaces.

As you know, in scientific knowledge and its functioning structures, they have always created a place so that women could have the "excellent" role of assistant. Hence, the meaning of the word "assistant" is inseparable from that of woman. Therefore, we don't know what came first, the egg or the chicken. Did women become assistants because they had no other choice, or was it simply that they were called assistants because they were women? All of this is what is being questioned. Women's authority has been publicly recognized, even though they don't have broad-based public recognition. For example, I'm talking about Margarita Salas, who has been well respected for many years, but even though she is a marvelous woman and an extraordinary scientist, she isn't the only one, not even in her generation. We only speak of a few of them, not all of them. My fight always is when we are in women's groups and we say, "Hey, let's propose names for a prize." Why don't we propose new women? [...] We should be sharing, because the community of expert women is very big; the number of them is enormous. This sometimes is reflected poorly. What we need is younger women to enter into the field, the universities, the research centers; we also need more money so that there are more young people who want to become scientists and want to enter into research communities. Then we will perhaps achieve this, so that more women join than before.

Of course, the social scene is dissuasive. Many things are happening at the same time. Younger women are having children; maternity is being mythologized. They are returning home while they raise their children because they think it is important, or because the neoliberal system forces them to have two roles when they only want to have one. Or maybe there is a culture of sending young mothers back home. We don't know what significance this is going to have in the reorganization of gender and in the hierarchies that are emerging from this new environment. All of these include consequences we cannot foresee. So, on the one hand, we have the vindication of maternity by women who are more progressive as a decision that should be made individually, and they do not feel obligated either to work or to be mothers. The women of my generation didn't want to stop working to have children. Well, I never wanted to occupy myself only with my children, even being, as they are, the thing that I love most in the world. We can't fall into the trap of deciding between working and taking care of our children. The trap was that we wanted to do the two things, and that sometimes ended by becoming a bit too stressful. Then of course, problems change. We don't know if these women, who are young, intelligent, and hardworking, will have these maternity leaves that will be different from the ones we had. And we don't know what that means and where it will take us.

It is very interesting because we are in the process of demanding recognition for women in all spheres. As feminist sociologist María Ángeles Durán has said, this work that women do in their homes can be, and thus should be, accounted for in the GDP statistics, because the input of the work that women do in their homes is enormous. It complements the deficit of a nation. Countries where family structures are very strong have survived the crisis better than countries that don't have such strong, supportive family structures. Governments take advantage of this in order to concede less budgeted monies and to avoid developing social policies for assistance. So we are in a difficult moment.

VLK: You have alluded to the role of the media and journalism in some of these cultural shifts, and we have noticed that you have published various articles in El País. We wanted to know why you have chosen what one might describe as the role of "citizen scientist," and what you hope to achieve with this type of publication that reaches the masses or has more diffusion?

MJS: Well, I've always been interested in journalism, probably in part because some of the men that I have most loved were journalists. My husband is a journalist, and it is a world that has always interested me. I even worked briefly in it, as a reporter. I didn't do much, but during a season I wrote an opinion column in a weekly magazine that no longer exists. It was because the director of the weekly publication was a friend of a friend. It was a good group of people, and I believe that I was amusing to them. They relied a little bit on my scientific training as a young person.

So, I started with the weekly column that was, more or less, an opinion column, and things would happen, and I would put those incidents in relation with my scientific knowledge in some way. That was the time period of the Iran-Iraq war, so I could blow off steam. It was a way in which I could express my opinion. At that time, everyone greatly needed to express opinions after so many years of the dictatorship. I grew up during Franco's dictatorship and became an adult during the transition to democracy. Therefore, the ability to speak freely about things was very important to everyone, and everybody liked to do it. But also, I have always liked to write, and I didn't have any other thing to do but that. [. . .] Later, when I began to work at other places, I stopped writing that way. And when I returned to research, well, there were other things in my daily life that were happening, those of which I would have liked to write about, but I haven't done so on a regular basis.

Every once in a while, when there are things that irritate me too much, I write because I get worried about things. I believe that people that want to study science need to study, sit down to study. [. . .] That's the way it is, and studying is something to which people need to dedicate themselves, and it takes a lot of time. It takes years, as is well known, for those who pursue university degrees. Studying should be taken seriously. And diffusion is a problem. If it is done well, and written in a clear way, then it is sometimes not really scientifically correct, because everything is very complex, and simplifying too much to make something clear can result in something misleading. So then, I don't write to divulge, to popularize scientific knowledge—I think I would not know how to do it, although I have seriously tried and finally got it when I collaborated with my husband, Antonio Calvo, and we published a biography of the British woman scientist Rosalind Franklin.[3] I collaborated on this biography for popularizing purposes. It was a nice experience. I have written these few short pieces for newspapers and online periodicals because for me there seem to be situations that are fixable, and there are things that I know personally. Then I write about them. I would like to write more about women . . .

DFM: Would you like to have a weekly column?

MJS: I would love it! Yes, yes, a weekly! Well, I don't know. I'm not sure about a weekly column, but something periodic, yes.

VLK: Do you have the time?

MJS: No, but sometimes I can write quickly. Writing books is a little bit more complicated, of course, but writing something like this, that has to do with something that you know about and that is happening, it requires very few lines, approximately a page, right? Then it's something that one can write rapidly with some passion. And then, somebody reads it for you and suggests changes, corrections, and you create the final version after receiving those comments. I have worked in a newspaper for a short time period, and I know how it is done. People read your piece, and many times they suggest corrections, changes, and you learn a lot, and then you revise it. [. . .] When I wrote the piece for *Femenismo/s* about gendered pronouns, I thought, "Hey, I could do something monthly." But honestly, I am not sure whether I would find the time—although definitely I would like to try. I have arrived at a moment in my life when I have to put my life in order and not be so ambitious. I have to learn how to rest, but it's very difficult to learn that.

DFM: Yes, it is very difficult to do that. Well, we have one last question. In one of your earlier responses, you commented about Franco's dictatorship and its influence in scientific fields. In our field of literary analysis, we often speak about the recuperation of historical memory. We wanted to ask you about your work related to historical memory and scientific research. Do you see a connection? Your work with the history of women seems to be in line with the contemporary project related to the recuperation of historical memory in Spain, which has been so dominant in the last twenty years. Is there a connection?

MJS: Well, this is a very specific thing about determined political and academic activities—research and publications—that have been accompanied by very interesting work from the anthropological perspective. From them, we have learned a lot. As you have formulated it, as the recuperation of historical memory, I don't feel that I participate in this movement. I would like to, but truthfully, I have to confess that I'm not a participant in this intellectual movement, and yet I believe that this work is very inspiring, and in a way I do not feel far from it. In other words, I have paid a lot of attention to what the women and the men with whom I was speaking were telling me, their remembrances, but conceptually and methodologically, I am not exactly dedicated to historical memory.

The historiography of Francoism is an ongoing issue. That is to say that there are great historians, and the historiographical programs that are most inspiring and revealing are those that talk about details related to how people lived, how they managed to cope with the norms imposed by the dictatorship, about real people, and lived concrete situations. Some citizens were in favor, and others were against the dictatorship without ever making it explicit because it was never permitted to be against. It has been considered a given fact that if one did not express oneself against it [the dictatorship], then one was in favor of it. That is not correct. There were interior exiles, about whom social historian Valentina Fernández Vargas and other historians have written.[4]

Regarding the history of science and research, there are outstanding issues, a historiographical debate, and in the case of the history of science it is very clear. I would like to participate in this discussion [about the historical memory related to science]. I believe that I have tried a couple of times with friends, but that hasn't been easy. Then for it to be a useful project, we really need to convene meetings and write books about it. It would have to take into account all points of view, as well as the places, cultural and symbolic, in addition to physical, in which we have situated the people that we study, the roles that we are assigning them, and the epistemological perspective we are using. It is a debate to which I would like to contribute. I think that there is an analysis of the history of the sciences in Spain to be completed, of course. Now, that is not easy. It is an unfinished issue, and the fact that we have not yet had such kinds of debate as a community of scholars who are dealing with the sciences, technologies, and medicine during the Franco regime produces a lot of melancholy in me. We will see what happens.

We have to arrive at agreements and consensus in order to reconstruct history. For that, well, we have to come together in a monographic manner to talk about it and write about it. The problem with the dictatorship is that there are a lot of emotional implications. There are many people who have fought against fascism, and along the way they have lost influence, suffered too much, or died, and all those people have fallen to the wayside. Not only the people who were killed, but also the suffering that was caused, that destroyed people's mental health. It goes beyond talking about the incidents [. . .], which were tragic. The history of the Transition is also dramatic, and sometimes also tragic—remember ETA terrorism. But, as democracy was being constructed, the general euphoria created great expectations and promoted creativity. We have a way of approaching Francoism that is emotional, very emotional. That is okay, because all approaches in history are emotional, and one shouldn't deny the reflexivity that historical research embodies. This allows us to make an explicit reflection. In other words, we are always committed observers, and also participants, which makes the historical project interesting.

NOTES

1. *Mujeres científicas en España 1940–1970: Profesionalización y modernización social.* Ministerio de Asuntos Sociales, 2000; *Entre Cajal y Ochoa: Ciencias biomédicas en la España de Franco, 1939–1975.* Consejo Superior de Investigaciones Científicas, 2001; "Towards Denaturalization: Women Scientists and Academics in Twentieth-Century Spain." *Feministische Studien*, vol. 1, no. 11, 2011, pp. 52–63; *Gendered Drugs and Medicine: Historical and Socio-Cultural Perspectives*, with Teresa Ortiz-Gómez, Routledge, 2014; and *The Circulation of Penicillin in Spain: Health, Wealth, and Authority*, Palgrave, 2018.
2. *Creative Couples in the Sciences.* Edited by Helena M. Pycior, Nancy G. Slack, and Pnina G. Abir-Am. Rutgers UP, 1996.
3. Santesmases, María Jesús, and Antonio Calvo Roy. Rosalind Franklin. PRISA, 2019.
4. Among Fernández Vargas's books, see *La resistencia interior en la España de Franco.* Istmo, 1981; *Las militares españolas: Un nuevo grupo profesional.* Biblioteca Nueva, 1997; *Memorias no vividas: Madrid qué bien resiste.* Alianza Editorial, 2002; and the edited *El Madrid de las mujeres: Avances hacia la visibilidad (1833–1931).* Comunidad de Madrid, 2007. 2 vols. See also Valentina Fernández Vargas's biography at www.csic.es/es/el-csic/mujeres-y-ciencia/mujeres-ilustres/valentina-fernandez-vargas (last access 9 Sept. 2020).

PART II

ON STE(A)M

Integrating Scientific Inquiry into the Cultural Realm

CHAPTER 6

Science in the Works of Clara Janés

A Poetics of Theoretical (Meta)physics

DEBRA FASZER-MCMAHON

In Clara Janés's June 2016 presentation upon being named to the Real Academia Española, she describes the title and focus of her talk using a combination of scientific and mystical terms: "el enigma ronda la escritura. Y este puede presentarse como azar, mediante estratos, ósmosis, fluctuaciones, cruces y selecciones naturales o intencionadas, mutaciones, confluencias" ("Una estrella" 13; enigma surrounds writing. And this can present itself as chance, through strata, osmosis, fluctuations, crossings, and natural or intentional selections, mutations, confluences).[1] It is fitting that Janés's speech upon accepting one of Spain's most important literary honors employs both scientific and mystical language to describe her poetic vocation. Throughout her publications, Janés has consistently shown an interest in the poetics of scientific discourse. Early in her career she produced poetic meditations on geology via works like *Lapidario* (1988; Lapidary), and by the late 1990s she began explicitly engaging with theoretical physics in books like *La indetenible quietud* (1998; Unstoppable stillness) and *El libro de los pájaros* (1999; Book of birds). Since that time she has published numerous collections that embrace the convergence of poetic and scientific experimentation via *Paralajes* (2002; Parallax), *Fractales* (2005; Fractals), *Los números oscuros* (2006; Dark numbers), *Fósiles* (2008; Fossils), *Variables ocultas* (2010; Hidden variables), *Orbes del sueño* (2013; Dream orbs), *Movimientos insomnes* (2014; Insomniac movements), and *Estructuras disipativas* (2017; Dissipative structures). This study analyzes Janés's determined combination of mystical and scientific discourse. It argues that, for Janés, the creative poetic process replicates and challenges the quest for order, disruption, and meaning found in scientific thought, and this process of interrogation and creation reveals the unstable boundaries between scientific inquiry and mystical poetic rumination. Janés links these disparate ways of knowing by explicitly incorporating scientific thinkers and their ideas into her mystical poetry and by highlighting scientific language as a historically embedded discourse.

This essay will first provide background on Janés's larger corpus and how her interest in bridging scientific and poetic thought connects with her larger poetic trajectory. Second, it will offer close readings from several works to highlight how Janés's interpretation of three specific scientific concepts serves to connect scientific and poetic discourse communities. Finally, the study will address why these efforts by Janés to bridge scientific and poetic worlds are important for Spain, for global women's issues, and for science today.

Janés's Poetic Trajectory and Scientific Thought

Janés's poetry demonstrates a sustained effort to bring diverse discourse communities into contact. She has proven herself adept at this endeavor throughout her career, particularly in the way her publications bring together distinct cultural and religious traditions.[2] Works like *Creciente fértil* (1989; Fertile crescent), *El hombre de Adén* (1991; The man from Aden), *Rosas de fuego* (1996; Roses of fire), and *Diván del ópalo del fuego* (1996; Fire opal divan) demonstrate a desire to connect Spanish poetic production with Islamic and Eastern cultural forms. This focus on bridging cultural divides remains present in Janés's contemporary work, but since the turn of the century she has shown a growing propensity for linking scientific and poetic discourse via metaphysical and mystical forms. Felix K. E. Schmelzer highlights this pattern in his 2018 article "Los versos pitagóricos de Clara Janés," noting that "la poesía tardía de Clara Janés está marcada por la recepción de las matemáticas" (Clara Janés's later poetry is marked by the reception of mathematics), and he argues that Janés follows a Pythagorean trajectory that "'poetiza' las matemáticas para insinuar el misterio inefable de la creación" (365; "poeticizes" mathematics in order to insinuate the ineffable mystery of creation).[3] Indeed, Janés's recent works clearly engage mathematical concepts, and they also deeply interrogate theoretical physics. Her poetry juxtaposes and blends scientific discourse with metaphysical ideas and mystical forms. This kind of bridging between different discourse communities is intriguing, particularly in the context of current discussions about how women and the arts have been relegated or distanced from the scientific realm.

Janés notes in the introduction to her 2014 work *La tentación del paraíso* (The temptation of paradise) that ever since childhood she has been fascinated by scientific discourse: "Quería ser astrónoma. Todavía quiero ser astrónoma. Este deseo volvía ahora, avanzados los años ochenta" (27–28; I wanted to be an astronomer. I still want to be an astronomer. This desire has returned now, after eighty years). In fact, in our 2014 personal interview about science and poetry, she repeated the notion two times, imitating herself as a child and saying, "Quiero ser astrónoma, quiero ser astrónoma" (I want to be an astronomer, I want to be an astronomer). One might ask why Janés did not in fact become an astronomer, considering this passionate recollection. While Janés does not directly address this question in her writings or interviews, it is well documented that in the 1940s and 1950s, when Janés was coming of age, finding a career in astronomy was not a very feasible goal for Spanish women, with overt government propaganda disparaging women's roles in the public sphere and encouraging them to become mothers and housewives.[4] Yet Janés's fascination with the stars was never far from her literary production. In her article "Las ecuaciones de la poesía" (Poetic equations) she recalls her childhood fascination with algebraic formulas, noting how they combined "por un lado, una certeza (se trataba de algo que era exacto), y, por otro, un enigma," (on the one hand, a certainty [it was something that was exact], and, on the other hand, an enigma), and she goes on to emphasize that "verdad y enigma son también dos características de la poesía, dos características importantes. Pero hay más cosas que aproximan la poesía a las ciencias" (4; truth and enigma are also two characteristics of poetry, two important characteristics. But there are more things that connect poetry to the sciences).

She similarly emphasized in the aforementioned personal interview how her scientific passion affected her writing, with her first major poetic publication, *Las estrellas vencidas* (1964; The conquered stars), indicating her astronomical interests in its title, followed by works like *Límite humano* (1973; Human limit) and then beginning in earnest with many poetic texts focused explicitly on scientific topics, such as *Paralajes*, *Fractales*, and *Variables ocultas*. Janés links the beginning of this shift toward scientific discourse with *El libro de los pájaros* and her interpretation of Chillida's works in *La indetenible quietud*, noting that her Chillida-inspired poems stemmed from exposure to wave theory and a range of scientific notions that she began to encounter via conferences and publications.[5]

When analyzing the scientific trajectory of her corpus, Janés references the work of several theoretical physicists like Basarab Nicolescu, Albert Einstein, Stephen Hawking, Ilya Prigogine, and Erwin Schrodinger, as well as the philosophical writings of diverse authors such as Derrida, Lacan, Nietzsche, Attar, and Camus. Janés employs these philosophers' and physicists' ideas in order to contextualize her efforts to challenge, incorporate, and disrupt the hierarchies and divides between poetic and scientific thought. She highlights poetic language and verses that "rondan en torno al anhelo de absoluto, es decir de eso que apetecemos y nos falta" (*La tentación* 12; hover around the longing for the absolute, that is to say that which we desire and lack). The absolute, for Janés, is not so much an object or a concept but rather a desire or an absence. She notes that many philosophical approaches, including the work of some of the famous figures mentioned above, have a tendency "definir las cosas a través de una negación" (to define things through negation), yet she insists that "yo, en vez de negar, de apartarme, voy rodeando algo, voy intentando enfocar algo de lo que tengo solo indicios, algo que se escapa cuanto más intento aproximarme" (*La tentación* 13; I, instead of negating, or distancing myself, circle around it, trying to focus on something of which I only have indications, something that escapes the closer that I try to get to it).

Janés's description sounds very much like a scientist who seeks to measure even when the very act of measuring makes a final answer impossible. Indeed, Janés finds that neither poets nor philosophers seem to offer the right metaphor for her experience, and that instead she has found deep insights from physicists such as Edwin Schrodinger, who notes that his scientific knowledge encompasses hundreds of millions of years of insight, but at the same time it only holds an infinitesimal fraction of the vastness of time and space: "yo, un pequeño punto en el inconmensurable tiempo, nada incluso en relación al número finito de millones de años que he aprendido a medir y a calcular—. ¿De dónde vengo y adónde voy? Esta es la gran cuestión insondable, la misma para cada uno de nosotros" (Schrodinger qtd. in Janés, *La tentación* 13; I, a small point in the immeasurability of time, nothing in fact in relation to the finite number of millions of years that I have learned to measure and calculate—Where do I come from and to where am I going? This is the great unfathomable question, the same one for each of us).

Janés approaches this impenetrable question by pointing to three key themes or perspectives that she argues theoretical physics and mystical poetry share: "un nivel fuera del tiempo, la unicidad y el 'saber del no saber.' El primero estaría representado por la teoría de la relatividad, el segundo por la función de onda y el tercero por el principio de incertidumbre" (*La tentación* 34–35; a level outside of time, uniqueness, and the "known unknown." The first would be represented by the theory of relativity, the second by the wave function, and the third by the principle of uncertainty). Janés is drawn to physics because it seems to explain, first, "a level" of experience or reality "outside of time," as in Einstein's theory of relativity; second, it offers a complex notion of "unicidad," or something unique that unifies all matter, as in Schrodinger's wave theory; and third, theoretical physics postulates a recognition of the limitations of knowledge, the "saber del no saber," or Heisenberg's uncertainty principle.

In order to understand how and why Janés's work incorporates these scientific ideas, the next section will highlight how scientific theories postulated by Einstein, Schrodinger, Heisenberg, and others both appear in Janés's poetry and clarify Janés's efforts to bridge scientific and poetic divides. Janés's poetry connects scientific and mystical discourse in part by offering explicit dialogue with scientists and their ideas. However, the connection also goes further and offers important subtlety. Janés's poetry not only connects with science explicitly, it also alludes to how scientific thought has developed over time, thus highlighting language and the role of history in mediating the discursive system. The following will provide close readings from four works in order to clarify the ways Janés's publications link scientific and mystical poetic thought.

Scientific and Mystical Discourse in Janés's Twenty-First Century Poetry

Janés's works bridge mystical and scientific discourse communities in part by offering explicit dialogue with scientists and their ideas. In *Orbes del sueño* (2013), for example, Janés incorporates quotations from a range of scientific thinkers from antiquity through the twenty-first century, including Heraclitus, Parmenides, Macrobius, Ilya Prigogine, Erwin Schrodinger, and Basarab Nicolescu. *Orbes* was published in 2013, but one can see this same propensity going back to 1999 with *El libro de los pájaros*, where she references Basarab Nicolescu, Albert Einstein, Stephen Hawking, and Werner Heisenberg, as well as a range of scientific theories like the "butterfly effect," "strange attractors," and the "laws of thermodynamics" (11). Some of the quotations that she selects from physicists demonstrate her interest in specific scientific theories, like the relativity of time, and how they relate to mystical and scientific blending, but the quotations also highlight how physicists themselves recognize a level of artistry and mystery in their work. For example, her quotation from Ilya Prigogine, a Nobel Prize-winning expert in thermodynamics, demonstrates the ongoing interrogation of time among scientific thinkers: "Cómo se imprime el tiempo en la materia? En definitiva esto es la vida, es el tiempo que se inscribe en la materia" (Prigogine qtd in Janés, *Orbes del sueño* 42; How does one imprint time in matter? In short, this is life, it is time inscribed in matter). Other quotations are less focused on specific issues like time and instead point more generally to science as a discourse. She quotes Erwin Schrodinger, for example, on the issue of science and the question of subjectivity: "Todo conocimiento científico se basa en los sentidos. Sujeto y objeto son una sola cosa" (Schrodinger qtd in Janés, *Orbes del sueño* 57; All scientific knowledge is based on the senses. Subject and object are one thing).

But Janés is not limited to simply referencing specific scientific insights or scientists in her work. Her original poetry also employs scientific ideas in intriguing ways. *Paralajes* offers some particularly compelling examples of how Janés incorporates and blends scientific and mystical thought. The title of the work references an important astronomical concept, the notion of parallax, or the seeming shift or displacement of an object based on the perspective of the viewer ("Parallax"). An everyday example would be the speedometer on a car. If you are seated in the driver's seat, you see the indicator pointing at a particular number, perhaps seventy, while the passenger seated to your right might see the speedometer, at the exact same moment, pointed at sixty-five. The speedometer has not shifted, but the viewpoint has. The implications of the parallax, however, go beyond the limitations of perspective. Parallax is used to determine unknown distances to new objects based on trigonometry and the known distance of other, closer objects. If an observer moves a particular known distance and then measures the change of angle for two objects, one closer and with a known distance, trigonometry can then be used to find the value of the unknown variable. Janés, as a mystical poet, is coming at science from a different angle, and the effects of her shifting perspective lead to new insights.

One poem in *Paralajes* addresses both the notion of parallax as well as Heisenberg's uncertainty principle, which Janés has identified as one of the key concepts she sees linking theoretical physics and mystical poetry. Heisenberg's uncertainty principle postulates that because particles are interacting so quickly and randomly at the subatomic level, one cannot measure an object's speed and its location at the same time without altering the results (Furuta; Jaeger). The act of measuring changes the motion, and thus one can measure either speed or location, but never both simultaneously. One of the implications of the uncertainty principle, which relates back to mystical thought, involves the notion not just of uncertainty, but that there might not actually be a stable truth to measure.

Some interpret Heisenberg's principle to mean that science can never really measure things accurately, and thus they argue that while science is limited by the principle, Heisenberg's notion does not really challenge the notion of scientific truth, since the act of measuring is what throws off the "true" result (Furuta). However, for Janés (and Heisenberg himself), the problem is in fact much deeper. The principle implies that subatomic particles do not actually have a particular value that could ever be defined, meaning that perhaps an intrinsic property of nature itself is this lack of a well-defined or core "being." Michael Raymer addresses this problem in *Quantum Physics: What Everyone Needs to Know*, noting that "a quantum particle doesn't *have* a position or momentum (that is, velocity) before it is measured" (127). The truth is in the constant shifting of reality. Janés plays with this concept, and its implications, in the fifth poem of the section called "Línea de paralajes" (Parallax line):

> Buscan asiento los signos,
> pero nómada es la forma
> en el desierto inasible
> y falaz la transparencia.
> El oleaje del reverbero
> aviva la incertidumbre.
> El sol ignora el movimiento del sol.
> El silencio multiplica su paralaje. (*Paralajes* 69)

> The signs look for a seat,
> but form is nomadic
> in the inaccessible desert
> and transparency is false.
> The wave of reverberation
> heightens the uncertainty.
> The sun is unaware of the sun's movement.
> Silence multiplies its parallax.

The poem begins by personifying signs, and in the opening verse signs are looking for a seat but are unable to settle: "Buscan asiento los signos / pero nómada es la forma." Signs are "nomads" wandering in an "inaccessible desert," constantly changing form. It is not difficult to connect this nomadism of signs with Heisenberg's uncertainty principle. The term "inasible" is intriguing, used alongside "nómada" to describe the signs. "Inasible" denotes the impossibility of accessing something, but it also implies from its root, "asir," that this thing or idea is not accessible in part because it is unmoored, unattached, or unanchored. It is not just that we cannot establish physical or mental contact, but that indeed its very presence in space/time is in question, much like Heisenberg's postulation. Indeed, as the poet continues in the fourth verse, it appears any seeming transparency or accessibility is misleading: "y falaz la transparencia."

In the second half of the poem Janés links these "signos" explicitly to the principle of uncertainty, creating a metaphor between mystical poetic form and scientific thought: "El oleaje del reverbero / aviva la incertidumbre" (69). The "wave" that reverberates with this constant motion of signs "heightens" or "intensifies" the uncertainty. With "oleaje," Janés's verse addresses not only Heisenberg's principle, but also the unifying notion of wave theory that she cites as another key connection between mystical and scientific thought. In 1925, the Austrian physicist Erwin Schrodinger created a mathematical equation to explain the movement of light that proved so powerful other physicists soon realized it could be applied to all matter (Klus). For Janés, what is crucial about wave theory is the notion of unification of matter, bringing disparate elements into a shared framework. Indeed, a recent study updating Schrodinger's work describes the ongoing importance of the theory by emphasizing its impact on all phenomena: "The time-dependent Schrodinger equation is a cornerstone of quantum physics and governs all phenomena of the microscopic world" (Schleich 5374). Janés extends this idea beyond the microscopic into the ontologic. In the middle of her poem "Línea de paralajes," the "oleaje" or "swell of waves" caused by the constant motion of "los signos" is said by the poet to "enliven the uncertainty" of form, both poetic and ontological, pulling together Heisenberg's uncertainty principle and Schrodinger's wave theory within Janés's own mystical poetic space.

Janés's desire to bridge varied perspectives or systems of thought appears again as the above poem draws to a close. It ends by asserting that the sun, the dominant force in our solar system, is unaware, indeed ignorant, of its own movement, and the poet links this blindness with the parallax view: "El sol ignora el movimiento del sol. / El silencio multiplica su paralaje" (69; The sun is unaware of the sun's movement / Silence multiplies its parallax). Even the most dominant force in our midst is affected by its own perspective, and like most sentient beings, is seemingly unaware of the contradictions (and possibilities) inherent in its own movement.[6]

The concept of parallax has received important attention in recent years. Slavoj Zizek, for example, published what he considers his most important work about precisely this concept in his 2006 *The Parallax View*. Zizek addresses all kinds of philosophical, scientific, and political issues in the work, but of particular interest is his critique of contemporary neuroscience, particularly the notion that all mental activity can be simply reduced to a set of predictable neural processes. Zizek argues that while ideas may be initiated by concrete chemical reactions, they are not determined by how they are initiated, and can "break away" in unpredictable fashion and in a way that preserves an ontological separation between chemical processes and knowing or being. Janés's poetry has long pointed to realities beyond the physical, and as the poem ends, this reference to parallax and to silence situates signs within a complex worldview that embraces the metaphysical: "El silencio multiplica su paralaje" (69; Silence multiplies its parallax). Silence is not the absence of signs but rather recognition of their constantly shifting form, a kind of awe for the wave that acknowledges and multiplies the sense of changing and immeasurable perspective. While the sign is nomadic and unreachable, the awe stemming from such constant creativity impels a silence that becomes the agent of awareness.

This same sense of movement toward the labyrinth that is mystical poetry, as toward scientific insight, comes across strongly in several of Janés's poems from other works, such as *El libro de los pájaros*. For example, poem 11 addresses the issue of artistic form and connects it with Einstein's important groundbreaking theories about space and time:

> Sigue,
> síguete en este avance
> que es hilo laberinto de belleza,
> de una belleza que no se detiene en forma,
> Y en ese tránsito el espacio es abolido
> el tiempo es abolido,
> y hasta el aliento
> está a punto de desmoronarse. (*El libro de los pájaros* 35)

> Continue,
> follow the advance
> that is a labyrinthine thread of beauty,
> a beauty that does not hold form,
> and in that crossing, space is abolished
> time is abolished
> even air
> is on the verge of collapsing.

The poem begins by pushing the reader to continue in the task of interpreting and creating signs, repeating the call to advance: "síguete en este avance / que es hilo laberinto de belleza." The complex process of artistic creation, like scientific discovery, is apparent from the beginning, since this advance is not so much a straight geometric line but rather a "hilo laberinto de belleza," a threaded labyrinth of beauty. This beauty is not explicitly poetry, since it does not stop to take a form ("no se detiene en forma"). What is key is not the form itself but the movement, a motion that redefines both space and time: "Y en ese tránsito el espacio es abolido / el tiempo es abolido" (35).

These references to overturning space and time recall another of the three scientific principles central to Janés's work, Einstein's famous theory of relativity, with its deep implications for being and time. Einstein's 1905 Special Theory of Relativity changed the way scientists understood space-time. The theory demonstrated that while space and time appear to be stable constructs, they are actually dependent upon the perspective and stability of the observer (Greene). This first theory, however, could not explain precisely how space-time was affected by accelerating objects, and in the above poem, the focus is clearly on acceleration and movement, with the poetic speaker commanding the reader to "síguete en este avance" with a beauty that "no se detiene." The poem seems to allude to Einstein's second major insight, the General Theory of Relativity (1915), which explains that the mass of objects in motion warps space-time, thus changing the definition of gravity from a stable force to a reaction involving acceleration caused by the warping of space and time (Ferreira).

For Janés, this means that a resistance toward space and time is not something simply desired in mystical thought but also evocative of a cosmic reality, the potential for space-time to bend depending on the forces working in its midst. It is through this movement, "en ese tránsito," that space and time are abolished. In fact, in Janés's poem, even breath seems threatened, with "aliento" on the point of collapse or breakdown. But the experience hangs just before the edge of collapse ("está a punto de desmoronarse"). It is this tentative hanging point, this balance, this uncertainty of time and space that Janés finds a fitting link between mystical poetry and scientific thought. The poetic advance continues into the labyrinth of beauty that is both theoretical physics and mystical poetry.

In addition to explicitly incorporating scientists and their ideas into her poetry, Janés also challenges the boundaries of scientific and mystical discourse by showing scientific language to be part of a genre that is historically determined. She reveals science as a discourse community via several of her poetic ruminations on scientific history. *Los secretos del bosque* (2002; The forest's secrets) is a prime instance of this effort. The work follows the stages of an alchemical experiment to ground a mystical experience, and it simultaneously highlights the tremendous changes that have occurred in scientific thought over time while also honoring some of the attraction of alchemy, particularly its connection to mystery, experimentation, and exploration of the unknown.[7]

In a subsequent work, *Los números oscuros* (2006), Janés continues to demonstrate the way scientific practice operates as a discourse. In the titular poem of the collection, occurring about halfway through the work, the poet uses an extended metaphor to treat the issue of dark numbers, recalling the notions of dark matter and dark energy, and employing scientific language throughout such as "oscilante," "eco," "candente," "vértigo," "ecuaciones," "orientación," "cero," "números," and "operación" (oscillating, echo, red-hot, vertigo, equations, orientation, zero, numbers, and operation). The poem, as all the other pieces in the collection, takes the form of prose, making it even more connected in a formal sense to typical scientific writing by the use of paragraph frames and strong descriptive components:

"Los números oscuros"

Desde la primera noche hubo un mensaje oscilante, que se mostraba y se ocultaba. Recogí su eco y lo guardé en un cofre: era el primer número oscuro que llegaba a mis manos.

Por entonces hubo también una respuesta; el segundo de aquellos números. Igualmente lo guardé. Ambos, además, eran candentes y no podían tocarse. No sumé ni resté, dejé que siguieran su curso. Luego llegaron otros. De vez en cuando abría el cofre y veía que habían aumentado y que se trenzaban y destrenzaban, de tal modo que me daba vértigo mirarlo.

Fuera del cofre las ecuaciones eran distintas y algún día pasaba todavía aquel pájaro que llevaba una flor en el pico y la depositaba en mi pelo.

Los números oscuros son cifra de lo incomunicable y a la vez ensanchan la propia visión. Aún no han despejado todas las incógnitas e incluso alguno se ha escapado del cofre, pero actúan como espejos.

Yo sigo sin tocarlos, respetando su orientación. Tampoco he despejado mi incógnita: mis números, que son distintos, se perdieron en el bosque de los secretos. (*Los números oscuros* 48)

"Dark numbers"

From the first night there was an oscillating message, that appeared and disappeared. I gathered its eco and I saved it in a chest: it was the first dark number that came into my hands.

At that time there was also a response; the second of those numbers. I saved it too. Both were also red-hot and could not be touched. I did not add or subtract, I allowed them to follow their path. Then the others arrived. From time to time I would open the chest and see that they had grown and were braiding and unbraiding themselves, in such a way that it gave me vertigo to watch.

Outside of the chest the equations were different and one day that same bird was passing by and carried a flower in its beak that it deposited in my hair.

Dark numbers are a cipher of what is incommunicable and at the same time they expand one's vision. The unknowns have still not become clear and some have even escaped from the chest, but they act like mirrors.

I keep going without touching them, respecting their orientation. My own mystery has not cleared up either: my numbers, which are distinct, were lost in the forest of secrets.

The poem reads like a detailed journal or recollection of a dream. The poetic speaker uses preterit verbs to describe the initial scene, not the "había una vez" of a fictional tale, but the "hubo" of a recorded moment in time. The first two paragraphs present a poetic speaker describing an encounter with a kind of Morse code, an "oscillating message" that is alternatively visible and hidden ("se mostraba y se ocultaba"), and that the speaker is only able to capture in the form of an echo and then hide away. This first "número oscuro" soon leads to replies, and each time a new message surfaces the speaker saves the echoes. They are burning and cannot be touched, alive in some way. The speaker does not try to stop the messages or interrupt them but lets them run their course, and more and more keep coming. Every once in a while, the speaker opens the chest and sees how the messages have become intertwined and braided, in the way Janés sees mystical and scientific thought. But looking at them too closely gives the speaker vertigo. The constant motion is too much to grasp.

In the third paragraph the perspective shifts, from the level of signs to physical experience. Outside the box the equations are different, and the poetic speaker links the dark numbers to the natural world. The speaker says that "Fuera del cofre las ecuaciones eran distintas y algún día pasaba todavía aquel pájaro que llevaba una flor en el pico y la depositaba en mi pelo" (Outside of the chest the equations were different and one day that same bird was passing by and carried a flower in its beak that it deposited in my hair). Here the poet implies that the equations (the signs and messages swirling and mixing in the box) are connected to an experience of the natural world, linked through one's perception of them, much like the quotations discussed above that Janés employs in *Orbes del sueño* from Schrodinger and Prigogine.

The next segment clarifies this linking between the scientific and mystical worlds. The speaker defines "dark numbers" as a cipher or code for what is not communicable and at the same time claims that they act as mirrors of something else, something deeper. The speaker continues without disrupting the blending of signs, respecting their creative, unpredictable movement: "Yo sigo sin tocarlos, respetando su orientación. Tampoco he despejado mi incógnita: mis números, que son distintos, se perdieron en el bosque de los secretos" (48; I keep going without touching them, respecting their orientation. My own mystery has not cleared up either: my numbers, which are distinct, were lost in the forest of secrets). The poem ends with a nod to the speaker's own numbers or signs, perhaps referencing poetic texts, and to "el bosque de los secretos," a clear reference to Janés's previous work, *Los secretos del bosque*, which highlights the discursive framework of science via the history of alchemical exploration and thought. Janés alludes to how her own work addresses science as a discourse via the trajectory of medieval alchemical processes and the labyrinth that is scientific experimentation and discovery. The reference to the "forest of secrets" also recalls Heidegger's work *Holzwege*, literally "Forest paths," which was translated as *Caminos del bosque* in Spanish and which Janés has referenced as having an important impact on her work.[8] The work addresses ontology with a particular focus on how poetry provides a privileged medium for deep existential questions related to being in the midst of the social, moral, and scientific complexities of the world. Janés's references here to the natural world, to Heidegger's poetic ruminations, and to the alchemical tradition in scientific thought all recall the ways both science and poetry function as parts of discursive communities. The speaker is not trying to disrupt either discourse—"Yo sigo sin tocarlos, respetando su orientación"—but also acknowledges that no one has uncovered the mystery, the unknown: "Aún no han despejado todas las incógnitas" (I continue without touching them, respecting their orientation; The unknowns have still not become clear). The cipher remains secret, lost in the forest of signs, the space where both science and poetry are seen as forms of discourse, and both are beautiful, powerful, artistic, and constantly changing.

Another example of how Janés connects disparate fields and highlights science as a discourse appears in a poem she dedicated to Basarab Nicolescu in the late 1990s after encountering his scientific work. At that time Janés had begun to explore a range of scientific theories, including Nicolescu's *Teoremas poéticos* (Poetic theorems), a work that was itself an effort to bridge scientific and poetic divides and to exemplify his notion of the "hidden third" (Nicolescu, *From Modernity*). Janés first published her dedication to Nicolescu, "Tercero oculto" (The hidden third), online in the early 2000s, but it appeared in print much later in *Estructuras disipativas* (2017).

Estructuras disipativas offers hints of Nicolescu's theories via the cover art: a mathematical formula and graph that represent the function of two vectors in multiple dimensions, demonstrating rates of change as time progresses. The "t" in the equation stands for time, and it appears on only one side, meaning that it is a hidden factor on the other side of the formula. The cover image connects implicitly with the titular section of the work, "Part IV: Estructuras disipativas," which opens with Janés's poem "Tercero oculto," dedicated to Nicolescu. The poem provides a response to Nicolescu's complex notion of a "hidden third," or an unexpected force that is created by the interaction between subjects or objects. Nicolescu is a theoretical physicist whose research focuses primarily on elementary particles, but he is also a major proponent of interdisciplinary dialogue between science and the humanities.

Nicolescu's theories of the hidden third are complex, but one central concept relates to the broader philosophical problem of how to interpret or represent "reality" and what that might mean about interactions between self and other. Nicolescu argues in his essay "Transdisciplinarity" that despite major scientific challenges to Newtonian conceptions of reality, most people continue to follow classical modes of thought and to consider reality through the lens of subject / object dualism (13). For Nicolescu, "reality" is all about resistance, and he has chosen this scientific term with care (15). For him, the nature of reality involves exploring the multilayered systems, experiences, and actors that resist or create nonresistance in any context or field, and these are always relational (21). In that sense, subject and object are linked together via a hidden third, which is a level of shared reality, or resistance. In Janés's 1999 work *La palabra y el secreto* (Word and secret) she cites Nicolescu's *Teoremas poéticas* and notes that his ideas offer a "lugar de encuentro entre la física cuántica, la Filosofía de la Naturaleza y experiencia interior" (Nicolescu qtd in Janés, *La palabra* 117; place of encounter between quantum physics, Natural Philosophy, and inner experience). Then Janés extends this to explain how she views poetry connecting to scientific discourse:

Esta experiencia — a la que se une la constatación científica — es la de la complejidad que invade todos los campos del saber, y es la que lleva a Nicolescu a definir la Naturaleza basándose en dos pilares: los "niveles de la Realidad" y la "lógica del tercero incluido." Ambos responden al hecho de que lo diverso en un nivel puede ser idéntico en otro, es decir, lo divergente puede converger, entrañar, pues, ese punto: el "tercero secretamente incluido." Por cientos se cuentan estos *Teoremas poéticos* que representan de modo excelente la aproximación de la ciencia a la poesía. (*La palabra y el secreto* 117)

This experience—which is connected to scientific verification—is that of the complexity that invades every field of knowledge, and it is what led Nicolescu to define Nature based on two pillars: the "levels of Reality" and the "logic of the included third." Both respond to the fact that what is diverse on one level can be identical on another; that is to say, the divergent can converge, become implicated, thus, that point: the "secret included third." They count in the hundreds these *Teoremas poéticas* that represent in an excellent way the approximation of science to poetry.

For Janés, Nicolescu's concept of the "hidden third" exemplifies the connections between poetry and scientific inquiry: both involve recognizing the complexity of all fields of knowledge and that these fields "converge" and become "entangled" with other forces.

Janés thus dedicates a poem to Nicolescu, titling it "Tercero oculto" (The hidden third):

> *a Basarab Nicolescu*
> Descansar en el verde
> de la selva,
> en el pájaro que llama al alfabeto,
> en las gotas de agua suspendidas,
> que son letras
> ajenas a concepto
> y se posan en las hojas,
> como un alado respirar
> que apacigua
> los turbiones oscuros
> de la palabra.

Vuelva a mí la virginal llamada
de una forma
pura resonancia,
que cruza el corazón
y lo llena de luz comunicante
anulando los límites
que establece el otro
al enunciarse. (Janés, "Basarab Nicolescu")

   to Basarab Nicolescu
Rest in the green
of the forest,
in the bird that calls to the alphabet,
in the suspended drops of water
that are letters
unconnected to concept
settling in the leaves,
like a winged breath
that pacifies
the dark showers
of the word.

Return to me the virginal call
of a form
pure resonance,
that crosses the heart
and fills it with communicating light
nullifying limits
established by the other
upon enunciation.

The natural world is immediately present in the imagery of the poem, operating on a scientific level not of quantum physics but of biology: *verde, selva, pájaro, gotas de agua, hojas, alado, turbiones, resonancia, corazón*, and *luz* (green, jungle, bird, drops of water, leaves, winged, downpours, resonance, heart, and light). In the first stanza, the poetic voice is resting in the lushness of a natural scene, and throughout each verse the natural world becomes more explicitly linked to poetic creation: a bird calling the alphabet ("el pájaro que llama al alfabeto"), water drops hanging in suspension and becoming the letters ("las gotas de agua suspendidas, / que son letras"), and a shower of words portrayed as dark torrential rains ("los turbiones oscuros / de la palabra"). This extended metaphor between the natural world and the poetic word comes to a culmination in the second stanza, where the poetic voice is no longer resting in a natural / linguistic space but is now directly commanding: "return to me" (vuelva a mí). The goal of this return or this longing involves breaking through the limits of self and other: "anulando los límites / que establece el otro / al enunciarse." The way the lines are formulated, it is unclear if the "other" is the one establishing limits, which the poet here calls to be annulled, or if in fact the limits, as in Nicolescu's theory, are what create access to the other, a force that enables the shift to announce oneself. The act of enunciation is an effort to break down barriers, to create a force or resistance that must be acknowledged, and via this extended nature metaphor Janés offers a poetic interpretation of Nicolescu's notion of the "hidden third."

Janés quotes Michel Camus in order to explain why Nicolescu's concept of the hidden third is key for understanding the value and role of literature. According to Camus, Nicolescu's theory focuses on the level of symbol, and the capacity for symbolic thought to "unite contradictions, embracing and forging contradictory things together" (qtd in Janés, "Basarab Nicolescu"). Metaphor offers precisely such a leap between conceptual space, but Janés does more than simply model the hidden third via metaphor. One can see in Janés's poem evidence of the drive toward complex unification, not simply through metaphor, but also through blending poetic language and scientific discourse, as in wave theory, where "pura resonancia" can cross the deepest subjective divides and annul the limits between self and other. Significantly, Janés uses terms from theoretical physics to describe this leap: "una forma," "resonancia," "luz," and "límites" (a form, resonance, light, and limits).

The wave structure, referenced by "resonance," helps to clarify the role of language in the second stanza. In physics, resonance refers to the way external forces or resistance can make other systems vibrate or oscillate, and these vibrations or movements are waves (Sekihara 2). Interactions between systems create oscillations, and in this case, the oscillations of poetic creation become a "communicating light" that allows connections between disparate forms. Indeed, it seems that what is breaking down barriers is not so much the "other" in the poem but rather the "llamada" in the form of pure resonance, which jolts the other into motion and breaches the divide between speaker and other. In this case, the speaker is also the "other," longing for the "llamada" to return to break down barriers. That kind of shifting subject position is intriguing, and it links the resonance in the poem to the natural world via the bird, the water droplets, and the leaves that symbolize the call of this larger resistance. It puts into motion a set of vibrations that can "cross the heart," fill it with "communicating light," and break down limits of reality. The natural world is both enmeshed in the word and juxtaposed to it. The water, suspended on the leaves, reminds readers how the natural world offers a pacifying space of nonresistance to the torrent of words, letters, and quests for communication and meaning. Janés's poem dedicated to Nicolescu ends with an affirmation of nature's potential, and thereby poetry's, to use resistance itself to bridge seemingly insurmountable divides.

Conclusion: Why Janés's Work Matters for Spain, Spanish Literature, and Global Women's Issues

What Janés finds fascinating about Nicolescu's "hidden third," and about theoretical physics in general, is its ability to completely reconceptualize reality, to offer new ways of seeing and being. Theoretical physicists insist that their discoveries must push thinkers to test new models for reality.[9] For example, in Nicolescu's 2014 work *From Modernity to Cosmodernity: Science, Culture, and Spirituality* he critiques public discourse about science, which he argues has focused on a "classical" scientific model which served to "reinforce our belief that we were living in a rational, deterministic, and mechanist world, destined to an endless progress" (1). He argues that all of the scientific breakthroughs of the twentieth century, including in quantum mechanics, biology, and information technology, should have shifted our notion of reality but in fact did not: "the triple revolution that spanned the twentieth century—the quantum revolution, the biological revolution, and the information revolution—should have thoroughly changed our view of reality. And yet, our mentalities remained unchanged" (1). Nicolescu also quotes Wolfgang Pauli, who won a Nobel Prize in physics and was one of the founders of quantum theory, noting that Pauli had stated in 1948 that "the formulation of a new idea of reality is the most important and most difficult task of our time" (Pauli qtd in Nicolescu 1). Nicolescu argues that now, "more than sixty years later, this task remains unfulfilled" (1). Janés is drawn to theoretical physicists and their efforts to reconceptualize reality, perhaps in part because it mirrors the way she herself has pushed gendered limitations and literary boundaries, at times using the resistance to forge new opportunities and relationships that cross artificial divides and that call for radically new ways of perceiving reality.

One might argue that not only in Spain, but indeed globally, a radical shift of perspective is needed in order to bridge the gendered divide in STEM fields. For Janés, transdisciplinarity, or STEAM, is at the heart of that effort. When Janés reflects on her early poetic interactions with scientific ideas, she emphasizes transdisciplinarity and its importance for her developing poetics (*La palabra y el secreto* 116–17). Indeed, it was her work with translation that first led her to the deeper exploration of scientific and poetic connections, as she describes in her online article for *Adamar*, "Basarab Nicolescu: Teoremas poéticos." She was captivated by a French poetry journal titled *Sources* that a contact from Belgium sent her that included readings excerpted from works like *The Tao of Physics* by Fritjof Capra, *A Brief History of Time* by Stephen Hawking, or *Science mathématique et poésie* by Fernand Verhesen. The journal's articles also included Michel Camus' study of Nicolescu and transdisciplinarity, and they inspired Janés's chapter "La aventura" in *La palabra y el secreto*. She notes that this began her journey into theoretical (meta) physics: "Los *Teoremas poéticos* me hicieron comprender, de otro modo, algunos títulos de las obras de Chillida, como *Rumor de límites, Espacios sonoros* o *Modulación del espacio*. Así nació el libro que hice con el escultor: *La indetenible quietud*. Y luego *El libro de los pájaros, Paralajes, Fractales, Variables ocultas* . . ." ("Basarab Nicolescu: Teoremas Poéticos"; *Poetic theorems* made me understand, in a different way, some of the titles of Chillida's works, like Rumor of limits, Sonorous spaces, or Modulation of space. That is how the book that I did with the sculptor was born: Unstoppable quiet. And then Book of birds, Parallax, Fractals, Hidden variables). This kind of interdisciplinarity and connection between art and scientific thought began to appear with increased frequency in Janés's corpus, perhaps because theoretical physics provided a complex framework capable of imagining radical forms of reality, bridging the divide between self and other and insisting, like poetry, on the need to form a deeper, more nuanced understanding of experience.

Janés has long been intrigued by the stars and by scientific ideas, but in her later poetry, readers can see a steady momentum toward more and more explicit integration of scientific discourse. One might wonder why this matters for Spain, for Spanish literature, or for global women's issues today. Both mystical poetry and quantum physics are two highly gendered discourse communities, historically associated with certain gender norms—mysticism and writing as a feminine pursuit and science and quantum theory as masculine terrain. Janés's poetry thus not only breaks down the discursive boundaries between these very different forms but also challenges the gendered norms associated with each.

Janés sees mystical practice as very much related to the process of scientific inquiry, both discourses straining to reach a deeper understanding of human ontology and epistemology. Mysticism has often been portrayed as antithetical to scientific reasoning. But for Janés, the connection could not be more obvious. Physicists inspire her poetry because, like Heisenberg, they understand the frustration and the joy of circling around something that can never be defined or explained; like Einstein, they grasp the notion of an experience that challenges accepted boundaries between space and time; and like Schrodinger, they envision a possible point of connection, a motion, that can bridge tremendous and unfathomable diversity.

As conversations about STEM become more focused on traversing gender boundaries, Janés offers a reminder of the role of STEAM, or the artistic endeavors connected with scientific pursuits. At one point in her ruminations on the development of her mystical and scientific work, Janés describes a series of poems that she titled "linces," or "lynxes," after an intellectual community in Rome in the 1600s that aimed to promote the understanding of natural science without being held back by any external authority. Janés notes that her own academy involved reading a range of great scientific thinkers and that these scientists were her muses: "Mis linces eran, de hecho para mí, verdaderas musas: Nicolás Copérnico, Galileo Galilei, Johannes Kepler, Isaac Newton, Max Plank, Albert Einstein, Erwin Schrodinger, Werner Heisenberg, Ilya Prigogine, Edward Lorenz y Benoit Mandelbrot" (33–34; My lynxes were, in fact for me, true muses: Nicholas Copernicus, Galileo Galilei, Johannes Kepler, Isaac Newton, Max Plank, Albert Einstein, Erwin Schrodinger, Werner Heisenberg, Ilya Prigogine, Edward Lorenz, and Benoit Mandelbrot). The list is diverse, covering a wide range of time periods, but one cannot help noting that the muses, including Copernicus, Galileo, Kepler, Newton, Plank, Einstein, and Heisenberg, are all male.[10] The gendered disparity in the list serves as a reminder of the way women have been relegated to the periphery of scientific thought for much of history. One is hopeful that Janés's efforts to bridge discourse communities will lead to many future muses of both genders.

There is good reason for hope. In February 2018 *El País* ran a special section dedicated to women in science. It was comprised of a website with images of and links to information about important female STEM leaders from all over the world, including Spanish women such as Josefina Castellví (biologist, oceanographer, and writer), María Wonenburger (mathematician), and Gabriela Morreale (chemist).[11] Clara Janés is not part of the STEM movement, but she is clearly engaged in advocating for STEAM, incorporating the arts in ways that challenge the artificial separation between scientific advances and humanistic endeavors. Her work demonstrates how mystical poetic thought and scientific experimentation each contribute in unique and important ways to the ongoing struggle for meaning inherent in contemporary cultural forms, whether those be mathematical formulas or literary texts. Such interdisciplinary conversations bode well for the future of women and STEM not only in Spain or in Spanish letters, but everywhere.

NOTES

1. Unless otherwise indicated, all translations are mine.
2. For book-length critical studies of Janés's corpus, see Faszer-McMahon (2010), Scaramuzza Vidoni (2012), and Mancilla (2016), as well as a 2014 collection edited by Mékouar-Hertzberg.
3. See also Robert Simon's "El misticismo geométrico y la poesía de Clara Janés en el siglo XXI" (Geometrical mysticism in Clara Janés's twenty-first-century poetry) for an earlier study of Janés's mathematical and mystical explorations.
4. For an important study of the postwar period and its impact on Janés and her generation's literary work, see Raquel Lanseros, whose 2017 study provides a personal interview with Janés and charts the postwar period's influence on her work and the development of her literary style.
5. For a detailed study of Janés's interdisciplinary work with Chillida, see Candelas Gala.
6. Just as humans for centuries incorrectly perceived the earth as stable with the sun orbiting our sphere, so this poem points out the complex layers of parallax regarding the sun. It might not recognize its own movement, but as noted by NASA, it is also flying rapidly through space. The sun rotates (different parts at different rates) as it orbits the center of the Milky Way, and it moves through space (along with our entire solar system) at around 450,000 miles per hour (NASA Science "Our Sun").
7. For more on *Los secretos del bosque* see Faszer-McMahon, *Cultural Encounters in Contemporary Spain: The Poetry of Clara Janés*.
8. See Faszer-McMahon, *Cultural Encounters*, for more on Janés's interest in Heidegger's

thought, particularly via his notions of poetry in *Caminos del bosque*. The German original, *Holzwege*, was translated into English as *Off the Beaten Track*.
9. Nicolescu and Janés met several times in Paris, and Nicolescu cites Janés in his books. Indeed, the poem analyzed above was printed in translation in one of Nicolescu's own books (*From Modernity to Cosmodernity* 237).
10. See Anna Casas Aguilar for more on Janés's autobiographical writings and what Casas Aguilar terms "una tradición literaria llena de nombres de hombres, a la que Janés incorpora su voz de mujer" (30; a literary tradition full of the names of men, into which Janés incorporates her woman's voice.
11. See *El País*'s weblink for details on the other female figures in STEM celebrated via the website: elpais.com/especiales/2018/mujeres-de-la-ciencia (Valdés).

WORKS CITED

Casas Aguilar, Anna. "La Barcelona de Clara Janés: Dolor, melancolía, herencia invisible." *Secretos y verdades en los textos de Clara Janés*, edited by Nadia Mékouar-Hertzberg, Peter Lang, 2014, pp. 29–48.

Faszer-McMahon, Debra. *Cultural Encounters in Contemporary Spain: The Poetry of Clara Janés*. Bucknell UP, 2010.

Ferreira, Pedro. "General Relativity." *New Scientist*, Oct. 2010, images.newscientist.com/wp-content/uploads/2010/10/instant_expert_1_-_general_relativity.pdf.

Furuta, Aya. "One Thing Is Certain: Heisenberg's Uncertainty Principle Is Not Dead." *Scientific American*, 8 March 2012, www.scientificamerican.com/article/heisenbergs-uncertainty-principle-is-not-dead.

Gala, Candelas. "Convergencia y transdisciplinaridad: Clara Janés y Eduardo Chillida." *Secretos y verdades en los textos de Clara Janés*, edited by Nadia Mékouar-Hertzberg, Peter Lang, 2014, pp. 49–70.

Greene, Brian. "Special Relativity in a Nutshell." *Nova*. PBS.org, 15 Sept. 2011, www.pbs.org/wgbh/nova/physics/special-relativity-nutshell.html.

Heidegger, Martin. *Holzwege*. Vittorio Klostermann, 1950.

Heidegger, Martin, et al. *Caminos del bosque*. Alianza, 2018.

——. et al. *Off the Beaten Track*. Cambridge UP, 2002.

Jaeger, Gregg. "Uncertainty Relations and Possible Experience." *Mathematics,* vol. 4, no. 2, 2016, pp. 40–48.

Janés, Clara. "Basarab Nicolescu: Teoremas poéticos." *Adamar*, vol. 40, 2008, adamar.org/ivepoca/node/1562.

——. *Creciente fértil*. Hiperión, 1989.

——. *Diván del ópalo de fuego (O la leyenda de Layla y Machnún)*. Regional de Murcia, 1996.

——. *El hombre de Adén*. Anagrama, 1991.

——. *El libro de los pájaros*. Pre-Textos, 1999.

———. *Estructuras disipativas*. Tusquets, 2017.

———. *Fósiles*. Twenty-one poems with nine engravings by Rosa Biadiu. 2nd ed. Z.I.P., 1985, 1987.

———. *Fractales*. Pre-Textos, 2005.

———. *La indetenible quietud*. Thirty-two poems with six engravings by Eduardo Chillida. Boza, 1998.

———. *La tentación del paraíso*. Poética y poesía, 29 Apr. 2014, Fundación Juan March, Madrid. www.march.es/events/100043. OCLC: 929680366.

———. *Lapidario*. Hiperión, 1988.

———. "Las ecuaciones de la poesía: Para una lectura de *Orbes del sueño*." *Secretos y verdades en los textos de Clara Janés*, edited by Nadia Mékouar-Hertzberg, Peter Lang, 2014, pp. 1–28.

———. *Las estrellas vencidas*. Agora, 1964.

———. *Límite humano*. Oriens, 1973.

———. *Los números oscuros*. Ciruela, 2006.

———. *Los secretos del bosque*. Visor Libros, 2002.

———. *Movimientos insomnes*. Del Centro, 2014.

———. *Orbes del sueño*. Vaso Roto, 2013.

———. *Paralajes*. Tusquets, 2002.

———. Personal Interview. Madrid, Spain. Clara Janés's home. 10 July 2014.

———. *Rosas de fuego*. Cátedra, 1996.

———. "Una estrella de puntas infinitas. En torno a Salomón y el 'Cantar de los Cantares.'" Discurso leído el 12 de junio de 2016 en su recepción pública, Real Academia Española, Madrid. RAE.es. 1 October 2016.

———. *Variables ocultas*. Vaso Roto, 2010.

Klus, Helen. "Schrödinger's Wave Equation." *How We Came to Know the Cosmos: Light and Matter*. The Star Garden, 18 Dec. 2017, thestargarden.co.uk.

Lanseros, Raquel. *Los poetas toman la palabra: La construcción de la educación literaria en los autores nacidos en posguerra*. Visor Libros, 2017.

Mancilla, Araceli. *Los astros subterráneos: Mito y poesía en Clara Janés*. Universidad Veracruzana, 2016.

Mékouar-Hertzberg, Nadia, editor. *Secretos y verdades en los textos de Clara Janés: Secrets and Truths in the Texts of Clara Janés*. Peter Lang, 2014.

NASA Science. "Our Sun." 19 Dec. 2019, solarsystem.nasa.gov/solar-system/sun/in-depth.

Nicolescu, Basarab. *From Modernity to Cosmodernity: Science, Culture, and Spirituality*. SUNY P, 2014.

———. "Transdisciplinarity: The Hidden Third, between the Subject and the Object." *Human and Social Studies. Research and Practice*, vol. 1, no. 1, 2012, pp. 13–28.

"Parallax." *Cambridge Online Dictionary*. Cambridge UP, 2018, dictionary.cambridge.org/us/dictionary/english/parallax. Accessed 24 March 2018.

Raymer, Michael G. *Quantum Physics: What Everyone Needs to Know*. Oxford UP, 2017.

Scaramuzza Vidoni, Mariarosa. *Compás de códigos en la poesía de Clara Janés*. Devenir, 2012.

Schleich W. P., et al. "Schrödinger Equation Revisited." *Proceedings of the National Academy of Sciences of the United States of America*, vol. 110, no. 14, 2013, pp. 5374–79.

Schmelzer, Felix K. E. "Los versos pitagóricos de Clara Janés." *RILCE*, vol. 34, no. 1, 2018, pp. 365–82.

Sekihara, Takayasu, et al. "Comprehensive Analysis of the Wave Function of a Hadronic Resonance and Its Compositeness." *Progress of Theoretical and Experimental Physics*, vol. 2015, no. 6, 2015, pp. 1–32.

Simon, Robert. "El misticismo geométrico y la poesía de Clara Janés en el siglo XXI." *Revista de Estudios Hispánicos*, vol. 34, no. 2, 2007, pp. 157–70.

Valdés, Isabel. "Especial: Mujeres de la ciencia." *El País*, 8 Feb. 2018, elpais.com/especiales/2018/mujeres-de-la-ciencia.

Zizek, Slavoj. *The Parallax View*. MIT, 2009.

CHAPTER 7

An Extension of Sympathy

Science and Posthumanism in the Paintings of Remedios Varo

MARTA DEL POZO ORTEA

Remedios Varo (Anglès, Gerona, 1908–Ciudad de México, 1963) was a visual artist whose work has most often been studied as a confluence of styles (namely fantasy, surrealism, and symbolism) in which alchemy, the kabbalah, Jungian psychology, and other esoteric and philosophical traditions converged. Nonetheless, the influence of science in Varo's oeuvre, although acknowledged by critics, has received little attention. This influence appears in the scientific personae that populate her paintings, such as the explorer, the geologist, the botanist, and the astronomer, but it is also evident in the omnipresent laboratory that Varo uses as a setting for many of her works, often including scientific apparatuses and even the depiction of invisible networks or molecular connections. This chapter looks beyond existing art-history studies of Varo's art in order to interrogate the role of gender and science in her paintings. Ideas stemming from quantum physics, such as the entangled nature of the observer and the observed, provide key insights into Varo's work, as do studies on the physical nature of light, gravity, time, and space vis-à-vis the theory of relativity. A consideration of posthumanist concepts will illuminate

Varo's visual ecologies within a scenario that ontologically and epistemologically decenters the human and advocates for an eminently relational and profoundly vital reality.

An Extension of Sympathy

> Evidently, I have no proof of the inner life of a neutrino.
> But strictly speaking, I also have no proof of the inner life of
> a bat or a cat, or indeed of another human being. This absence
> of proof is unavoidable, given the spectral nature of inner,
> private experience. But because I nevertheless acknowledge
> and respect the inner lives and values of other human beings,
> I can potentially do the same with other entities of all sorts.
> What's needed, perhaps, is an extension of sympathy.
>
> STEVEN SHAVIRO

Remedios Varo's painting *Sympathy* (1955; see fig. 7.1) depicts the invisible affective connections between a human figure and cat while the latter is being stroked. The cat has jumped onto the table, spilling a glass of milk all over the floor. Sparks like fireballs fly out and hover above the two subjects, configuring an "invisible" network of threads that connect the cat's spine to the figure's own neck, head, and hands, suggesting the neurological responses to the physical gestures of affect on both sides. The bodies of the cat—arching half in the air—and the human figure—slouched forward in a chair—form a visual ellipse that suggests a reciprocal circulation of energy. There also appears to be another cat under the table, hiding beneath the figure's gown. Its feet, with the same color fur as the cat on the table, are jutting out from beneath the gown, creating the illusion that they are the figure's feet, further connecting the cat on the table with the figure in the chair, or even implying that the figure is turning into a cat him/herself. Furthermore, the use of the same color palette, the feline complexion of the person, and the spiky hair on both figures (apparently caused by the emotional rapport), emphasize a mirroring relationship and a common constitution beyond their differences as species. Finally, the connection established through their gazes, as the punctum of the painting, implies what the title of this work calls for: a strong sense of sympathy.

In general, art critics have explored Remedios Varo's visual universe through three main vectors: gender,[1] her association to Mexican and Latin

FIGURE 7.1. *Sympathy* (1955) by Remedios Varo. © 2020 Artists Rights Society (ARS), New York / VEGAP, Madrid

American surrealism and magical-realism,[2] and the alchemical-Jungian prism.[3] Regarding the question of gender, *Sympathy* reveals a striking and pervasive characteristic of Varo's universe: the destabilization of gender categories. Figures generally described by critics as "women" or "female" are closer to humanoid or androgynous subjects. Such hybridity often transcends gender and racial divides, presenting subjects that appear to become inter-species, inter-objectual, or cyborg. Posthumanly deprived of a fixed and "pure" humanness or gender specificity, they forecast what Donna Haraway would enunciate in her *Cyborg Manifesto* almost thirty years later: "creature[s] in a post-gender world" (126). From such a post-humanist stance, Varo's paintings emphasize the relational properties of her often-solitary subjects with entities of all kinds: beings, things, tools, architectures, vehicles, etc. This conceptualization of entities has yielded, for example, subjects merging architectonically with a range of objects, such as with floors in *The Useless Science of the Alchemist*, with furniture in *Mimesis*, and with fantastic vehicles in *Twisted Roads* and *To Women's Happiness*. The pattern of such merging makes them appear like a new species, as in *Homo Rodans*, which develops wheels instead of legs as a new mode of locomotion. In other cases, such as in *Troubadour, The Escape*, and *Exploration of the Sources of the Orinoco River*, Varo's adventurous explorers embark on spiritual journeys in impossible vehicles that, like mandorlas, seem to be appendages of their own bodies.[4] In other works, like *Sympathy* (above), *Microcosm*, and *Three Destinies*, "invisible" threads create an interconnected universe within the painting. Thus, by reinforcing the hybridity, interconnectedness, inner vitality, and shared agency of her figures, Remedios Varo's universe announces a posthuman and post-gender world. To relegate her work to the constraints of surrealist and/or fantastic trends ignores the more complex implications that these narratives suggest, especially in light of contemporaneous scientific discoveries and current posthumanist theories.

Still, the Catalan-born painter's association with surrealism warrants mention. Such influence is evident via Varo's self-declared affinities for the poetics of André Breton's circle,[5] which she joined in 1937 while exiled in Paris during the Spanish Civil War.[6] As the critic Gloria Durán notes, Varo's Jungian-inspired paintings connect "[the] affirmative side of surrealism . . . [with the] legacy in the works of 'magical realism' by leading Spanish American writers" (309). Indeed, Varo's work stands in stark contrast to the surrealism of her Catalan male counterpart, Salvador Dalí, whose Freudian-based "critical-paranoic" method would in time come to represent surrealism itself. Varo's surrealism remains more related to the magical-realist movement in Latin America than to the European-based surrealist movement. This Latin American link is strengthened by the fact that, due to the Nazi invasion of France in 1941, the painter exiled herself to Mexico, where she lived until her death in 1963. Her association with Latin American artistic heritage is such that her most famous biographer, Janet Kaplan, goes so far as to consider her a Mexican artist.[7]

In the critical corpus dedicated to the painter, the presence of the alchemical "soft" sciences in her pictorial universe has been emphasized over more "hard" science-centered perspectives. Nonetheless, Kaplan reminds us that both influences equally marked Varo's upbringing: "Still very much the engineer's daughter, she read science as avidly as metaphysics, and her personal journey was propelled as much by her interest in scientific phenomena as by the study of the mystics. Turning to the sciences, she recognized in the newest developments in medicine, biology, chemistry, physics, astronomy and botany infinite possibilities for further exploration" (*Unexpected Journeys* 172). Another critic, José Luis Antequera Lucas, argues for the central role of scientific thought: "Natural phenomena were transcendent to Remedios (eclipses, telluric forces . . .). She was fascinated by the logics of cause-effect phenomena. . . . Accustomed ever since she was a little girl to machines and calculus, she would adapt her knowledge to her spiritual journeys" (356).[8] Furthermore, Luis-Martín Lozano notes that "Varo's work shows a fascination with the idea of different logics, different sense of time, and multiple understandings of causal networks" (46–47). This well-noted interest in science appears through Remedios Varo's work in a variety of ways: the scientific personae that populate her paintings, the omnipresent laboratory or studio as the setting for many of these characters' introspective explorations, the presence of scientific apparatuses throughout her canvases, and the already-mentioned depiction of the invisible thread of networks and connections between beings of all sorts that, perhaps, suggests a physical reality that is invisible to the naked eye.[9]

By studying the above-mentioned characteristics through the lens of posthumanist theories, this study expands the critical corpus of Varo's work via the perspective of science, particularly notions from quantum physics and relativity. Such analysis of Varo's expansive visual universe will help transcend any single feminist, fantastic, or surrealist label. It will relocate Varo's visual ecologies within the contemporary scenario, one anticipating twenty-first century scientific thought by ontologically and epistemologically decentering the human, as well as advocating for a relational and post-gendered reality.

On the Complementarity Principle and the Perseverance of a Rose

In alignment with contemporary feminist and ecofeminist readings of culture and technoscience, Varo actively participated in fostering a new cultural and scientific paradigm. Her works replace ideas of instrumentality, domination, and profit with more democratic values such as wonder, harmony, and connectivity. Kaplan believes that Varo saw scientific inquiry as analogous to a spiritual pursuit: "science must adopt the role not of domination but of harmony of natural forces. . . . She made careful distinction . . . between the kinds of scientific practice that she trusted and those that she did not, warning in a number of her paintings against manipulative abuses of authority, myopic belief in facts, infatuation with gadgetry, and misguided attempts to conquer nature" (*Unexpected Journeys* 172). In this way, Varo not only forecasts the affective turn, but she also aligns herself with the intuition of brilliant minds of the past, such as Goethe: "My perception is itself a thinking, and my thinking a perceiving"(Miller 39) and Emerson: "never did any science originate, but by a poetic perception" (304). Varo's subjects are inextricably bound in her pictorial narratives through a nonbinary generic ontology (hybrid physical characteristics), and a hybrid episteme (intuition/reason, imagination/thinking).

Such hybridity is evident in her work *Unsubmissive Plant* (1961; see fig. 7.2), which analyzes the instrumental uses of science, particularly the employment of scientific knowledge to control and manipulate reality. The work undermines a sense of scientific control by highlighting the omnipresence of unexpected occurrences in scientific experiments. In this painting, the botanist is sitting at his/her workstation outdoors, trying to clone plant specimens in an array of test tubes. Curiously, the shapes of the leaves of those plants, upon closer examination, take the form of mathematical formulae, as do the botanist's own clothing and Einsteinian hair. The scientist stares with suspicion and disbelief at the subject of the painting: an insurgent rose that has unexpectedly grown from the strictly calculated experiments. In Varo's reflection upon the painting, she states: "this scientist experiments with diverse plants and vegetables. [He/she] is perplexed because there is a rebellious plant; the rest are all growing their branches in the shape of figures and formulae, except for one that insists on growing a flower, and the only mathematical leaf that [he/she] planted at

FIGURE 7.2. *Unsubmissive Plant* (1961) by Remedios Varo. © 2020 Artists Rights Society (ARS), New York / VEGAP, Madrid

the beginning, and which falls upon the table, was very weak and withered, and also erred since one can read on it: Two plus two equals almost four" (Varo and Castell 238).[10] It is interesting to note that Varo herself signals a scientific concept in her painting that is nearly imperceptible to the naked eye. The possible inequality in the mathematical equation raised by the word "almost" evokes Heisenberg's principle of uncertainty. The limited precision offered by mathematics, or in this case botany, invites viewers to question their known world.

While immersed in a mental and mechanistic universe, and despite being outdoors, the scientist has failed to really "see" nature for what it is, to acknowledge its own agential powers. Instead of a predictable or humanly dominated natural order, chance has mutated the experiment, resulting in a vigorous act of emergence. This disconnection and alienation from nature's own self-regulating processes renders the scientist awe-struck by what he/she cannot understand. In her book *Vibrant Matter,* Jane Bennett calls for a similar re-engagement of today's vital materialism, highlighting the role of "critical vitalists" from the early twentieth century. Bennett notes that "nature was not, for Bergson and Driesch, a machine, and matter was not in principle calculable: something always escaped quantification, prediction and control. They named that something *élan vital* (Bergson) and entelechy (Driesch). Their efforts to remain scientific while acknowledging some incalculability to things is . . . . exemplary" (63).

Like Varo's painting, Bennett's perspective of a vital materialism recognizes a non-anthropocentric universe in which nature is endowed with agency and drive; it is active rather than passive—unpredictable, unaccountable, and self-regulating. Karen Barad correspondingly promulgates an agential realism, a posthumanist approach to understanding techno science that recognizes matter's dynamism. Moreover, this dynamism is, according to Barad, "intra-active": "matter and meaning come into existence, are iteratively reconfigured through each intra-action, thereby making it impossible to differentiate in any absolute sense between creation and renewal, beginning and returning, continuity and discontinuity, here and there, past and future" (ix). One notices the insights from quantum physics in the concept of intra-action, and specifically that of *entanglement*. For the posthumanist critic, matter and mind are part of a dynamic, material, and discursive process of entanglement. This process takes into account Niels Bohr's complementarity principle, which states that there is no inherent separation between the knower and the known, that there is an ouroboric quality inherent in the act of perception whereby the old passive category of the observer gives way to that of the *participant* or *creator*.[11] Therefore, they are "entangled," implying that the categories of subject and object are undone in the act of perception, giving way to a non-dualist ontology.

This idea of entanglement is visually presented in Remedios Varo's painting through the analogy between the shape of the scientist's hair and that of the plants (both are tangled), as well as in the form of the scientist's robe, which not only resembles the grass around him/her but also its dynamic quality. Soo Moll registers the irony that this self-transformative process takes place beyond the scientist's own perceived sense of control: "Varo's suggestion, I would argue, is that the observer and observed are merging characteristics or identities, but not through the intentional action of the subject (scientist). Since the scientist presumably cannot see her/his own hair or the back side of the clothing, the observer and the observed seem to be merging identities through processes of which the observer is unaware" (41). In other words, there is no perception from the point of view of the scientist of such porosity between the inside (the inner mental space of the botanist) and the outside (the natural world). Such realization would require a new way of seeing and thinking that transcends the human-centered ontological and epistemological worldview of mechanistic natural philosophy. In Varo's painting, human control is replaced by chance, shared agency, entanglement, and intra-action, key components of a new non-formulaic reality.

On Gravity: One Foot in Science, One in Intuition

In addition to addressing notions from quantum physics like entanglement, Varo explores concepts of gravity in works like *Phenomenon of Weightlessness* (1963; see fig. 7.3). In this case another scientific figure, an astronomer, appears in the scientist's studio, which contains an open window through which the influences of celestial bodies are made visible. Other critics have noted a "vertical relationship between man and the stars" (Durán 308) in Remedios Varo's paintings. However, rather than focusing on the passive and hierarchical connotations of such verticality (in the sense of receiving light or grace), Varo appears to present a more exploratory or co-creative experience with nature. In this sense, the connection is visually more "oblique" than "vertical" and calls attention to the light or force

FIGURE 7.3. *Phenomenon of Weightlessness* (1963) by Remedios Varo. © 2020 Artists Rights Society (ARS), New York / VEGAP, Madrid

that appears to fall on the character's hands. Varo's use of light and the allusions to gravity evident in the work's title offer an intriguing visual narrative in this painting, which focuses on the scientist's scrutiny of complex interconnected forces. These forces, as was the case in *Sympathy*, are not clearly visible. However, they are made distinguishable by the painter in the form of a thin white line, a reminder that, albeit imperceptible to the naked eye, unseen forces are *real* and affect the actors in the painting.

The *Phenomenon of Weightlessness* moves into the speculative realm in that it depicts a hypothetical scenario: the lack of the force of gravity. In it, an astronomer in eighteenth-century attire (possibly suggesting a Newtonian worldview), reacts with perplexity to the abnormality taking place in the studio: one of the models of the solar system has separated from its pedestal and thereby broken free from its presumed and static gravitational force. It floats in the air in the exact center of the painting, perhaps implying an ironic comment on earth-centric views of the universe. This unexpected movement of the model has provoked a change of the room's dimensional space, creating a new 3D plane visually juxtaposed with the first. The astronomer, with one foot on each plane, has no option but to find balance in both worlds, literally and metaphorically.

The painting points to multiple dimensions of reality, and the scene creates scientific conundrums that emanate from the dispositions adopted by the painter. For example, Varo places a celestial body in the window space, thus provoking an optical illusion. It is difficult to ascertain whether this figure is part of the model floating in the air, a representation of an actual star (the sun), another planetary satellite (the moon), or all things at once. The object's disproportionate size and apparent gravitational force raise questions that depend on mass and distance. Nonetheless, there is one certainty about the physics in the painting: that of the curvature of the light reflected in the tilting of the room, which recalls Einstein's theory of relativity, namely that space-time is curved and light is bent by gravitational fields. Stephen Hawking illustrates this concept with the example of a distant star whose light passes near the sun. He notes that in such a case, the star's light "would be deflected through a small angle, causing the star to appear in a different position. . . . It is very difficult to see this effect, because the light from the sun makes it impossible to observe stars that appear near to the sun in the sky. However, it is possible to do so during an eclipse of the sun, when the sun's light is blocked by the moon" (32–33). If Varo's painting depicted a solar eclipse (making both stars physically coincide), then the deflection of light would be made visible, causing the space of the room to bend.

Varo succeeds in visually portraying the lack of gravitation implied in the title of her work, in part by recalling the experience of weightlessness in space. Kaplan reminds us that this "painting earned praise from one of the North American astronauts, who contacted Varo to applaud it as an excellent visualization of the weightlessness he had experienced in space. It is also a sufficiently accurate depiction of the phenomenon to have been chosen as the cover illustration for *The Riddle of Gravitation*, a scientific text written by the physicist Peter Bergmann, an associate of Einstein" (*Unexpected Journeys* 177). Yet Varo's work was not just scientific, it was prescient, in that it was painted in 1963, six years before the first images were broadcast of a man walking on the moon. Varo's knowledge of scientific phenomena and her intuitive artistic vision coalesced in her art, thus anticipating a physical phenomenon yet to be seen by humanity.

A New and Vertiginous Idea of Time

In addition to interrogating the complexities of gravity, Varo's paintings demonstrate an artistic exploration of time. Her work *Revelation or the Clockmaker* (1955; see fig. 7.4) provides an intriguing counterpoint to perceptions of time offered by other artists such as Salvador Dalí, whose preoccupation with time was explored in his enduring painting *The Persistence of Memory* (1931). Whereas Dalí's famous work portrays a highly subjective and deformed world of melting solids set on an oneiric coastline, *Revelation or the Clockmaker* offers more familiar symbols, presenting an internal logic. Its symbols, rather than being personal, pertain to the collective unconscious, therefore establishing it as more Jungian than Freudian. The work's general setting and figures are familiar, as once again there is a studio, a scientist-watchmaker sitting at a desk, and an open window through which the scientist receives an idea or revelation on the nature of time. In Varo's words, "the watchmaker represents our ordinary time, but through the window a 'revelation' appears and suddenly he understands many things. . . . Around him there are a number of clocks that strike the same hour, but within each is the same character in different periods, and each clock has a window with bars as in a jail" (Varo and Castell 174). Like the astronomer in the previous painting, the watchmaker

FIGURE 7.4. *Revelation or the Clockmaker* (1955) by Remedios Varo. © 2020 Artists Rights Society (ARS), New York / VEGAP, Madrid

perceives this phenomenon with awe and concentration, so rapt that the cogwheels he handles in his mechanical task fall and roll to the ground. The world-view of Newtonian physics is being questioned via the row of clocks that surrounds the scientist. Time, according to Newton, was absolute, unchanging, and independent of space, but in this painting, viewers can see how humankind has metaphorically fallen prisoner to such a fixed epistemological conception.

The painting emphasizes one of Einstein's key insights, described here by Stephen Hawking: "[T]he theory of relativity put an end to the idea of absolute time! It appeared that each observer must have his own measure of time, as recorded by a clock carried with him, and that identical clocks carried by different observers would not necessarily agree" (22).[12] Varo's painting captures Einstein's insight through the jailed clocks and the human figures trapped at their base, guarding multiple timepieces in an attempt to seize time. The work recalls the notion that all time exists simultaneously, just as the human figures in each clock are perhaps from different eras, yet in the space-time continuum, they all exist at the same moment, indicated by the 12:15 time of each clock.

Other symbols in the painting, such as the cat and the circling nebula, connect this work to Varo's corpus and integrate her visual universe. First, there is the presence of the feline, formerly encountered in *Sympathy*, which symbolizes an openness and psychic rapport between human and other non-human realities. Varo was fascinated with cats, which she saw, according to Kaplan, as mysterious and personal allies (*Unexpected Journeys* 123). Yet here, the presence of the black cat also recalls its symbolism in ancient Egypt, where its aloofness and deified status allowed it to overcome time. While the cat faces away, the clockmaker and the many human figures at the base of the clocks still try to observe (from within the clocks) or conquer (from the worktable) time. The visual triangulation between the clockmaker, the cat, and the whirlpooling form penetrating the room implies a kind of psychic dynamism, elasticity, or receptivity toward a new concept of time suddenly entering the clockmaker's mind. Indeed, even the ceiling of this room, evocative of Dalí's melting solids in *The Persistence of Memory*, has begun to bubble or bend, reminiscent of the idea that space-time is not linear but curved. In other words, the three-dimensional world of solids in which this scientist has always operated is giving way to a four-dimensional world that accounts for the inextricability of space-time.

This unpredictability in the clockmaker's psyche has yet another visual projection: at the intersection of the wall and the floor, some uninvited plants and weeds are growing, indicating once again nature's external agency, imposing itself in unpredictable and uncontrolled ways. The scientist perceives the spiraling nebula, perhaps pondering the notion of an expanding universe. While the circling nebula is how Varo chooses to depict a new concept of time, the image links her work to the circularity of time as seen in the works of Latin American writers such as Borges.[13] Yet critics like Octavio Paz insist that it is precisely non-time that Remedios Varo has always painted: "She doesn't paint time but the instants in which time rests. In her world of stopped clocks, we hear the flow of substances, the circulation of shadow and light: time ripens in it. . . . She surprises us because she painted in a state of surprise" (10). Wonder, the initiation of the scientific enterprise, stands at the emotional center of Remedios Varo's visual universe. Her works visualize what experimentation, whether scientific or artistic, reveals about our conceptions of nature and form, letting familiar symbols and new ideas penetrate the apparent walls of the mind, giving shape to a visual and exploratory co-creation of new meanings.

When Matter and Antimatter Collide

Varo's paintings regularly foreground scientific insights, and they often connect theoretical notions with everyday experiences from the physical universe. *Still Life Resuscitating*, also known by the title *Reviving* (1963; see fig. 7.5), demonstrates this intriguing combination. It was Remedios Varo's last painting and rare for her corpus in that no human figures appear. Yet this still life of a set dinner table is no ordinary still life. A vortex above the table spirals around a lit candle, thus making the tablecloth, plates, and fruits spin in the air. The collision of the orbiting fruits spills seeds to the ground, which germinate on the concrete floor of what seems to be a chapel. The religious connotations of the scene and the title, implying resurrection, or life beyond physical death, are evident. There are, however, additional science-laden metaphors elicited by the scene, such as the big bang theory, the solar system, or the atomic structure, with a candlelit nucleus around which other particles orbit. Fireflies surround the scene as well as insurgent plants growing on the floor. The work defamiliarizes the still-life scene, inviting viewers to explore the scientific and symbolic meanings behind everyday objects.

Critics have long recognized the symbolic and archetypal influence of the collective unconscious throughout Varo's work, and this painting connects those concerns to her scientific explorations. For example, the work foregrounds not only the element of fire but also the round table, which is, according to Gloria Durán, "a well-known symbol of wholeness and plays a role in mythology—for instance, King Arthur's round table, which itself is an image derived from the table of the Last Supper" (307). The scientific scenario taking place in the painting also recalls Jung's own scientific explorations and exchanges with the Nobel Laureate in Physics, Wolfgang Pauli. Their conversations addressed correspondences between neuroscience (or the nature of consciousness) and the atomic structure, and these ideas were published in an article from 1955, "The Interpretation of Nature and the Psyche." Such ideas, contemporary to Varo, may have inspired this final painting, which appears to represent the physicist's

FIGURE 7.5. *Still Life Resuscitating* (1963) by Remedios Varo. © 2020 Artists Rights Society (ARS), New York / VEGAP, Madrid

and the psychologist's established analogy between "the atom, consisting of a nucleus and shell, and the human personality, consisting of 'nucleus' (or self) and 'Ego'" (Jung and Pauli, *Atom and Archetype* 92).

Moreover, the collision of orbiting fruits recalls, in the realm of science, the image of the collision of particles with their corresponding antiparticles, an idea linked by Jung to the union of opposites. The psychological analogy of the collision of matter and antimatter is the Jungian concept of *conjuntium oppositorum*, or the union of pairs of personality opposites. Just as the physical collision of particles gives way to annihilation and energy release, so does the *conjuntium oppositorum*, fostering psychological freedom from pairs of opposites (dualisms) to create oneness. Varo's corpus, when considered from such a framework, might be interpreted as an exploration and critique of dyadic form: science/intuition, self/otherness, human/non-human, observer/observed, man/nature, linear time/circular time, etc. From this dyadic perspective, *Reviving* can be seen to fuse the physical and the psychological, resulting in freedom from pairs of opposites via a resurrection that yields the attainment of a mysterious and complex oneness. Moving beyond binary oppositions requires seeing a deeper, more complex vision of reality. Such interpretations might also help contextualize the nonbinary or androgynous gender of many of Varo's subjects. Just like the alchemist, and in Jungian fashion, Remedios Varo was trying to integrate the feminine and the masculine components of the psyche (the anima and the animus), thus symbolizing the myth of the lost androgyne.

One could further argue that *Reviving* represents the element for which Jung and Pauli searched, a "psychophysical monism," an actual requisite for conjunction: "the double vision or perception of both the external sensory aspect of the world and also its hidden depths and meaning" (*Atom and Archetype* 157). In the painting, this monism is found in the mandala-like structure around which all elements revolve, and are thus integrated into what might be described as an atomic light of consciousness. Additionally, Varo's painting elicits a "double vision," where the ideas on the nature of the life-death cycle are found within the sensory and the domestic worlds represented by the fruits set on the table. In this sense, Varo reanimates the iconic art history motif of the *still life* (in Spanish, literally, "dead nature"—*naturaleza muerta*) if only to subvert it pictorially and conceptually in order to bring it back to life. If, as Brad Epps argues, "Fusion—or indeed synthesis, union, or harmony—is one of the most conspicuous objectives of Remedios Varo's art" (185), then the title of this painting not only symbolizes the *conjunctium oppositorum* of life and death, but also announces the result of this conjunction by appealing to a third and synthetic element: *resucitando*, or reviving.

As Stephen Shaviro points out in this chapter's epigraph, there is no proof of the inner life of objects, but, as he argues, "because I nevertheless acknowledge and respect the inner lives and values of other human beings, I can potentially do the same with other entities of all sorts" (40). An extension of sympathy is what Shaviro promotes, and such ideas and values underlie Remedios Varo's universe. In each of her pictorial narratives, one can discern an articulation of an extension of sympathy toward otherness: animals, flowers, gravity, space, time, or, as the last analysis suggests, the male component of her own psyche. The investigative interactions of her subjects with such otherness is precisely Varo's formula for creating new meanings. This entails a democratization of affects and announces a genuine recognition of a shared and distributive agency. With the decentering of human and epistemological methods, Remedios Varo forecasted new post-human ecologies and paved an ontological turn, one that advocates for an ultimately fluid, unknown, and interconnected reality consisting of potentiality, rather than of certainty, as science has been postulating in the last decades.

To approach the work of Remedios Varo from these new perspectives is to open one's eyes to a more encompassing way of perceiving the work of an artist whose vision transcended the subjectivity of the surrealists, toward the creation of a postgender and posthumanist reality, which decenters not only *man* but also *the human* and announces, in alignment with the latest scientific discoveries, an eminently relational reality. Moreover, Varos could also be, from a Jungian point of view, all these personae or masks (the botanist, the astronomer, the clockmaker) with their respective investigative and entangled interactions. It is evident that through her visual work, Varo was trying to both intuitively and scientifically (imaginatively and logically) come to terms with the unknown, with what humans call reality but which is being constantly co-created. Thus, it is not coincidental that *Still Life Resuscitating*, if interpreted as the entanglement of the artist with her ultimate reality, would be her last work.

NOTES

1. See Chadwick; Petersen and Wilson; Kaplan, "Remedios Varo (An Art Essay)"; Lauter; and Stellweg.

2. See Rodríguez Rampolini; Martín; and González Madrid.
3. See Andrade; Anzures; Crespo de la Serna; and Gónzález Madrid.
4. In medieval art, a mandorla is an oval-shaped framework that encircles some religious images.
5. Nevertheless, and despite her regular attendance at the group's meetings in Paris during the 1930s, she was not recognized as anything more than the lover of Benjamin Péret, the surrealist writer (Durán 299).
6. Remedios Varo had previously participated in the Exposición Logicofobista (Logicophobists Exhibition) held in Barcelona in 1936. Together with painters like Esteban Francés, Antoni Clavé, and Joan Massanet, the group daringly declared war on logic. Also, her strong friendship with the British painter Leonora Carrington is evidenced by their pictorial affinities and the influence of Carl Gustav Jung's theories.
7. Despite her many years in Mexico, Janet Kaplan has noted the absence of Mexican landscapes in Varo's work. Apparently, Varo used to return through her imagination to the Catalan landscape of Anglés and also to North Africa, where she spent several years due to her father's work (Kaplan, *Unexpected Journeys* 24).
8. My translation from the original Spanish.
9. Ellen Alyce Soo Moll's chapter of her doctoral dissertation, *Unlikely Comparison and the Transdisciplinarity of Comparative Literature: The Boundaries of Gender, Technoscience, Literature and Visual Culture* (2011), explores the otherwise very likely intersections of Varo's work with the ideas of the cultural critic from the Santa Fe group, Karen Barad. Soo Moll takes into account matters of technoscience and concepts such as "agential realism" or "intra-action," which in Barad's theoretical framework connect quantum physics and philosophy.
10. My translation from the original Spanish.
11. Niels Bohr introduced the concept of "complementarity" in his famous 1927 Como Lecture, reproduced in Niels Bohr's *Atomic Theory and the Description of Nature*.
12. A further explanation of the consequence of relativity on time: "In Newton's theory, if a pulse of light is sent from one place to another, different observers would agree on the time that the journey took (since time is absolute), but will not always agree on how far the light traveled (since space is not absolute). Since the speed of light is just the distance it has traveled divided by the time it has taken, different observers would measure different speeds for the light. In relativity, on the other hand, all observers must agree on how fast light travels. They still, however, do not agree on the distance the light has traveled, so they must therefore now also disagree over the time it has taken" (Hawking 22).
13. See, for example, "Las ruinas circulares" (The circular ruins) by Borges.

WORKS CITED

Andrade, Lourdes. "Remedios Varo y la alquimia." *México en el arte*, no. 14, 1986, pp. 67–69.

Antequera Lucas, José Luis. "Remedios Varo (1908–1963): El viaje interior." *Espacio, Tiempo y Forma. Serie VII. Historia del Arte*, no. 20–21, 2007–2008, pp. 341–61.

Anzures, Rafael. "Poesía, magia y surrealismo: Leonora Carrington. Remedios Varo." *Cuadernos médicos*, vol. 2, Dec. 1956, pp. 29–40.

Barad, Karen. *Meeting the Universe Halfway: Quantum Physics and the Entanglement of Matter and Meaning*. Duke UP, 2007.

Bennett, Jane. *Vibrant Matter. A Political Ecology of Things*. Duke UP, 2010.

Bohr, Niels. *Atomic Theory and the Description of Nature*. Cambridge UP, 1934.

Borges, Jorge Luis. "Las ruinas circulares." *Ficciones*. Editorial Sur, 1944.

Chadwick, Whitney. *Women Artists and the Surrealist Movement*. Thames and Hudson, 1985.

Crespo de la Serna, Jorge Juan. "Hechizo de Remedios Varo." *Novedades*, 27 May 1956, *México en la cultura*, p. 4.

Durán, Gloria. "The Antipodes of Surrealism: Salvador Dalí and Remedios Varo." *Symposium*, vol. 42, no. 3, 1988, pp. 297–311.

Emerson, Ralph Waldo. *The Later Lectures of Ralph Waldo Emerson, 1843–1871*, Vol. 1, edited by Ronald A. Bosco and Joel Myerson, U of Georgia P, 2010.

Epps, Brad. "The Texture of the Face: Logic, Narration, and Figurative Details in Remedios Varo." *Journal of Spanish Cultural Studies*, vol. 4, no. 2, 2003, pp. 185–203.

Haraway, Donna. "The Cyborg Manifesto." *Nature: Documents of Contemporary Art*, edited by Jeffrey Kastner, MIT P, 2012, pp. 124–29.

Hawking, Stephen. *A Brief History of Time*. Bantam, 1996.

Gónzález Madrid, Maria José. *Surrealismo y saberes mágicos en la obra de Remedios Varo*. 2013. Universitat de Barcelona, PhD dissertation.

Jung, C. G. and W. Pauli. *Atom and Archetype*, edited by C. A. Meier, Princeton UP, 2000.

———. *The Interpretation of Nature and the Psyche. I, Synchronicity: An Acausal Connecting Principle*. Vol. 51. Bollingen Series. University of Michigan: Pantheon, 1955.

Kaplan, Janet. *Unexpected Journeys: The Art and Life of Remedios Varo*. Abbeville P, 1988.

———. "Remedios Varo (An Art Essay)." *Feminist Studies*, vol. 13, Spr. 1987, pp. 38–48.

Lauter, Estella. "The Creative Woman and the Female Quest: The Paintings of Remedios Varo." *Soundings*, vol. 63, Sum. 1980, pp. 113–34.

Lozano, Luis-Martín. *The Magic of Remedios Varo*. Washington, DC, National Museum of Women in the Arts, 2000.

Martín, Martín Fernando. "Remedios Varo, pintora de lo maravilloso." *Lápiz. Revista mensual de arte*, vol. 3, Jan. 1985, pp. 52–56.

Miller, Douglas. *Johann Wolfgang von Goethe—Scientific studies*. Suhrkamp, 1988.

Paz, Octavio. *Corriente alterna*. Siglo Veintiuno, 2000.

Petersen, Karen, and J. J. Wilson. *Women Artists: Recognition and Reappraisal. From the Early Middle Ages to the Twentieth Century*. Harper and Row, 1976.

Rodríguez Rampolini, Ida. *El surrealismo y el arte fantástico de México*. Instituto de Investigaciones Estéticas, Universidad Nacional Autónoma de México, 1969.

Shaviro, Stephen. "The Consequences of Panpsychism." *The Nonhuman Turn*, edited by Richard Grusin, U of Minnesota P, 2015.

Soo Moll, Ellen Alyce. *Unlikely Comparison and the Transdisciplinarity of Comparative Literature: The Boundaries of Gender, Technoscience, Literature and Visual Culture*. 2011. University of Maryland, PhD dissertation.

Stellweg, Carla. "Feminine Sensibility, as Viewed in Mexico by Some Artists, Art Critics, and Art Historians." *Artes visuales*, vol. 9, Jan-Mar. 1976, pp. 53–59.

Varo, Remedios, and Isabel Castell. *Cartas, sueños y otros textos*. Era, 2002.

CHAPTER 8

Subversive, Combative, Corrective

Carmen de Burgos's Interventionist Translation of Möbius's Über den physiologischen Schwachsinn des Weibes *(The mental inferiority of women)*

LESLIE ANNE MERCED

Throughout the nineteenth and early twentieth centuries, German literature and science found their way into Spain through the work of translators and travelers who visited various German cities. "Viajeras" such as Emilia Pardo Bazán and Carmen de Burgos traveled to Germany and/or Switzerland, bringing back to Spain not only material for travelogues and narratives, but also an interest in debating German literature and scientific theories in their essays and medical translations. There was a need for translators of medical texts to serve the reading public's interest in debated health topics of the day.

As early as 1861, one can find newspapers such as *El Debate Médico*, solely dedicated to homeopathic medicine, or other better-known publications such as *La Abeja*, with medicine and hygiene sections in every issue. *La Abeja* was pivotal in the dissemination of German thought since it included both literary and non-literary German authors as well as many Spanish translators when the journal appeared in Spanish during the years of 1862 to 1866 and 1870. At the time, no other journal was as focused on German scientific texts, and indeed the title page emphasized that aspect of the publication: "La abeja. Revista científica y literaria ilustrada. Principalmente extractada de los buenos escritores alemanes por una sociedad literaria" (Bergues de las Casas 1; The bee. Illustrated scientific and literary magazine. Mainly extracted from the good German writers by a literary society). Editor Antonio Bergues de las Casas explained the topics readers might expect from the journal, including "reseñas de las obras de mayor importancia que vayan saliendo a luz en alemán . . . biografías de hombres célebres, [y] los bellos productos de las investigaciones científicas" (Bergues de las Casas 1; reviews of the most important works that come to light in German . . . biographies of famous men, [and] the beautiful products of scientific research), emphasizing the coverage of German scientific research. *La Abeja* had also published the biographies of three important German translation theorists: Humboldt, Goethe, and Schleiermacher. It is significant that these three German theorists of translation methodologies were known to Spanish audiences. Indeed, as Lawrence Venuti states, "In the West, from Antiquity to the late nineteenth century [. . .] the most frequently cited theorists comprised a fairly limited group [. . .] Goethe, Schleiermacher and Nietzsche" (*The Translation Studies Reader* 4). The aforementioned translation theorists established the parameters of the field as a professional endeavor.

It is in this milieu that Carmen de Burgos decided to undertake the translation of Möbius's misogynistic study *Über den physiologischen Schwachsinn des Weibes* (*The mental inferiority of women*). Burgos joins other Spanish translators such as Sofía Casanova and Francisca A. de la Barella as women who were educated, multilingual, and who understood the profession of translation to be as important as journalism or literary writing. But Burgos's work offers something Casanova and Barella's translations did not. Her translation of Möbius's text challenges translational norms, revealing Burgos as a subversive and proto-feminist translator using innovative strategies to critique gendered hierarchies and scientific exclusions.

While Carmen de Burgos is well known today for her varied and prolific authorship, including a large number of short stories, hygiene books, newspaper articles, editorial projects, war correspondence, and pseudonymic publications, her work with translation has received less attention. Alongside Burgos's extensive corpus of literary works, the literary historian María Simón Palmer lists thirty-five different translations from French, German, Russian, Italian, and English attributed to Burgos (*Escritoras españolas del siglo XIX* 140). In 1904 alone, Burgos translated Hellen Keller, Paul Moebius, Geza Mattchich, Ernesto Renan, and Leo Tolstoy.[1] Ana Martínez Marín writes about the importance of the year 1904 as one in which Burgos starts her translation work, but Marín's study focuses on literary translations, "a través de las cuales va a tomar contacto con la obra de los vanguardistas extranjeros" (17–18; through which she will have contact with the work of the foreign avant-garde). These 1904 translations are thematically diverse and present the modern reader with questions about Burgos's methods and motivations for translating such varied texts. Recent biographers such as Marcia Castillo Martín acknowledge Burgos's translations ("publica y traduce sin descanso") (9; publishes and translates without stopping) but present them as a bulk list, not even identified by year. Burgos's best known biographer, Elizabeth Starčević, does not even mention translations when she defines Burgos's body of work in the context of gender roles: "Posteriormente, la visión de la mujer retratada por Carmen de Burgos en sus novelas, ensayos, cuentos y artículos nos demuestra que las ideas sobre la inferioridad de la mujer seguían vigentes" (22; Subsequently, the vision of the woman portrayed by Carmen de Burgos in her novels, essays, stories, and articles shows us that ideas about the inferiority of women were still held).

Sylvia Truxa's chapter on Carmen de Burgos's life and literary and non-literary writings, which appeared in German, does contain a paragraph on Burgos as translator. Yet even there, the translational work is mentioned in a brief aside focused on her "other activity," namely her literary accomplishments and her political candidacy of 1932: "Und *nebenbei* schreibt sie. Fertigt fast dreißig große Übertsetzungen an, in denen sie vor allem neure Autoren von Hellen Keller über Max Nordau bis zu John Ruskin in Spanien bekannt macht und selbst vor Moebius nicht zurück schreckt" (200; And besides all this other activity, she writes. She completed thirty large new author's translations in which she made known Hellen Keller, Max Nordau, and John Ruskin in Spain and was not scared of even doing a translation of Möbius).[2]

María Simón Palmer wrote the first article that deals in depth with Burgos's translations, focusing on her relationships with the editorial world. Simón Palmer notes an important point, namely that most of the translations penned by Burgos stem from the Sempere editing house, all except the Möbius translation: "Antes de que figure como pie de imprenta Sempere, Carmen va a publicar en la Imprenta del Pueblo, propiedad de Blasco Ibáñez, una obra que estallará la polémica. [. . .] Su autor, Paul Julius Möbius" ("Carmen de Burgos, traductora" 160; Before Sempere appears as printer, Carmen will publish with Pueblo Press, owned by Blasco Ibáñez, a work that will fuel the controversy. [. . .] Its author, Paul Julius Möbius). This is an important detail because Blasco Ibáñez later bought the Sempere editorial house and became a very important player in the publication of translations in Spain. Burgos's translation of Möbius, the National Gallery edition, thus involves both editorial houses and shows the beginning of Burgos's long connection to Sempere, her main publisher for her later translations.

Another critic, Lola Sánchez, notes other unique aspects of Burgos's translation of Möbius, offering a thorough account of all the commentaries made by Burgos and her contemporaries about the translation, and concluding that "reading or rereading the heterogeneous documents, dispersed notes, seemingly anecdotal or insignificant, has led me to rethink how she performed her translation as an act of militancy and public agitation" (75). Sánchez's insight raises the question of Burgos's own theory of translation. Did Burgos reject the norm of the invisible translator? This study argues not only that her work with Möbius's misogynistic text demonstrates an innovative approach to translation but also that her subtle subversion of translational norms demands a reevaluation of her translation on its own merits. To view Burgos's translations as mere economic vehicles serving to relieve her from an impoverished state (as Simón Palmer and Sánchez argue in the aforementioned articles) would be a diminishment and a misreading of Burgos, the translator.

When approaching Burgos's translation of a text as polemically titled as *The Mental Inferiority of Women*, readers familiar with Burgos's corpus might rightly be puzzled about what to expect. The challenges involve bridging not only gendered and linguistic divides, but also discursive ones, as the original was presented as a medical treatise. Burgos's task was thus unusually complex and involved a range of methodological choices, including to employ or reject dominant translation theories such as Schleiermacher's German model of reader-centered versus author-centered translation, or the French tradition that valued "the close reproduction of the style of the source text" (Baker 408). When one compares Burgos's source text with her translated text, it becomes clear that her translation rejected what Lawrence Venuti calls "the translator's invisibility" (*The Translation Studies Reader* 1). This study analyzes how Burgos dealt with the culturally complex and politically controversial issues of Möbius's text via radical translation choices. She modeled a new approach to translation theory, one that redefined the role of the translator as activist and offered a space for subversion, ultimately demonstrating in yet another genre how Burgos was a "Mujer adelantada a su tiempo" (Simón Palmer, *Escritoras españolas del siglo XIX* 130; A woman ahead of her time).

The Interventionist and Visible Translator: New Additions to the Translated Text[3]

In his *Methoden des Übersetzens* (On the different methods of translating*)*, Friedrich Schleiermacher calls upon the translator "to prepare an imitation, a whole composed from parts noticeably different from the original, but which nevertheless comes as close in its effect to that original whole as the difference in material permits" (41).[4] Schleiermacher's choice of the word "imitation" to define the translation process denotes a particular stance toward how translators should approach a foreign text and one that limits the interpretative aspect of that work. For Schleiermacher, the translated text must follow the foreign parameters of the source text, even when the translation is presented to an audience culturally different from the original. Yet contemporary translation theorists like Lawrence Venuti have convincingly argued that such foreignization proves problematic in that it completely disregards "the relative autonomy of translation" (*The Translation Studies Reader* 5). The translator following Schleiermacher's method cannot change or transform the translated text because the foreignness needs to be preserved along with the linguistic characteristics of the source text. The only transformation that can result from this type of translation is one that closely resembles the original text, highlighting once again its foreignness. Neither new interpretations, nor any translator's liberties with the source text, are allowed in this type of translation methodology.

Schleiermacher's approach would likely trouble many translators envisioning their task, but in the case of Burgos's translation of Möbius, the admonition to embrace imitation would have been particularly difficult given the misogynistic content and tone of the original. As Burgos embarked on her translation of *Über den physiologischen Schwachsinn des Weibes* or *The Mental Inferiority of Women,* the first vital challenge is evident in the title itself, as the German noun phrase "das Schwachsinn" (Möbius 3), means lack of intelligence or mental deficiency due to underdevelopment, "Mangel an Intelligenz durch angeborene Unterentwicklung" (Wahrig 3336; Lack of intelligence due to congenital underdevelopment). While Möbius focused on the word "deficiency" in his title, Burgos's translation highlights the other meaning of the word, underdevelopment, as a form of inferiority, in her translated title "La inferioridad mental de la mujer" (4).

Burgos also makes a subtle yet important change that greatly affects the message of the source text when she switches the main title to a subtitle and includes her new title as the main, radically altering the final text. This choice is explained by Burgos in 1909, four years after the translation was published, in an article penned by her alter ego and pseudonym, Colombine, in her *Femeninas* column in the *Heraldo de Madrid* newspaper. Colombine comments on her translation process and the book itself, stating: "Es un libro con suerte, porque tuvo que la de que sus traductores le cambiamos el título y aparece con el sugestivo y batallador de *La inferioridad mental de la mujer,* cuando Moebius le llamó sencillamente *Deficiencia mental de la mujer*" (Colombine 4; It is a book with luck, because it had translators change the title with the suggestive and combative *The mental inferiority of women,* when Moebius simply called it *Woman's mental deficiency*).[5]

Burgos's explanation offers a clear demotion of the original German title as one that is not representative of the arguments made by Möbius in his book, but also most importantly, it establishes Burgos's translation as an authority over Möbius very early in the process, as early as the title itself. By moving the original title to a secondary position on the cover page, the translation grants precedence to Burgos's choice for a more specific, topic-connected title above the one selected by the original author. By acknowledging herself as the translator and imposing a deliberate change to the title, as she states in her *Femeninas* column above, Carmen de Burgos implies that from the beginning she understood herself as an interventionist translator, one who will not simply "imitate" but rather interpret the original source text. Reader be warned: this translator will not be "the modest, self-effacing translator, who produces a smooth, readable target language version of the original" (Flotow, "Feminist translation" 76). The effect of such a move is more than a translation strategy by Burgos. Indeed, it can be viewed as the beginnings of what the feminist translation methodology would become later in the twentieth century, that is, a performance of "actively participating in the creation of meaning" (Godard 50). The new meaning is thus the one offered by the translator and not the one originally intended by the author. Indeed, this activist stance will become increasingly clear throughout the rest of Burgos's translation. Burgos's work results in a translated text that has become a new text, "a mediated form of it, a 'representation'" (Venuti, *Translation Changes Everything* 200), and one that addresses the cultural parameters of Burgos's own situation and that of her ideal Spanish readers.

In addition to creating a new title for the work, Burgos adds a new section to the book titled "Prólogo de la traductora" (Translator's prologue) that consists of a five-page introduction, which precedes Möbius's own prologue. Burgos's prologue functions in two ways: first, it offers a personal explanation as to why she decided to tackle this particular translation, and second, it provides a more procedural discussion where she explains the task of translating a misogynist text. Burgos's ambivalence at undertaking such a scandalous translation is weighed against her curiosity for analyzing the text in order to, as Burgos explains: "penetrar seriamente en las deducciones atrevidas del neurópata" (5; seriously penetrate the bold deductions of the neuropath). She assures her readers of her procedure to prevent a tainted translation: "quise despojarme de todos los prejuicios; si alguna idea nublaba mi espíritu, interrumpía la lectura para volver a reanudarla pasada la ráfaga de protesta" (5–6; I wanted to shed all prejudices; if any idea clouded my spirit, I interrupted the reading to resume it again once the burst of protest passed). As a translator, Burgos anticipates the attacks she will receive as a woman translating a scientific text, particularly one on this topic. Claiming for her readers that she would provide a pure, objective, and untainted translation can be seen as a strategy to gain trust, particularly from male readers, and to further convince them to read the translation and, in that process, also to continue reading her comments, criticism, and refutations of Möbius's theories.[6] This clarification also reflects Burgos's preoccupation with her female readers, perhaps for those who were wondering why a woman would want to get involved in translating a misogynistic text. Appealing to their sense of public morality, Burgos connects to them as follows: "Cuando comencé la traducción de este libro, un temor instintivo detuvo mi pluma" (5; When I began the translation of this book, an instinctive fear stopped my pen). What Burgos deems as a pure translation is what Lawrence Venuti describes as what happens to the translator during the process of translating a foreign text. In *The Translator's Invisibility*, Venuti defines the experience as follows: "A translated text [...] is judged acceptable by most publishers, reviewers, and readers when it reads fluently, when the absence of any linguistic or stylistic peculiarities makes it seem transparent, giving the appearance that it reflects the foreign writer's personality or intention or the essential meaning of the foreign text—the appearance, in other words, that the translation is not in fact a translation, but the 'original'" (1). By doing his or her task so perfectly, so purely, the translator becomes "invisible," resulting in, as Venuti continues to argue, "a weird self-annihilation, a way of practicing translation that undoubtedly reinforces its marginal status" (7). Since the most important aspect of the end product is the original author and not the translator of the foreign text, the translator's voice is erased from the process of translation, and the translator is praised for producing a translated text as close as possible to the original. However, in this instance, the voice of the presumed "invisible" translator, Burgos, becomes visible and is heard in the process of translation. Burgos warns her readers of a change, becoming not only the visible translator but an interventionist and feminist translator as well.

Already in the prologue, which does not exist in the source text, Burgos establishes a counterargument to Möbius via the disclosures of the translator, where she states what the reader can expect regarding the task at hand, namely a translation that is greatly influenced by diverse arguments in the realms of science, feminism, and religion. In this prologue, Burgos attacks what she describes as "feminismo militante" (6; militant feminism), which for her is directly connected to the religious fervor among conservative upper-class women, with its aim of making women slaves to a cause. She insists that poor women are anarchists and not feminists, because they ask for human equality among the sexes and not for absolute identity among the sexes (7–8). On whether Burgos was a feminist or not, her biographers are split. Concepción Núñez Rey states that "negar el feminismo parece una táctica para Carmen" (110; denying feminism seems to be a tactic for Carmen), and María del Carmen Simón Palmer asserts of Möbius's book's scandalous content: "causará la indignación de las feministas, como sin duda lo era la propia Carmen" ("Carmen de Burgos, traductora" 160; it will cause the outrage of feminists, as Carmen undoubtedly was herself). Detailed studies of Spanish feminisms, such as Roberta Johnson and

Olga Castro's "First-Wave Spanish Feminism Takes Flight," clearly identify Burgos as part of Spanish first-wave feminism and clarify some of Burgos's ambivalence about the movement, noting that "Burgos rues the fact that the different facets of feminism (many of them with class origins) are often at cross-purposes when women should be working together for the common good" (230). The ideology espoused in Burgos's translation does point toward a feminist stance, but the connotation of the word in Spain, as Núñez Rey argues, made it a matter of what was accepted publicly and by which social group.[7] Burgos was not unique in not wanting to call herself a feminist in public; other women writers, such as Emilia Pardo Bazán and Concepción Gimeno, did the same.

Burgos uses her translator's prologue to anticipate her later attacks on Möbius, whose theories of women's intellectual deficiency originated from his perception of "un degenerado físico" (a physical degeneration) as well as his "atinadas observaciones" (wise observations) on women's education and motherhood (9). As a translator, Burgos attacks Möbius's use of long prologues, three to be exact. Burgos makes a choice to include the third prologue, from his fourth edition, after her own original prologue in her translation. Burgos defends her translation of Möbius at the same time that she debunks his main thesis in the book, namely that women are intellectually inferior to men: "La diversa aptitud de los dos sexos no indica inferioridad en ninguno de ellos, sino modalidades diferentes, armónicas y necesarias para la marcha de la humanidad" (9; The diverse aptitude of the two sexes does not indicate inferiority in either of them, but different, harmonious, and necessary modalities for the march of humanity). She insists that Möbius's book needs to be read, because it "hace pensar" (11; makes one think), even though stylistically it is "seco, árido, descarnado, frío, y violento en ocasiones" (10; dry, arid, stark, cold, and violent on occasion). She defends her translation because it provides the reader with a continuum regarding the development of scientific discourse within the history of Spain, particularly a view of the Spanish woman such that she "tiene un papel activo de excepcional importancia, admirablemente determinado dentro de su sexo ... pues no podemos suponer que evolucione en sentido inverso" (11; has an active role of exceptional importance, admirably determined within her sex ... because we cannot assume that it evolves in the opposite direction). Finally, Burgos plants doubts not only about Möbius's theories, but also about the scientific apparatus itself when it comes to the study of the intellectual capabilities of the sexes: "la antropología y la biología no se hallan adelantadas para sentar principios absolutos" (10; anthropology and biology are not advanced enough to establish absolute principles). Even though in many of her writings she publicly stated that she believed in the progress science would bring, one can see via this translation that she also attacked the misogyny that existed among the men of science.

The Subversive Translator

Burgos's transformation of her translated text is greatly enhanced by her subversive use of "Nota de la T. [Traductora]" (30; Note of the translator). These notes exemplify what one translation critic describes as "explicit and direct statements of the translator's voice for the purpose of annotating the text" (Toledano-Buendía 150). Defined as "paratexts" by Gerard Genette, these translator's notes are seen as "liminal devices and conventions within the book that mediate the book to the reader: title and subtitles, forewords, [and] notes" (xviii). There are two distinguishable kinds of translator's notes. The first type are those deemed "explanatory notes," which "provide additional information that is considered necessary in order to achieve a perfect understanding of the source text and reproduce all the effects of the original texts in the target language" (Toledano-Buendía 156). The other type of notes, called "discursive notes," function "not to provide the implicit reader with objective information or details . . . the translator does not only say something, but also comments on something and expresses an opinion about it . . . the communicative intention of the paratextual messages is not restricted" (Toledano-Buendía 158). Burgos does not use her translator's notes as explanatory notes but rather as discursive notes in order to argue with the source text author and to undermine the authorial voice by pointing out the illogical and unprovable aspects of Möbius's theory.

Burgos's translator's notes thus often appear in critical moments of the discussion by Möbius, such as when he is attempting to connect mental inferiority and physical inferiority. Möbius places this argument within a discussion of the challenge of defining scientific terms, arguing that "Mental deficiency is defined as a state in between imbecility and normality, but making a demarcation between them is very difficult. What is clear is that an intelligent man is one that can discern or think well, while a stupid man is one that cannot do so" (Burgos 29).[8] In scientific terms, stupidity is seen by Möbius "as a morbose anomaly, as an enormous reduction or weakness of discernment, whereas physical deficiency, as seen in children and later on, in elderly adults, becomes an illness that cannot be stopped and ends up producing the diminishment of mental faculties" (Burgos 30). At this point Burgos turns her note into a contestatory remark that points at the illogical argument of Möbius, who sees illness everywhere, instead of seeing aging as a natural occurrence in human beings, "La misma vejez es ya una enfermedad—(N.de la T.)" (30; Even old age is now seen as a sickness—[Translator's note]).

Burgos's translator's notes continue to serve as contestatory remarks on the source text, and they often help clarify instances in which her translation differs from the original. For example, one of Möbius's most extensive arguments is based on head size, whereby Möbius argues that men have larger brains and women smaller ones, and relative size determines intellect. While the German source text states "Ein kleiner Kopf umschliesst natürlich auch ein kleines Gehirn" (Möbius 6; A small head naturally encloses a small brain), Burgos translates the German term "kopf" as, "Un cráneo pequeño encierra evidentemente un cerebro pequeño" (Burgos 36; A small cranium evidently encloses a small brain/mind). At the time of publication in 1883, the German noun "kopf" literally meant "head," but it also figuratively meant "mind" (Wesseln 278). Burgos's translation thus makes explicit the bias of such associations. Her choice of terms is significant because the continuation of Möbius's argument is aimed toward disproving Bischoff's theory of brain weight, embracing instead the anatomical observations of Rüdinger's comparisons of the size of male and female brains. This is key because Möbius bases his argument about women's mental inferiority on Rüdinger's conclusions: that women's mental inferiority is due to a less developed physical composition of their brains, beginning at birth, as compared to a more fully developed brain composition in males.

It is at this point in the argument, literally at the brain itself and not at the head, that Burgos aims her attack, as she states in her note: "El cerebro de Voltaire es de los más pequeños que se conocen y basta para contener un mundo" (36; Voltaire's brain is one of the smallest known, and it was enough to contain a world). Burgos anticipates the faulty argument and makes her own comparison between Voltaire's brain and a woman's brain, in an attempt to show how prejuiced Möbius's arguments were. Burgos's note does not clarify any technical aspect of the translation; instead, it opens the door to scrutinize and counterargue with Möbius by juxtaposing Voltaire, a male cultural icon, with Mobius's anatomical argument. In this instance, Burgos achieves what Venuti calls "the violence wrecked by translation" (*The Translator's Invisibility* 15), for she has effectively shown herself to be a subversive translator by portraying Möbius's argument as the one that needs to be confronted, criticized, and transgressed. Furthermore, Burgos also anticipates what Luise von Flotow calls a mandate of feminist translators "[to] correct texts they translate in the name of feminist truths" (*Translation and Gender* 24). This correcting and rewriting process transposes the original text into a new text by counterarguing, highlighting the absurd comments made in the source text. Such translational work adds a new perspective to the way the reader perceives the text. With this bold and aggressive strategy, Burgos appropriates Möbius's text, corrects its errors, revises and edits it, with the purpose of redefining it as an invalidated text for the reader. This endeavor reflects a feminist point of view that critiques the arguments in the source text, but it also asserts Burgos's authorial voice as an expert translator who, five years later in her newspaper column entitled "El secreto de Moebius" (Moebius's secret), further reduces Möbius to a confused man peddling not even a book but a worthless "folleto" (Colombine 4; brochure).

Burgos's notes also become both historically insightful and combative, as she shows on the one hand her knowledge of how Möbius's treatise was received in Germany and on the other how she plans to combat its premises. Möbius's argument about women as instinctive beings provides his most misogynist postulate in the source text: "instinct makes women the same as animals" (9). It is probably in reference to this part of the book that Burgos commented in her prologue about the scandalous nature of this text in Germany. Burgos's reaction to Möbius's assertions is shown in her translator's note as follows: "Admitiendo desde luego sin reserva, como el autor lo hace, la existencia de los instintos en el ser humano, en vez de los actos reflejos, es evidente que tanto el hombre como la mujer dominados por ellos [los instintos], han de parecerse a la bestia, pero ni el instinto obra solo, ni puede dar ningún atractivo." (43; Admitting of course without reservations, as the author does, the existence of instincts in human beings, instead of instinctive actions, it is evident that men, like women, are dominated by them [instincts], and thus both must be animals, but instincts do not work alone, nor are they able to offer a special appeal). For Burgos, the difference between animals and human beings is clear. By choosing the Spanish verb "obrar" to demonstrate an act of individual and intellectual choice, the translator explicitly mounts a contrary argument to the source text: human beings are not dominated by instincts as beasts are, because both sexes use their intellect and free will to discern. In Burgos's counter-argument, she highlights that women are not overly dominated by instincts and that both sexes have them. The final critique of this line of Möbius's argument comes when Burgos demonstrates an important contradiction, criticizing Möbius's claims that women are dominated by instincts like animals by noting that such women would never be able to fulfill even Möbius's vision for their mission in life. Burgos counterargues that women use their intellect to think logically, insisting that "El instinto no domina en ella, como en los animales, casi aislado de todo, pero está sujeta al pensamiento individual" (43; Instinct does not rule in her, as it does in animals, almost isolated from everything, but is subjected to individual thinking).

Möbius makes several attempts to define women as overly dependent on feelings and motivated only by their social responsibility as mothers, portraying them as without any interest in matters beyond children or the home. He repeats the gender bias common among his peers, labeling all women as private creatures defined by physical limitations and traditional social norms. Burgos reacts forcefully to such claims in her note, attacking Möbius as a bad psychologist who does not understand the social pressures that women face: "Inexacto, pues, fanatizadas, prefieren, más que nada, los intereses de la religión y del sacerdote, a los del hogar, con un ahínco que se llama heroísmo entre los hombres" (Burgos 44; Inexact, because, fanaticized, they prefer, more than anything, the interests of religion and of the priest to those of the home, with a determination that is called heroism in men). Burgos posits that the representation of women offered by Möbius's pseudo-science is not only uninformed but also reinforces a discriminatory practice against women that keeps them uneducated and under male dominance and traditional religious hierarchies. Burgos will not let such ideas stand without interrogation. As Massadier-Kenny argues, "translating is an exercise in interrogating the complex ways in which gender becomes bound up with language, and consequently with translation" (55). In this case, Burgos has become the interrogator or inquisitor of Möbius's ideas.

By questioning Möbius's definition of women in terms of biological determinism grounded in social obligations, Burgos uses the source text itself to challenge its own logic. One example involves Möbius's discussion of the problem of multiple simultaneous mental activities, or what today might be termed multi-tasking. He argues that such distraction is one reason women are mentally inferior (15–16). Burgos reflects sarcastically in her translator's notes as she deliberately misstates the title of Möbius's book, focusing the reader's attention on the irony of Möbius's claim by using italics for emphasis: "entonces este libro debe llamarse no la *inferioridad*, sino la *diversidad* de la mujer" (Burgos 58; then this book should not be titled *inferiority*, but *diversity* of women). With this note, Burgos implies that, far from proving women's mental inferiority, as Möbius claims, the necessity and multiplicity of cerebral tasks performed by women prove their mental agility. For Burgos, the argument against diversity of thought is simply not logical, and thus it undermines all claims about women's mental inferiority, to the point of renaming the book. For Burgos, mental agility is a clear strength in women and a factor that forcefully contradicts Möbius's claim of their inferiority.

Yet while for Burgos multitasking is a positive characteristic of women, for Möbius it is a negative one and an impediment to her maternal calling: "Kraft und Drang ins Weite, Phantasie und Verlangen nach Erkenntniss würde das Weib nur unruhig machen und in ihrem Mutterberufe hindern, also gab Sie die Natur nur ein kleinen Dosen"(15; Strength and urgency, imagination, and craving for knowledge would make a woman restless and hinder her maternal obligations, so nature gave her only a small dose). Möbius contradicts himself with these claims, when he has just argued that women are inferior because of the distractions of their varied and diverse interests. His conclusion demonstrates a similar logical fallacy: "Nach alledem ist der weibliche Schwachsinn nicht nur vorhanden, sondern auch notwendig" (16; Therefore, women's intellectual inferiority not only exists, but is necessary). Whereas for Burgos, women's intellectual difference is seen as "diversidad" (58), a positive characteristic for women, for Möbius, difference is synonymous with inferiority to her male counterpart. Burgos inverts this weak argument by stating and highlighting via italics the word "inferiority" in her following note: "si es necesaria [la diferencia] no hay tal *inferioridad*" (58; if it is necessary [the difference], then there is no such inferiority). In her translator's aside, Burgos once again critiques a controversial and contradictory claim from the source text, and she does so from a feminist translational stance. As Johnson and Castro note in their study of first-wave Spanish feminisms, "Burgos's rational style of argument, her cool level-headedness, and ability to go straight to the heart of the matter with incontrovertible evidence align her with the long rational equality feminist tradition that began in Spain with Benito Jerónimo Feijóo" (226).

Burgos's resistance throughout the notes to the argument contained in the text redefines the role of translator in feminist terms, as one of challenging supposedly "expert" scientific claims (in this case, those of a German doctor), by showing their illogical conclusions.

The Translator's Decisions and Tasks: Strategic Insertions and Substitutions

Burgos's translation undermines Möbius's authority as a German scientific arbiter not only by demonstrating the source text's logical fallacies but also by including insertions, Spanish colloquial expressions, and substitutions that serve to facilitate her audience's understanding of the foreign text. In some ways, one might argue that this would be true for nearly any non-literal translation and thus not specifically an innovative or feminist move on Burgos's part. Yet the specific choices made by translators deeply affect the way a text is received by the reading public and, most importantly, how the translated text is understood by a different audience than that of the original author. Indeed, translated texts demonstrate "a decision process: a series of a certain number of consecutive situations [or] moves, as in a game, imposing on the translator the necessity of choosing among a certain (and very often exactly definable) number of alternatives" (Levý 148). In Burgos's case, the decisions involved not only employing an activist voice and an innovative approach but also taking great care with word choice, particularly via the translation of linguistic and colloquial phrases, as well as the inclusion and substitution of various target language forms.

In her translation of Möbius's prologue, Burgos inserts Spanish colloquial phrases that help the reader visualize the lively discussion that Möbius had with his psychologist friend, who injects doubts into the German doctor's assertions about women's mental inferiority to men. For example, Burgos inserts the phrase, "¡Ay de mí" (Oh my!), not existent in the source text, as Möbius defends himself: "Pero, contra cuántos prejuicios ¡Ay de mí! Me veo precisado a luchar" (14; But, against so many prejudices, Oh my! I find myself obligated to fight). According to the online dictionary of the Real Academia Española, "¡Ay de mí!" can signify a range of emotional responses, including "pena, temor, conmiseración, amenaza, suspiro, [o] quejido" (sadness, fear, commiseration, threat, sigh, [or] moan). Burgos's added phrase is thus a Spanish colloquialism used when the speaker wants others to hear a lament that creates pity in the reader. Burgos uses this as a way of familiarizing the text to her Spanish readers, thus making the text sound more like a Spanish text rather than a foreign one. In another example from the prologue, Burgos inserts another word with the expected effect on the reader of demonstrating how unmovable Möbius is with his theories; not even his friend can convince him of his flawed assumptions: "-[Amigo] Existen con toda certeza cualidades más desarrolladas en la mujer. -[Möbius] Tontería. ¿Cuáles son, si no, esas cualidades?" (14; –[Friend] There are certainly more developed qualities in women. – [Möbius] Nonsense. What, if not, are those qualities?). The word "tontería" (foolish remark) does not appear in the source text and is intended to negate the argument of the speaker, making it unimportant or without any value. This insertion sways the reader to view Möbius negatively.

In another instance, Burgos expands her commentary to a complete phrase, as she inserts words in Spanish meant to make very clear where Möbius stands on the societal need for women writers and artists, or those who believed that women had to self-emancipate by working and contributing to society: "A mi juicio, no son absolutamente indispensables más que las actrices y las cantantes. Ninguna persona sensata querrá sostener que son necesarias las pintoras, las escultoras, las doctoras" (23; In my judgment, they are not absolutely indispensable any more so than actresses and singers. No sensible person will want to argue that painters, sculptors, doctors are necessary). The beginning phrase "a mi juicio" (to my mind / in my judgment), is a Spanish phrase that, because it employs the word "judgment," communicates that this is a subjective opinion on the part of the speaker. This inserted phrase implies for readers that with such judgment, there would also seemingly be no need for women intellectuals, and much less emancipated ones, and thus leaves them with the impression of Möbius as a misogynist with faulty judgment rather than a respectable German neurologist.

Another way that Burgos connects to her readers through her translation is by using idiomatic expressions in Spanish, a significant translational strategy that helps readers better understand a source text via familiar Spanish forms. Such idiomatic expressions are called equivalences and are defined as "fixed and belong[ing] to the phraseological repertoire of idioms, proverbs" (Vinay and Darbelnet 90). Burgos uses idiomatic expressions to help her readers understand Möbius's text and to clarify his meaning in Spanish. Möbius argues that the married woman undergoes a process of physical "decay" after multiple pregnancies and births that, he asserts, destroy her physical beauty and bodily strength, thus making her androgynous (72). Möbius finds in these losses a further reason for her declining mental faculties. The translator chooses a Spanish idiomatic expression known to her readers and writes it in italics, substituting it for the German one: "y la mujer, como suele decirse, *chochea*" (72; and women, as one commonly says, become *feeble minded*). The verb "chochear" means "mostrar debilitadas las facultades mentales por efecto de la edad" ("Chochear"; to show weakened mental faculties due to aging). The Spanish phrase is not as strong as the German "die Frauen versimpeln, wie es populär heisst" (Möbius 22; women become dumb, as the popular saying goes). Burgos's translation choice of "mental weakness," instead of the more literal meaning of portraying women as becoming "dumb" in their old age, is a way of evoking her resistance to the text, much like her modified title for the work. Burgos selects terms that serve to highlight the scandalous arguments made by Möbius in the source text. Her translation offers the perspective of a feminist translator, exerting her authority over Möbius and challenging his brand of science itself.

Burgos not only employs idiomatic expressions and colloquial phrases to help her readers critique the foreign text; she also alters or modifies concepts, substituting them with other words or ideas that might prove insightful for her readers. For example, in one of Möbius's conclusions, he states that excessive brain activity makes women strange, sick, and bad mothers, preventing them from fulfilling their so-called "natural" tasks. When it comes to defining those women (the androgynous feminists who do not follow strict sexual rules for social separation), he refers to them as "Die modernen Närrinen" (16; the modern women-fools). Burgos translates the phrase as "Las exaltadas *modern-style*" (59; the *modern-style* exalted), including an adjective and making a clear reference to the English suffragettes by using their English description as "modern" women. Burgos does not dismiss the argument as a foolish remark with the word "tontería" as she had done earlier. Instead, in this case she changes and, most importantly, substitutes the negative connotation given by Möbius with a positive one of the suffragettes through a connection known to her readers and to those who were fighting for women's rights.

Conclusion

Contrary to the accepted norms of her time, Carmen de Burgos did not follow the accepted translational practices in her work with Möbius's polemical text. Rather than following the stylistic norms of the French tradition or the German standards championed by Schleiermacher and others, she adopted an activist stance that was innovative and indicative of her first-wave feminist role. Burgos effectively redefines the role of the translator by becoming interventionist, subversive, combative, and corrective. Through the use of translator's notes, she implements a secondary text that tells a different story and that clashes with Möbius's treatise. Burgos does not accept the invisibility rule of translators but instead continually asserts her agency by becoming textually present. Already by 1905 when this translation was published, she was interpreting source texts in feminist terms, including a translator's prologue before the source text author's prologue that spoke directly to her readers and anticipated their concerns related to such a misogynist text. She inserted phrases and concepts that facilitated the reader's critical stance, and most importantly, she offered a socio-linguistic guide to alert readers to code words used by Möbius, highlighting their problematic status via translator's notes and italics. Within the text, she established her authorial voice as she confronted the German doctor, whose ideology was clearly opposed to the progressive philosophy she espoused. Her work carefully, deliberately, and convincingly substituted and redirected the German text toward her interpretation and feminist translation. Burgos was not only a ground-breaking writer and journalist but also a translator ahead of her time.

NOTES

1. Burgos's 1904 translations included Hellen Keller's *Historia de mi vida* (The story of my life), Geza Mattachich's *Loca por razón de estado. La princesa Luisa de Bélgica* (Demented by reason of state. Princess Louise from Belgium. Unpublished memoirs by Count Mattachich), Ernesto Renan's *Los Evangelios y la segunda generación cristiana*, (The Gospels and the second Christian generation) and Leo Tolstoy's *La guerra ruso-japonesa* (The Russian-Japanese War).
2. All German-to-English and Spanish-to-English translations are mine.
3. I borrow the term "interventionist" from Luise von Flotow's seminal book *Translation and Gender: Translating in the "Era of Feminism."* The invisibility designation was developed by Lawrence Venuti's *The Translator's Invisibility: A History of Translation*.

Lane-Mercier's article "Translating the Untranslatable" also informs this study.
4. This treatise was originally read at the Royal Academy of Sciences in Berlin on June 24, 1813.
5. All my translations of Burgos's *La inferioridad mental de la mujer* follow the 1905 edition by *Imprenta del Pueblo*.
6. Burgos demonstrates her knowledge and argumentative power in many other non-scientific texts that follow Möbius's translation, but her last essay, "Las mujeres de ciencia" (232) brings her back to the topic of medicine and presents her with the opportunity to respond to Möbius once again.
7. For a comprehensive study of Iberian feminisms, see Bermúdez and Johnson, eds.
8. Note that this quotation, and the following attributed to Möbius, is my English translation of Burgos's Spanish translation of the original German source.

WORKS CITED

"Ay de mí." *Diccionario de la lengua española*. Real Academia Española, dle.rae.es/ay. Accessed 6 September 2020.

Baker, Mona. *Routledge Encyclopedia of Translation Studies*. Routledge, 1998.

Bergues de las Casas, Antonio, editor. *La Abeja*, Tomo I, Barcelona, Librería de Don Juan de Oliveres, 1862.

Bermúdez, Silvia, and Roberta Johnson, editors. *A New History of Iberian Feminisms*. U of Toronto P, 2018.

Burgos, Carmen de. *La inferioridad mental de la mujer. (La deficiencia mental psicológica de la mujer)*. Imprenta del Pueblo, 1905.

"Chochear." *Diccionario de la lengua española*. Real Academia Española, dle.rae.es/?id=8w-JVwM4. Accessed 20 July 2018.

Colombine. "El secreto de Moebius." *El Heraldo de Madrid, Femeninas*, 11 Apr. 1909, p. 4.

Flotow, Luise von. "Feminist Translation: Contexts, Practices, Theories." *Traduction, Terminologie et Rédaction*, vol. 4, no. 2, Jan. 1991, pp. 69–84.

———. *Translation and Gender: Translating in the "Era of Feminism."* St. Jerome Publishing, 1997.

Genette, Gerard. *Paratexts: Thresholds on Interpretations*. Cambridge UP, 1997.

Godard, Barbara. Preface. *Lovhers*, by Nicole Brossard. Translated by Godard, Guernica P, 1988.

Johnson, Roberta, and Olga Castro. "First-Wave Spanish Feminism Takes Flight in Castillian-, Catalan-, and Galician-Speaking Spain." *A New History of Iberian Feminisms*, edited by Silvia Bermúdez and Roberta Johnson, U of Toronto P, 2018, pp. 221–35.

Lane-Mercier, C. "Translating the Untranslatable: The Translator's Aesthetic. Ideological and Political Responsibility." *Target*, vol. 9, no. 1, 1997, pp. 43–68.

Levý, Jiří. "Translation as a Decision Process." *The Translation Studies Reader*, edited by Lawrence Venuti and Mona Baker, Routledge, 2000, pp. 148–59.

Marín, Ana Martínez. *Carmen de Burgos (Colombine). Mis mejores cuentos.* Editoriales Andaluzas Unidas, 1986.

Martín, Marcia Castillo. *Carmen de Burgos 1867–1932.* Ediciones del Orto, 2003.

Massadier-Kenny, F. "Towards a Redefinition of Feminist Translation Practice." *The Translator,* vol. 3, no. 1, 1997, pp. 55–69.

Möbius, P. J. *Über den physiologischen Schwachsinn des Weibes.* Halle a. S., Carl Marhold, 1900.

Núñez Rey, Concepción. *Carmen de Burgos "Colombine" en la Edad de Plata de la literatura española.* Fundación José Manuel Lara, 2005.

Sánchez, Lola. "Productive Paradoxes of a Feminist Translator: Carmen de Burgos and Her Translation of Möbius' Treatise, The Mental Inferiority of Woman (Spain, 1904)." *Women's Studies International Forum,* 2014, pp. 68–76.

Schleiermacher, Friedrich. "Methoden des Übersetzens (On the Different Methods of Translating)." *Theories of Translation. An Anthology of Essays from Dryden to Derrida,* edited by Rainer Schulte and John Biguenet, U of Chicago P, 1992, pp. 36–59.

Simón Palmer, María del Carmen. "Carmen de Burgos, Traductora." *ARBOR Ciencia, Pensamiento y Cultura,* vol. 186, June 2010, pp. 157–68.

———. *Escritoras españolas del siglo XIX.* Castalia, 1991.

Starčević, Elizabeth. *Carmen de Burgos. Defensora de la mujer.* Cajal, 1976.

Toledano-Buendía, Carmen. "Listening to the Voice of the Translator: A Description of the Translator's Notes as Paratextual Comments." *Translation and Interpreting,* vol. 5, no. 2, 2013, pp. 149–62.

Truxa, Sylvia. "Carmen de Burgos." *Ein Raum zum Schreiben. Schreibende Frauen in Spanien von 16. bis ins 20. Jahrhundert,* edited by Ute Frackowiak, Verlag Walther Frey, 1988, pp. 198–219.

Venuti, Lawrence. *The Translation Studies Reader.* Routledge, 2000.

———. *The Translator's Invisibility.* Routledge, 2008.

———. *Translation Changes Everything.* Routledge, 2013.

Vinay, Jean-Paul, and Jean Darbelnet. "A Methodology for Translation." *The Translation Studies Reader,* edited by Lawrence Venuti and Mona Baker, Routledge, 2000, pp. 84–93.

Wahrig, Gerhard. *Deutsches Wörterbuch.* Mosaik, 1981.

Wesseln, Ignaz Emanuel. *Thieme-Preußer: Neues Vollständiges Kritisches Wörterbuch der Englischen und Deutschen Sprache.* Haendke and Lehmkuhl, 1883.

CHAPTER 9

Contrasting Images of Women Scientists in the Early Postwar Period (1940–1945) and the Novel *María Elena, ingeniero de caminos* by Mercedes Ballesteros

MIGUEL SOLER GALLO

This study analyzes how the official rhetoric in Spain portrayed women scientists and female intellectuals in the period directly following the Spanish Civil War, between 1940 and 1945. These years have special relevance because they constitute the so-called five-year blue period (*quinquenio azul*) of Francoism; that is to say, those most influenced by the Falange, the Mussolini-inspired movement founded in October of 1933 by José Antonio Primo de Rivera and subsequently employed by Franco to ideologically mask, under the auspices of a political movement, what was in fact a military dictatorship.[1] The discourse of the regime, following the lead of both the Catholic Church and the Falange, was deeply restrictive of women's rights and focused on the indoctrination of women according to the vision of society supported by the dictatorship.[2] Most publications disseminated at the time were written by men linked to the Franco regime, and representations of women tended to focus on appropriate conduct and traditional feminine roles, relegating women to the private sphere and excluding them from public life and thus from scientific participation. Consider for example the titles of several texts published anonymously in magazines for women such as *Revista para la mujer* (Women's magazine) and *Medina, Semanario Oficial de la Sección Femenina* (Medina, Official weekly of the Women's Section): "Lo que las armas victoriosas traen, mujer" (1939; What victorious weapons bring, woman), "Carreras para la mujer" (1941; Professions for women), "Destino de la mujer Falangista" (1941; Fate of the Falangist woman), "La mujer, alma del hogar, preparada por la Falange" (1942; The woman, soul of the home, prepared by the Falange), "Las terribles intelectuales" (1943; The terrible female intellectuals), and "Diez mandamientos de la universitaria" (1945; The ten commandments of the coed). These and many other works published in the early Franco period sought to inculcate the parameters of societal gender norms, asserting the primacy of the male over the female and confirming traditional military, educational, and religious hierarchies.[3]

Within this context of systemic political and social oppression, Mercedes Ballesteros published the novel *María Elena, ingeniero de caminos* (1940; María Elena, civil engineer). Several things were unusual about the text, not simply its scientific content. To begin, most works that were published in the immediate postwar period that were geared toward women's issues appeared anonymously, as evident in the samples listed above. They also tended to be published in journals with a predominantly female readership, controlled by the Falange, and for women between twenty and thirty years of age.[4] The goal of official discourse was to influence the consciences of young female readers so that they would not deviate from the dictatorship's intended mission, focused on developing women's roles as wives and mothers in accordance with traditional ideals of the female sex. Indeed, in official texts, when a woman gave advice or speculated on her own intellectual capacity, the narrative voice employed was often the first-person plural. In this way, the female collective was made to appear unified, and a type of feigned solidarity was offered, cloaking the manipulation of readers and the inculcation of traditional femininity, so central to Falangist doctrine.

However, in the midst of this atmosphere of restrictions against women's individual liberties, there were discordant voices, such as that of Mercedes Ballesteros, who were opposed to the official discourse and dared to reflect the ability of women to pursue professions based on their academic interests and preparation. *María Elena, ingeniero de caminos*, published in 1940, offers one such example.[5] The work, previously unstudied by critics, is itself unique within the author's artistic trajectory, not based on genre (Ballesteros published nearly twenty other novels of similar style between 1938 and 1943) but rather on scientific focus and content.

Ballesteros did not gain status and prestige as an author until after the 1950s, but this early work, despite its formal simplicity, reveals the progressive spirit that characterized Ballesteros's writing and that permitted her to confront themes considered taboo in the first years of Francoism. Ballesteros's novel departs from the standard approach to women's relationships with the worlds of education and science, offering a fictional testimony that is nonetheless a faithful reflection of the difficulties that women faced pursuing a professional future. The work thus offers a counterpoint to the dominant discourse, given that it advocates for educational and labor equality between the sexes, and it does so using the world of STEM (via engineering) as its focus.

The "Terrible Women Intellectuals": Negative Mirrors for Women Via the Sección Femenina

Before looking more carefully at Ballesteros's own background and the content of *María Elena, ingeniero de caminos*, it is important to contextualize Ballesteros's contribution within the discourse of the time, particularly the representations offered by official sources regarding women intellectuals. The years 1940–1945 were marked by a regression in women's access to intellectual and professional development, particularly in light of the reforms and freedoms Spanish women had experienced under the previous democratic government operating during the Second Republic (1931–1939). Yet despite the setbacks for women in the postwar period, there existed many who possessed a vision resonant with the atmosphere prior to the Civil War, that is, with Republican Spain, and who rejected the model of women that was offered by Francoism and the Sección Femenina de Falange.[6]

The period of the Second Republic, within the frame of the Constitution of 1931, had been very fruitful for women's intellectual and professional development. On the question of female education, the *Institución*

Libre de Enseñanza played a fundamental role, stipulating that the physical and intellectual formation of each individual was essential for society's evolution and thereby advocating for the coeducation of boys and girls (Jiménez-Landi 33–39; Vázquez Ramil 51–65). Even earlier, in 1910, the Minister of Public Education, Julio Burell, who was a supporter of the *Institución Libre de Enseñanza*, promoted the Burell Law during the reign of Alfonso XIII, which established women's free access to university without having to obtain permission from the Education Minister (Flecha García, "Las mujeres" 216). Indeed, beginning in the nineteenth century, successful Spanish women writers offered important works focused on the equal treatment of both sexes in educational settings, arguing that access to education was the only possible way for society to develop and that the country needed women in higher education and the labor market.[7]

Despite these important arguments and clear progress for women's education in the late nineteenth and early twentieth century, there was still strong resistance to scientific study for the female population—especially in those fields derived from the natural sciences, such as medicine (Formica Corsi 172). The case of Dolores Aleu Riera demonstrates how this resistance was linked to the belief that such environments should be reserved for men. Aleu Riera was the first woman to graduate with a Bachelor's degree in medicine in 1882 from the Universidad de Barcelona, and she specialized in gynecology and pediatrics after obtaining her doctorate.[8] She dealt with continuous obstacles that were mounted to limit her freedom and to keep her from fulfilling her desire to be a doctor. As one example, she had to be escorted to the medical school in order to avoid being assaulted (Flecha García, *Las primeras universitarias en España* 95–98). Nevertheless, in the midst of this hostile environment, she succeeded with her medical career, and there were important figures who supported her goals, despite resistance from most of the students, faculty, and administration. For example, Dr. Antonio Formica Corsi, who offered classes in anatomy and dissection (and from whom Aleu Riera obtained a grade of A+), noted that despite Aleu Riera's intellectual brilliance, she was experiencing a tirade that had been unleashed contesting her ability to practice medicine. Dr. Formica Corsi felt the need to publicly come to her defense:

> No crea usted [se refiere a una voz anónima crítica con la estudiante], sin embargo, que los conocimientos expuestos fueran espontáneos, fueron aprendidos en los libros, en los atlas, de boca de los señores catedráticos y

de la mía, si no le sabe mal, ya que los conocimientos no son congénitos, sino adquiridos por el estudio. [. . .] ¿Quién es usted para medir la "cantidad de ciencia" de ninguna señorita? (Formica Corsi, *Las primeras universitarias en España* 172).

Do not believe XXX [referring to an anonymous voice critical of the student], however, that the knowledge displayed was spontaneous; it was learned from books, from atlases, from the mouths of university professors and from my own, which should not seem problematic for you, given that knowledge is not congenital, but rather acquired through study. [. . .] Who are you to measure "the quantity of science" of any young woman?

Formica Corsi's defense, published in *La Independencia Médica* in 1878 under the title "Fuera escrúpulos" (Out with scruples), demonstrates the depths of resistance to women's participation in scientific fields and the ways women, and their capacity for science, were measured. Indeed, in this case, the faculty member had to challenge the way female students were being judged based on suspect notions of women's "cantidad de ciencia" (172; amount of science). Given that society as a whole was not very comfortable with women's access to university studies, let alone to scientific disciplines, it was extremely difficult for women to pursue STEM fields. However, there were women like Aleu Riera and others who successfully pursued scientific studies; particularly during the Second Republic, there were even institutions that supported women's participation. For example, the Instituto Nacional de Física y Química (National Institute of Physics and Chemistry), which functioned from 1931 until 1937, employed 158 scientists, 36 of whom were women (Magallón Portolés 111–139). Among this distinguished group of female scientists were Felisa Martín, who earned Spain's first female doctorate in physics; Jenara Vicenta Arnal, the first female doctorate in chemistry; as well as Teresa Salazar, Piedad de la Cierva, and the sisters Dorotea, Petra, and Adela Barnés González (Muñoz Páez).

After World War I, women's participation in university life, and later in professional life, became increasingly common. Consider for example the following women who played exceptional roles in their fields at that time: Victoria Kent, the first woman to enter the *Colegio de Abogados de Madrid* in 1925; Matilde Huici Navaz, teacher, lawyer and pedagogist; and Clara Campoamor, the second woman to enter the *Colegio de Abogados de Madrid* (these three women founded the *Instituto Internacional de Uniones Intelectuales*); María de Maeztu, pedagogist and humanist, proponent of the *Residencia de Señoritas* between 1915 and 1936 and president of the *Lyceum Club Femenino* (1926–1939); María Zambrano, philosopher and essayist; María Cegarra Salcedo, the first female bench chemist, after obtaining her degree in 1928; Pilar Careaga, the first female industrial engineer in 1929; and Matilde Ucelay, the first degree-earning female architect in 1936. Legislation during this period did not make intellectual work impossible for Spanish women. While their access to upper-level instruction was hindered by customs and traditional principles linked to patriarchal culture, such limitations were not yet established law (Formica 11).

With the Civil War and the Francoist dictatorship that followed, the world of opportunities for women disintegrated. On March 9, 1938, the *Fuero del Trabajo* (Work Jurisdiction), one of the eight fundamental laws of Francoism, was approved. In the context of women's participation in the labor force, it prohibited their work at night and emphasized that "se libertará a la casada del taller y de la fábrica" (*Fuero del Trabajo*; married women will be free from the workshop and the factory). The law sought to relegate women to domestic functions and to promulgate theories regarding the perniciousness of work and its supposedly detrimental impact on familial tranquility, arguing that work limited the capacity of women to offer healthy children to the fatherland. A priori, nothing was stated in the new law that outright prohibited university study among women, but the restrictive atmosphere that the law promoted regarding women's freedom created a deep impact beyond the rules or legal measures. Conceptions of women's roles and their equation with the feminine became a focus of the regime's propaganda channels as well as speeches, not only those offered by the *Jefe del Estado* (Head of State), Francisco Franco, but also by the national "jefe" of the Sección Femenina, Pilar Primo de Rivera.

Two periodicals in particular mark this first period of Francoism: *Y* and *Medina*.[9] The first gave a glowing review of the benefits that the *Fuero del Trabajo* would provide for women via a 1939 article titled "Lo que las armas victoriosas traen, mujer" (What victorious weapons bring, woman): "Son para ti, mujer, estos principios el rescate de tu feminidad. Tú no naciste para luchar; la lucha es condición del hombre y tu misión excelsa de mujer está en el hogar" (12; For you, woman, these principles are the saviors of your femininity. You were not born to fight; fighting is man's condition, and your exalted mission as a woman is in the home). The effort to destroy the model of the modern and independent woman that had been consolidated during the Second Republic is a constant theme in publications of the period. In the 1940 inaugural editorial of the magazine *Medina*, titled "Destino de la mujer falangista," one can see the explicit way Francoist rhetoric attacked the notion of the modern, independent woman, offering instead an ideal of passivity, seclusion, and religiosity: "Lo que nos resistimos a entender es el nuevo concepto, un poco libre, independiente y suelto, de la mujer moderna. Amamos a la mujer que nos espera pasiva, dulce, detrás de una cortina, junto a sus labores y sus rezos." (3; What we resist understanding is the new concept, a bit free, independent, and loose, of the modern woman. We love the woman that waits for us, passively, sweetly, behind a curtain, along with her work and her prayers). Modern women in Francoist discourse are not only rejected as not worthy of "understanding," but they are also labeled as loose and unattractive, while the ideal Falange woman waits passively at home ready to meet patriarchal expectations.

Clearly, the early Francoist period produced a large amount of propaganda attempting to redefine what a woman should be and how women should behave in the new political sphere, and part of that rhetoric, particularly for the Sección Femenina de Falange, involved confronting the dilemma of how to reconcile women's access to university. Higher education was associated with the concept of the modern woman, and thus Falangist educational advocates needed to ensure that such access would not undermine Falangist ideals of femininity. Within the Falange, there existed the Sindicato Español Universitario (SEU; Spanish University Union), founded in November 1933, and it had a feminine branch beginning in April 1935. While the Sección Femenina alluded to the role of the woman intellectual via other outlets, including speeches and publications, it was the aforementioned organization that concerned itself with determining the modes of behavior for university women, particularly after the Civil War. In a 1939 article from *Haz* (the magazine of the SEU), entitled "La mujer y la preparación intelectual" (Women's intellectual preparation), readers were reminded that "la misión eterna de la mujer era el hogar, pero tenía también el deber de dirigir ese hogar con la suficiente preparación intelectual para ser educadora de sus hijos y compañera de su marido" (40; the eternal mission of women was to be in the home, but she also needed to be able to manage the home with sufficient intellectual preparation in order to be a teacher for her children and a companion for her husband).

Thus, education could be justified, but only insofar as it served the "eternal mission" of serving the family. The same article also alludes to economic necessity as a possible justification for university studies so that the wife could work outside the home. The rationale was clearly not feminist but rather focused on finances and explained via an analogy of inequality, which confirmed women's lesser status even while acknowledging their usefulness: "así como la mujer no es igual al hombre, tampoco las carreras y profesiones son igualmente propias para unas que para otros" (41; just as a woman is not a man's equal, neither are careers or professions equally appropriate for all women).

Despite this grudging acknowledgment of the necessity of university education for some women, early postwar discourse clearly discouraged university study, and particularly study in STEM fields. For example, in a 1941 article titled "Consultorio" (Doctor's office) and published in the magazine *Medina*, the author disparages women's interest in education, particularly the hard sciences: "para empezar, y para ser sincera, hemos de reconocer que a ninguna muchacha le ha gustado muy en serio estudiar. Sobre todo, aquellas ciencias exactas y ásperas que nada tienen que ver con las inquietudes innatas de la mujer" (21; to begin with, and to be sincere, we must recognize that no girl has ever really seriously liked studying. Especially those exact and precise sciences that have nothing to do with the innate concerns of women). This quote appears as part of an advice column in *Medina* where readers were encouraged to send questions in order to receive expert recommendations. On this occasion, the response was intended for a young woman seeking advice about choosing a university major. The advice ended in the following way: "Me parece que tú prefieres la carrera matrimonial. Es la más femenina; pero ten en cuenta que también requiere muchos conocimientos y una gran dosis de abnegación" (21; It seems to me that you prefer the matrimonial major. It is the most feminine; but keep in mind that it also requires much knowledge and a large dose of self-sacrifice). The reference to the "conocimientos" (knowledge) that women should acquire was clearly not a promotion of formal education but rather an effort to promote marriage as the highest ideal for women, advocating self-exclusion from higher education as a preferable and more feminine educational option and thus elevating marriage to the same level as any other profession while relegating women to that one option alone.

Nonetheless, it seems clear that Falangist authors, including women, were forced to acknowledge their contemporaries' ongoing interests in higher education, particularly in STEM fields. For example, in María Gabriela Corcuera's 1942 article titled "La mujer universitaria" (The coed), she outlines the university majors most often pursued by female students of the period, listing them in order of preference: philosophy and letters, medicine, biology, chemical sciences, mathematics, and lastly, those centered on the artistic disciplines, such as music (24–25). Corcuera's article comments on the appropriateness of each of these specialties for women, arguing that certain fields are just too difficult for female students. In the case of a degree in medicine, for example, she claims that due to the subject's elevated complexity and the intellectual and physical fragility of women, "en el tercero o cuarto curso flaquean y la abandonan" (24; in the third or fourth course they weaken and abandon it). Not only does Corcuera argue that a degree in medicine is too difficult for women, but she also claims that even if a woman were to graduate successfully and become a doctor, her skills in the field would be less the result of her academic efforts and more based on an innate feminine capacity for caregiving, leading to effective healing "por su extrema sensibilidad y la intuición puramente femeninas que le harán ganarse la confianza de sus pacientes" (24; due to extreme sensitivity and a purely feminine intuition that will enable the female doctor to gain the trust of her patients).

In reference to majors in biology and chemistry, Corcuera's analysis focuses on the potential usefulness of those majors for homemaking, presenting the fields in a somewhat frivolous manner, as objects for mere decoration. For example, she describes biology as a "ciencia llena de colorido y sugestión" (24; science full of color and suggestion) that allows female students who earn degrees to dispose of true "museos naturalistas en sus casas" (24; naturalist museums in their homes). For chemistry, she notes that it allows women to create "filtros amorosos" (24; love potions) or to practice "fórmulas extrañas" (24; strange formulas).

It is evident that, far from analyzing the challenges that university-educated women might encounter in these disciplines, Corcuera attempts to link scientific knowledge with more generally accepted feminine pursuits, such as maintaining the home. Her descriptions never offer the possibility of employing scientific learning in order to work as a professional in the field. Indeed, Corcuera makes it clear that women cannot compete professionally in her discussion of the exact sciences (mathematics and physics), noting her disapproval for women who enroll in such programs since "Aquí vencen los varones" (25; Here, men prevail). She notes that if one were to insist on choosing such studies, the ideal would be to opt for astronomy, but only out of a purely contemplative desire to stargaze: "para mirar las estrellas y la luna" (25; to gaze at the stars and the moon). Finally, Corcuera praises women who decide to study fine arts, like, for example, music, because that field is considered "un gran terreno para la mujer" (25; a great field for a woman). As Corcuera's article on "La mujer universitaria" makes clear, while Falangist discourse did acknowledge the desire of women in the early postwar period to continue with university education, even the women writing to support that goal worked to undermine women's access to STEM fields and encouraged a narrow conceptualization of the purpose and use of such degrees in order to make them appear more socially acceptable and sufficiently feminine.

This type of discourse, restrictive of women's freedom and intellectual potential, proliferated within Falangist media, particularly at key time periods such as right before final exams or in the days prior to the beginning of the academic year. These were moments when the visibility of women in classrooms would increase, and thus the Falange focused on reminding women at those crucial and stressful times that their educational goals should be geared toward becoming better wives and mothers. Such arguments are evident in the article "Muchachas en la Universidad" (Girls at the University), published in *Medina* in June 1942, right around final exams. The piece was published anonymously and was thus likely written by a Falange official, but it is voiced as if it were written by a group of female university students, and it describes the aim of female education as enhancing women's temperamental natures and creating superior offspring for the Nation: "Nosotras queremos dotar de inteligencia a nuestras características temperamentales y eternas. Revalorizadas por la inteligencia, daremos a la Patria hijos de una preparación superior" (18; We want to endow our temperamental and eternal characteristics with intelligence. Revalued by this intelligence, we will give the Fatherland children of superior preparation). The same article also implies that women don't have the stamina to manage intense programs, noting that in the context of law school, many drop out before finishing: "acuden para emprender derroteros de intensa labor . . . pero se cansan y a los últimos cursos llegan pocas" (19; they go to take on tasks of intense labor . . . but they tire and few make it to the final courses). The article thus reasserts the inferiority of women, who lack the strength or will to finish their studies.

Another article from later that same year, also published in *Medina* (August 1942) and titled "La educación de la mujer" (The education of women), acknowledges that advanced studies can constitute a positive aspect of a woman's life as long as those studies "sean para que con dignidad afronte su gravísima responsabilidad moral como educadora de sus hijos" (14; serve to help her take on the grave moral responsibility of educating her children with dignity). Other articles appearing at the beginning of the school term—such as "Retornos a las aulas" (Returns to the classrooms) from October 1944—emphasize the importance of hygiene and that, at all times, those women who pursue degrees should be sure to focus on being clean and well-groomed because "el estudio no ha de enturbiar para nada la feminidad" (34; study should not obscure one's femininity in any way).

A primary goal of the Sección Femenina was to distance itself from any feminist initiatives, and indeed, for that same reason leaders wanted to extol femininity, which was considered a trait natural to women and the opposite of feminism. There was a desire to encourage occupations traditionally associated with femininity, as seen in the 1942 article "La mujer, alma del hogar, preparada por la Falange" (Woman, soul of the home, prepared by the Falange). The article charts the journey through a woman's education in order to prove that her mission, by her very feminine condition, should be no other than that of caring for her husband and children.[10]

Intellectual or scientific women from the Republican era, who had become architects, engineers, chemists, and mathematicians, were characterized as "virago," "guarra," "hombruna," "monstruo," "cómica," or "prostituta" (tomboy, filthy, manly, monstrous, comic, or prostitute), that is, models of antifemininity. Indeed, according to this critique, intellectual women were responsible for having made the stability of the Christian home lose its balance by dedicating themselves to functions not appropriate to their sex. The Catholic Church supported such ideals, arguing that the cultivated and free woman was less virtuous and less moral. The church preferred women to remain intellectually inferior if such lack of formation would protect them from circumstances leading to sin and waves of passion. The intellectually formed woman, particularly one steeped in the disciplinary practice of the sciences, might establish hypotheses that would reject rites and beliefs and lead to what the church saw as destructive social formations. For example, María Pilar Morales, in her 1944 book titled *Mujeres (Orientación femenina)* (Women [Feminine orientation]), with a prologue by Pilar Primo de Rivera, argued that scientific women might use intellectual reflection to disentangle themselves from their roles as wife and mother, rejecting submission to male dominance and thus rejecting what was right and natural (74–76). The scientific or intellectual woman was thus a type of feminist trend, or, as expressed in the article "Las terribles intelectuales" (The terrible intellectuals), which appeared in *Medina* in 1943, "criaturas perversas masculinizadas" (15; perverse, masculinized creatures).

This Falangist ideology divided reality into two spheres and asserted that men and women should be divided and educated only in the one sphere in which they were deemed born to operate. Thus, the intellectual woman's desire to develop professionally and to prepare intellectually was condemned as a process of masculinization from which it would be difficult to recover, given that femininity was considered a divine grace for a woman, essential for the nation, and irreconcilable with intellectual pursuits.

As a final example of the cultural milieu surrounding women's intellectual opportunities during the early period, consider the "Diez mandamientos de la universitaria" (Ten commandments of the coed), published by *Medina* in 1945. The article likely had a strong impact on female readership, as the magazine enjoyed success and had a large number of subscribers. The commandments evidence the Falangist fear that women might deviate from the path of femininity, attempting to lift themselves up in ways that recall the independent spirits promoted during the Second Republic, and thus they reject all influence of feminist currents:

1. In your order of values place always first womanhood; afterwards, being a student.
2. Be, at all times, the best and most selfless comrade of your male classmates.
3. Never be stingy with your time, help, and advice to those who need them.
4. Accustom yourself to giving a light, pleasant, and amiable tone to those things that, in your feminine personality, can seem too rigid and academic.
5. Consider your studies as a means and not an end. With respect to man, think that you are his complement, not his rival in battle.
6. Always see to it that your knowledge does not stand out too much.
7. Learn to be flexible and diverse, without letting it change you. Monotony is the worst enemy of happiness.
8. Learn to nuance your reactions. Many times, success depends upon the quality of a nuanced reaction.
9. Do not become a slave to the knowledge you acquire; better yet, oblige that knowledge to become your slave.
10. And flee from the scientific woman who wears spectacles even on her spirit,

who analyzes everything, who wants to dissect everything, even the most intimate, most sacred, most profound feelings. (18)

The commandments begin with the relegation of the student role behind that of "womanhood," implying that the two cannot coexist, and they end with the denigration of the "scientific woman who wears spectacles on her spirit, who analyzes everything, who wants to dissect everything" as if being scientific, studious, and analytic were the worst offenses and could not coexist with emotion or "profound feelings." Clearly, within Falangist discourse, the bounds of women's intellectual and scientific formation were becoming increasingly limited. However, despite official discourse, which offered little margin for resistance, there were discordant voices that offered other perspectives on feminine experience. Indeed, in the case of Mercedes Ballesteros, the author of the novel analyzed below, the theme of the scientific woman is treated most favorably, in sharp contrast to the time period's official discourse.

María Elena, ingeniero de caminos: A Case of Subversive Fiction Concerning Francoism's Female Prototype

Mercedes Ballesteros (1913–1995) grew up the daughter of two distinguished intellectuals, Antonio Ballesteros Beretta and Mercedes Gaibrois Riaño (both historians), and the sibling of Manuel Ballesteros, university professor, historian, and eminent archaeologist.[11] As a young girl, Ballesteros was introduced to many countries and cultures, given the professional tasks of her nuclear family members, who were all passionate about the humanities and historical research. Ballesteros began university studies and chose philosophy and letters as a major, but she abandoned her program with only a few courses remaining toward her degree. In 1932, she married Claudio de la Torre, who had a degree in law and was a two-time winner of the *Premio Nacional de Literatura* (National Literary Prize). Ballesteros published actively, contributing to newspapers and magazines such as *ABC* and *Ya*, as well as to the humor magazine *La Codorniz*, in which she signed her writings with the pen name "La Baronesa Alberta" (The Baroness Alberta). She was known for cultivating humor in her narrations, an unusual trait in women's narrative of the post-Civil War, and her versatility can be noted in genres as diverse as historical biography, theater, short story, and the novella. Her work was translated into English and German, and it saw adaptations to theater and television. Some of her most well-known titles include *Las mariposas cantan* (1952; The butterflies sing), *Eclipse de tierra* (1954; Earth eclipse), *La cometa y el eco* (1956; The comet and the echo), *Invierno* (1959; Winter), *Verano* (1959; Summer), *Taller* (1960; Workshop), *El chico* (1967; The boy), and *Pasaron por aquí* (1985; They passed through here). From among her humoristic novels, key texts include *Así es la vida* (1953; That's life), *El perro del extraño rabo* (1953; The dog with the strange tail), and *Este mundo* (1955; This world).

María Elena, ingeniero de caminos came to the market as part of the literary collection "La novela ideal" (The ideal novel), which had been founded in 1938 by Claudio de la Torre. Via this collection, Ballesteros published numerous titles created in the mold of the *novela rosa* (romance novel), and she used the pseudonym Sylvia Visconti. She also wrote detective novels, for which she used the pen name Rocq Morris.[12] "La novela ideal" was in operation until 1943, and the catalog of authors did not extend beyond the familial circle.[13] Between 1938 and 1943, Ballesteros published some twenty titles between the two types of popular genres, romance and detective fiction. The novels did not surpass 120 pages in length, as was logical for a collection aiming to offer entertainment and attempting to make readers fans of the collection and its authors without requiring too much effort and time to read the works.

While it is true that this type of literary genre, the novela rosa in particular, has not enjoyed high esteem among critics, there have also been scholars who have recognized the important use of diverse literary genres, particularly by female authors, to address challenging social issues.[14] In many cases, female authors like Ballesteros have taken advantage of a seemingly innocuous framework like the romance novel, using the plot lines to challenge existing ideology while making the emotions, modes of conduct, and representations of society recognizable to readers. In just that way, Ballesteros published *María Elena, ingeniero de caminos* (1940), selecting the profession of civil engineering for her protagonist in order to highlight the challenges faced by young intellectual women, who saw their desires to develop as professionals frustrated in the postwar period. Indeed, Ballesteros was likely acutely aware that under Francoism, the Escuela Especial de Ingenieros de Caminos (Special School of Civil Engineers) made it a matter of principle to not admit women into its classrooms. Ballesteros's novel, which had a six-thousand-copy run, unfolds in thirty-three chapters of scarcely two or three pages each, and the plot reflects extensively on the contemporary time in which it was published.

María Elena is an orphaned young woman from Madrid, having first lost her father as a child and later her mother, who dies following a lengthy illness before María Elena completes her studies. The protagonist's life unfolds alongside that of her foreign tutor, Miss Harris, who has concerned herself with the young woman's education and assists María Elena with daily chores in the Madrid home where they live. Between the two women there exists a generational conflict that permits the reader to contemplate reality from two different perspectives: Miss Harris, fifty-nine years old, represents tradition, and María Elena, at twenty-three, modernity. However, María Elena's modernity does not correspond to the morality for women being disseminated from official sources. Instead, Miss Harris represents the type of woman promoted under the Franco regime, and María Elena appears to represent the remnants of the Second Republic.

The protagonist has no friendships other than a couple, formed by Marta, a childhood friend, and her husband Carlos, who symbolize the exemplary family favored by Francoism: Marta, a housewife, tends to the home and the couple's small son, and Carlos, head of the family, sustains the home by exercising his profession of sculptor. María Elena rejects this type of life, especially after having obtained her engineering degree. Her highest aspiration is to be able to practice her profession and feel herself realized as a professional woman. Around her friend Marta, María Elena appears brave, strong, and determined, but even with these qualities, her friend's husband, Carlos, who represents traditional masculinity, insists on the need for protecting what he considers the weaker sex, warning María Elena of the dangers she faces in life due to her feminine and single condition: "Es muy difícil la vida para una muchacha joven y de talento cuando tiene que luchar sola" (6; Life is very difficult for a young, talented woman when she has to make her way alone).

While this study is focused on the scientific representation related to the novel, the physical descriptions of María Elena are intriguing and seem linked to this notion of the feminine condition. The author opted to give the protagonist a physical appearance that distances her from the model feminine woman: "Era más bien alta, delgada, de tez morena y ojos muy claros y penetrantes. Su cabello negro, peinado en cortísima melena, le daba cierto aire de chico, pero de chico serio" (6; She was fairly tall, thin, with tan skin and very blue, piercing eyes. Her black hair, set in a very short style, gave her a certain boyish air, but of a serious boy). Ballesteros appears to "masculinize" the protagonist, with the goal of showing through María Elena's physical appearance the protagonist's disobedient spirit and her desire to be accepted among her male colleagues. Indeed, in one intriguing passage, analyzed by Zemon Davis, the masculinized female protagonist, whose appearance and actions dissent from the discourse of femininity and of the self-denying woman, attempts to subvert patriarchy via a reflection on the defining schemes of sexual differentiation, questioning them directly.[15] The author employs the protagonist in this particular scene to voice the idea that while gender does not matter for the practice of a profession, what does in fact matter is educational training, an opportunity that, for women, was nearly prohibited.

The challenge of completing a scientific degree and then finding work within the context of a discriminatory patriarchal culture becomes evident as a central theme early in the work. It has taken María Elena eight years to finish her civil engineering degree, and she feels proud of her effort and accomplishment in what she considers a relatively short time frame. However, a male colleague who also recently received his degree expresses surprise that María Elena would emphasize the time that she has employed in culminating her studies, given that he finished in the same amount of time. The narrative voice responds with one of the most animated rejections of the patriarchal system that appears in the work: "¡La vida de los hombres! ¡Lleváis siglos, hijo mío, lleváis siglos siendo lo que sois! Y cuando una mujer quiere alcanzar algo nuevo ha de costarle un esfuerzo mayor . . . Sí, eso es" (9; The life of men! You've had centuries, my son, you've had centuries being what you are! And when a woman wants to achieve something new it costs her a greater effort . . . Yes, that's it). The use of exclamation points and the pejorative "my son" emphasize the anger of the narrator and the willful blindness of the young male engineer. In the quote, the speaker implies that society has consistently identified men as the active subjects of history, leaving women on the margins. María Elena knows that through her behavior she is distancing herself from the model of the traditional woman. Indeed, by studying engineering, she has become the object of insults from other women, from those who consider themselves "true women," those who have not lost their femininity: "¡Una mujer ingeniero, un marimacho, una desgraciada!" (9; A woman engineer, a dyke, a wretch!). The author's choice of the word "marimacho" at this point in the narration captures the sense of a society imbued in a patriarchal discourse that encourages attacks on women who deviate from the "natural" role of mother and wife. A woman who seeks to transgress the social and gender norms by studying civil engineering must be somehow less than female, a "dyke," and thereby a dishonorable "wretch" worthy of anti-feminist attacks.

The discrimination María Elena faces goes beyond mere verbal abuse, leading to exclusion from the workforce. In the days following the earning of her academic degree, her male colleagues (she is the only woman in the cohort) debate the uselessness of the degree for a woman, since they contend that as soon as she marries that will put an end to her intellectual life. Nevertheless, María Elena is certain: "Quizás me case alguna vez. Pero mi marido ha de resignarse a que su mujer sea ingeniero" (10; Perhaps I will marry some time. But my husband has to resign himself to his wife being an engineer). Still, the protagonist begins to lose hope upon learning that many of her male colleagues have already found work. At this point Miss Harris sees the opportunity to recommend that María Elena abandon her way of thinking and opt for marrying a colleague who is in love with her: "¡Pensar con cabeza! ¡Casarse, formar un hogar, tener niños rubios a los que ella les enseñaría el inglés . . .! Por fin, Elena, empezaría a dar señales de discurrir" (12; Think with her head! Get married, make a home, have blonde little children whom she could teach English . . .! Finally, Elena, you would begin to set a good example). While María Elena aims to participate fully in her field, the narrative voice uses both the protagonist's fellow engineers as well as Miss Harris to demonstrate the utter rejection from the broader society and the lack of consideration that existed for women with intellectual ambitions.

Amid this tension surrounding work, marriage, and women's intellectual pursuits, the romance theme surfaces when María Elena visits the home outside Madrid where her married friends live. Following the typical narrative arc of la novela rosa, María Elena has a chance encounter with a young man, Roberto, who lives bitterly with his two daughters. While taking a canoe ride, María Elena notices in front of her a man suffering from head wounds sustained while diving, and he is about to lose consciousness. During this first encounter, the protagonist takes it upon herself to return Roberto to his house, she waits for him to recover, and afterwards she leaves. Thus, while there is limited plot development in this first interaction, it is significant that from the first moment of contact between the two characters, it is the woman who exercises an active role. Indeed,

María Elena will always take the lead in this relationship. Once again, the author appears to distance herself from the archetypal model that would traditionally offer a similar scene, but one in which the woman would need to be saved. Indeed, in Ballesteros's version, it is the male character who is presented as melancholy, physically weak, and emotionally unstable, the opposite of the idea of traditional masculinity. The brave and bold attitude of the female character, in contrast to the male romantic lead, will continue to play a significant role in the remainder of the work.

If the love theme has already begun to sound through the described encounter, the other major motif concerns María Elena's professional goals, which present further challenges when she returns home and prepares to continue seeking work. She accepts the assistance of several professors, but the job ads include the requirement to "be male," and so, one by one, the doors to the labor market close. The figure of the university chair proves decisive in María Elena's future, as he informs her of the existence of an engineering position in Equatorial Guinea, at that time a Spanish African colony. Initially the chair tries to dissuade María Elena from the post, arguing that it would be better to look for other options given the demanding conditions in the colonies, which were considered unfavorable for women, particularly the physical effort required and the high temperatures of the zone. However, the chair decides to support María Elena's candidacy, and the narrative justifies his support by explaining his opposition to the type of life led by his own daughter, who is apparently given to excessive leisure and frivolity. His own negative experiences with his daughter, and María Elena's example, make him believe that exercising one's intellect is in fact healthier for women than walking around aimlessly between celebrations, flirtations, and shopping sprees. The author provides the foil of the department chair's daughter in order to offer a moralizing message emphasizing that a scientific woman's work appears to even the highest intellectual authorities as less harmful than devotion to leisure and indolence. In her novel, Ballesteros does not support traditional ideas about appropriate feminine pursuits, but instead chooses to advocate for women as agents who make decisions and control their own destiny. María Elena thus ultimately accepts the challenge and moves forward to fulfill her desire to work as a scientist, not on the peninsula, but in Spain's colonial territory.

The author's decision to address issues of patriarchy and women's access to scientific careers within the colonial context is particularly intriguing. The mere fact of placing a woman civil engineer as a novel's protagonist was a subversive act, particularly during the early post-Civil War period, but Ballesteros chooses to further challenge the dominant discourse by revealing the patriarchal schemes that predominated in the colonial space. The novel thus provides important material for analyzing Francoist colonialism, sexism, and even racism. Although racism is not alluded to directly in the novel, the choice to place the novel's action in the colonial theatre demonstrates that the interplay between gendered discourses and other forms of oppression did not escape Ballesteros's notice.[16] Once María Elena arrives in Equatorial Guinea, she confronts a challenging reception, with a difficult environment and colleagues not pleased to have a woman in her role. María Elena finds herself happy in her work, but she describes the challenges of interacting with her boss, who never stops lamenting having to work with a female colleague: "Tengo la mala suerte de que el jefe de todos nosotros es un viejo antipático, antifeminista, que se pasa el día metiéndose conmigo y diciéndome que este es un trabajo de hombres y que es ridículo verme a mí haciéndolo" (29; I have the misfortune that our boss is an old, mean, antifeminist man who spends the day getting into it with me and telling me that this is men's work and that it is ridiculous to see me doing it). By referencing this conflict between María Elena and her boss, Ballesteros highlights the difficult situation in which women scientists found themselves even in the colonial context, attempting to make their way forward as intellectuals in a world controlled by European men.

However, while the novel offers a clear critique of gendered discourse and of the gendered ideology of the Falange, it is less effective in critiquing the way the colonial context also served to further expansionist and racist Francoist ideals. Indeed, as Carmen Domingo's *Coser y cantar* and Castro Rodríguez's "Fascismo, colonialismo y masculinidad" maintain, the colonization of Equatorial Guinea provided ideological support for Francoist imperialist desire and symbolized the achievement of important imperial ideals, to which the Falange aspired and had supported consistently via Africanist interventionism during the dictatorship of Miguel Primo de Rivera. Expansionist aims served to salve the loss of other overseas colonies, including Cuba, the Philippines, and Puerto Rico, and postwar discourse was very interested in imposing Spanish identity on newly colonized communities. Thus, while the gendered challenges that María Elena encounters in Equatorial Guinea are an extension of what she was living in Spain, they are not equal, and the work does not fully explore that difference.

The criticism of Falangist discourse that the novel proffers moves in two directions, but it only explores the former: first, a woman may not freely develop herself professionally, not on the Peninsula, where it is prohibited at the administrative level (with the prerequisite of "being male"), nor in the colony, due to the prevailing machismo. Second, the problem does not exist equally in the colonial space but becomes something different in the translation of gender structures from one colonial space to another. Antonio M. Carrasco says in this regard: "Es como si en Guinea no existiese nadie normal y todos acudieran a la colonia a saldar las culpas del pasado y buscar la salvación social" (242; It is as if in Equatorial Guinea there did not exist anyone who was normal, and everyone came to the colony in order to vacate past responsibilities and to look for social salvation). While María Elena clearly fits Carrasco's description as "nadie normal," her attempts to find social salvation in the colony are quickly thwarted. Shortly after arriving in the colony, she falls ill, and although she attempts to hide her illness in order to avoid showing signs of what her boss and others might view as feminine weakness, the fever climbs, and she eventually must request sick leave and return to Madrid. Her boss sees this as exonerating his machoist perspective: "Se dará ahora cuenta de que yo tenía razón al decirle que esto no era un trabajo para niñas sino para hombres" (31; Now she will realize that I was right when I told her that this was no job for girls but for men). In the end, given the impossibility of practicing her profession, the protagonist is replaced by a male engineer, and given her precarious situation, María Elena decides to seek work as a tutor, a traditional role for women. Destiny, however, always helpful in the novela rosa, determines that she will fulfill her new role in the house of the man whose life she had saved.

Roberto, her new employer, is also an engineer, but readers will note how Ballesteros continues to present him in non-traditional gendered roles that allow María Elena to take initiative and demonstrate her intellectual and engineering skills. Roberto lives with his daughters and feels dissatisfied because his wife has abandoned them in order to move to England to pursue mundane pleasures and spend the money that Roberto sends her. The representation of Roberto's difficult marriage weakens his masculine status in light of patriarchal norms, as his wife manages him in a denigrating way. Roberto's wife also provides a foil for the virtues and personality of the new tutor, who captivates the girls and provokes Roberto's interest, given that he finds in her manner the attractive qualities that his own wife is lacking. The plot turns of the novela rosa lead to the wife's death in an automobile accident, and Roberto, upon learning the news, falls ill. Ballesteros's depiction of Roberto's weak constitution is no accident. The "feminization" of Roberto's character constitutes a highly subversive element, not only toward the novela rosa itself, but toward the patriarchal ideal, as it is out of Roberto's fragility that María Elena is able to demonstrate her own strength, distinguishing herself by her actions and decisions over those of the man.

María's opportunity as an engineer comes about through Roberto's illness, and she is able to overcome gendered obstacles through exercising her profession. The bridgework that Roberto oversees suffers a delay because he is unable to place himself in charge. María Elena, after confessing to the secretary that she is an engineer and capable of doing the job, substitutes for Roberto, with the condition that this intervention never be known. María Elena once again receives insults and derision from her male coworkers, who laugh shamelessly at the idea that a woman is going to direct the works. She is called, sarcastically, "la señorita ingeniero" (65; little miss engineer). However, the teasing attitude is transformed into admiration at the moment when María Elena demonstrates her ability and skill in the fulfillment of her profession: "Trabajo le costó a Elena convencer a Benito y hacerse respetar como jefe. Tuvo que echar mano de su flamante título. Bien es cierto también que, en cuanto el capataz y los obreros se convencieron de que aquella chica mona era su jefe, y además un jefe enterado y competente, su asombro y su veneración no conocieron límites" (65; It was difficult for Elena to convince Benito and become respected as boss. She had to take hold of her new title. It is also certainly true that, as soon as the foreman and the workers were convinced that that pretty girl was their boss, and what is more a well-informed and competent boss, their astonishment and veneration knew no bounds). While María initially struggles to earn respect as an engineer, taking initiative, claiming her title, and most importantly demonstrating her skills allow her to win over skeptical male colleagues.

The denouement of the novela rosa proves less satisfying, as readers see the author, via María Elena, succumb not only to the norms of the "ideal woman" but perhaps more crucially to the limitations of the Franco-era censors. As the novel draws to a close, Roberto learns that the engineer who has been substituting for him has been none other than his beloved. Shocked, he proposes that she change from tutor to collaborator, and, finally, from collaborator to wife. In this way, María Elena becomes a married civil engineer, yet although she will not stop being connected to her profession, she does take on a less direct role. While the novel as a whole contains many passages that deviate from the official rhetoric, censorship was fully in force, and perhaps for that reason the end of the novel accommodates itself to the morality and gendered limits of the time. María Elena's transgressive episodes as a civil engineer dissolve in the instant she decides to become a wife and her husband's collaborator. She will no longer perform as a professional but rather as a type of advisor, which connects the work with the message transmitted by Falangist ideology: a woman's knowledge should serve to perfect her mission in the home.

Despite this unsatisfactory ending, perhaps driven by Ballesteros's need to address the exigencies of censorship, the author's larger objective is still accomplished: the novel demonstrates that being competent in a STEM field is not a matter of sex but rather of the intellectual and professional preparation one has achieved. In fact, with this idea in mind, the novel's denouement seems forced and unnatural, since it is not logical that, after all the obstacles that the protagonist has had to overcome in order to be able to develop herself professionally and to achieve the respect of her environment, she would abandon her aspirations. In fact, the novel's cover reveals María Elena's impetus for achieving the objective of working as a scientist: "Desde que era una niña soñaba con su puente, con poder hacer un día uno de esos tremendos puentes que luego tienen un nombre famoso. Y he aquí que ya tenía el título de ingeniero. [. . .] La vida empezaba" (7; Since she was a girl she had dreamed of her bridge, of being able to one day make one of those tremendous bridges that bear a famous name. And now that she had her engineering degree. [. . .] Her life was beginning).

On the cover readers can see, in the foreground, a man with his back to the viewer, observing a young woman who stands in profile in the center of the scene, analyzing a plan of the work that she is creating (see fig. 9.1). It is none other than a bridge, the same bridge that appears in the background. The young woman appears traditionally feminine, with her hair tied back, leaving a slender neckline exposed. She wears a long dress, her arms exposed to the elbow and revealing part of her back. In her hands, she holds a pencil and appears to note something on the plan. The book cover reinforces that femininity need not be a stranger to intelligence. However, the conservative denouement belies the cover's visual optimism. In reality, the ending for the protagonist could not be so radical and still be published in the immediate postwar period. Indeed, as José María Fernández Gutié-

FIGURE 9.1. Cover of the novel *María Elena, ingeniero de caminos* (1940).

rrez argues after analyzing a corpus of brief novels by male and female authors of the 1950s, the repetition of conservative endings might have had the function of softening the episodes of rebellion that had previously taken place, thus making the works easier to publish: "Estamos convencidos de que son así para contrarrestar los relativos excesos y la falta de vida moderada y sin demasiado orden que llevan algunos protagonistas de las historias narradas en ellas" (34; We are convinced that they are this way in order to counteract the relative excesses and the lack of moderate and relatively orderly living that some protagonists experience in the stories narrated in them). While the censorship of the early postwar period may not have been the only factor in Ballesteros's choices for the novel's ending, the denouement certainly softens the critique of Falangist discourse, and in that sense belies many of the novel's other more intriguing critiques.

This study demonstrates the internal conflict within the novel, from the protagonist's subversive episodes with the official discourse to the conservative ending typical of the period. Yet despite its final romantic ending, *María Elena, ingeniero de caminos* presents an image of the intellectual woman that challenges the Franco-era norms and encourages women to act in assertive ways, cultivating their intellect and seeking access to labor markets. While Francoism attempted to eradicate the feminist perspectives that had become a part of Spanish life under the Second Republic, it was not successful in eradicating the convictions of educated women, both scientists and novelists, who sought greater equality and participation in STEM fields. In the case of Mercedes Ballesteros, she had seen the life of a female intellectual modeled in her own family through her mother, who demonstrated great erudition; she had gone to university, where she received the benefits of the *Institución Libre de Enseñanza*, which always championed higher education for women, and she had also experienced the cultural and intellectual environment of distant lands via the Canary Islands, where she moved with her husband during the postwar period. While Ballesteros's novel is not without limitations, it demonstrates the value of recovering early postwar writing by women and analyzing how women writers responded to the restrictive Francoist discourse with respect to women and science. In *María Elena, ingeniero de caminos*, Ballesteros offers a protagonist who resists patriarchal limitations on intellectual formation and scientific work, and she demonstrates that it is social and relational barriers that hold women back from STEM fields, not any inherent characteristics of their sex.

NOTES

1. Francoism's so-called blue period lasted until 1945, the year in which international fascism fell and the influence of Catholic doctrine became more appreciable in Spain. Franco shifted after that date toward National-Catholicism, affirming the close link between Church and State via control of education, culture, and all other aspects of life.
2. For analyses of this kind of repressive gendered discourse, see for example Connell; Bourdieu; Adams and Savran; or Sussman.
3. For a recent analysis of the broader cultural movements in the post-Franco period and their impact on literature, see Gajić's *Paradoxes of Stasis: Literature, Politics, and Thought in Francoist Spain* (2019).
4. For details about women and literature during this period, see Gallego Méndez; Sánchez López; Domingo; and Richmond. For an overview of Fascist discourse from the period, see Rodríguez-Puértolas.
5. While contemporary Spanish employs the widely accepted feminine form to refer to a woman who practices engineering (*ingeniera*), in this essay the masculine form will be used. In addition to its being the form used by the author, it is what was said at the time, given that engineering was a profession classified as one practiced solely by men, in the same way that other professions, such as *abogado*, *médico*, *juez*, or *arquitecto* (lawyer, doctor, judge, or architect), were limited by gender, both grammatically and practically, at the time.
6. Beginning in 1934, the Sección Femenina was the organization in charge of the indoctrination of Falangist women (and of Spanish women generally), initially important

as the Francoist regime was accumulating possessions during the uprising, and subsequently a center of power as an organization integrated into the State from 1940 until 1977. For more information about the Sección Femenina de Falange, consult the studies of Gallego Méndez (1983), Sánchez López (1990), Richmond (2004), and Domingo (2007).

7. For example, Concepción Arenal and Emilia Pardo Bazán advocated for women's educational access in diverse texts (Scanlon 61–80). Concepción Arenal's novels *La mujer del porvenir* (1869) and *La mujer de su casa* (1883) offer arguments contesting the dominant discourse of traditional education for women. Emilia Pardo Bazán's articles "La mujer española" and "La educación del hombre y de la mujer. Sus relaciones y sus diferencias" attempt to persuade male readers that education would make women better mothers and wives. Both Arenal and Pardo Bazán are considered proponents of a "conservative feminism," given that, as women marked by their time, they did not try to claim whole equality between the sexes but rather tried to convince society of the value of saving women from ignorance, arguing that women's education was in no way counterproductive for their sex.

8. Aleu Riera was the second woman to obtain her doctorate in Spain; the first was Martina Castells Ballespí, also a medical doctor, who defended her thesis just four days earlier (Flecha García, *Las primeras universitarias en España* 95–98).

9. The title of the first magazine, *Y*, which was published monthly, alludes to the first letter in the name of Queen Isabel the Catholic, a feminine model idolized by the female branch of the Falange. It was first published in February 1938 and ended in December 1945. It carried as a subtitle *Revista de las mujeres nacional sindicalistas* (National trade union women's magazine) until its third issue (April 1938), when it became simply entitled *Y. Revista para la mujer* (Y. Magazine for women). The other publication, *Medina*, which came out weekly, gave homage via its title to the provincial Valladolid locale of Medina del Campo, the place where Isabel the Catholic lived, confessed, and died, and thus offered an additional tribute to Queen Isabel. *Medina* was in circulation from March 1941 until December 1945. Along with Isabel the Catholic, the other feminine model considered exemplary for the Falange feminine ideal and for the Franco regime in general was Teresa de Jesús, the Saint of Ávila, who was offered as an example of a religious Spanish woman. Although since Santa Teresa was a Doctor of the Church she could be considered a protofeminist icon, this aspect of the saint's life was not extolled by the Falange but rather her religiosity, her virtuousness, and her constancy in bringing about the Discalced Order of Carmelite Nuns. These publications became the official channels of communication for the Falangist organization, created in an effort to mold the spirit of Spanish women.

10. Professions for women that were defended by the Sección Femenina, according to *Y* in a 1944 article entitled "Carreras para la mujer," were secretary, seamstress, salesperson or representative, teacher, practitioner, aesthetician, or telephone operator. Such activities were considered traditionally feminine, in part because excessive intellectual development was not considered necessary for these positions.

11. Mercedes Gaibrois, Ballesteros's mother, was also the first woman to occupy a seat in the Spanish Royal Academy of History.

12. The practice of using pseudonyms was very common among authors who wrote "popular" novels, a term under which one would include both la novela rosa as well as the detective novel. The use of pseudonyms might have been demanded by the collection editor or publisher, wishing to offer an international flair to the works, given that many of the names were foreign, as was the case for Ballesteros. However, the use of pseudonyms also constituted a recourse employed by authors to publish in diverse genres for higher remuneration without sacrificing name or prestige, protecting future literary reputations. In Mercedes Ballesteros's case, there is high confidence among scholars that all of the works she produced under her adopted pseudonyms have been identified.
13. In addition to Ballesteros, two other family members contributed: Josefina de la Torre, Claudio's sister, who wrote under the pseudonym Laura de Cominges, and Bernardo de la Torre Barceló, Claudio's nephew.
14. See for example Radway; López; Johnson; and Núñez Puente.
15. For more information and deeper analysis of this passage, see Zemon Davis (15).
16. It is evident that these concepts deserve deeper study than is possible here, given that it would diverge from the thematic line of this work, but the intersectionality between gendered discourses and other forms of oppression are apparent in Ballesteros's novel. For studies addressing the colonial context in greater depth, see McClintock; Carrasco; Medina-Doménech; Martín-Márquez; and Castro Rodríguez.

WORKS CITED

Adams, Rachel, and David Savran, editors. *The Masculinity Studies Reader*. Blackwell Publishers, 2002.

Arenal, Concepción. *La mujer del porvenir; La mujer de su casa*. 1869, 1883. Orbis, 1989.

Ballesteros, Mercedes (Sylvia Visconti, pseud.). *Así es la vida*. J. Janés, 1953.

———. *Eclipse de tierra*. Editorial Tecnos, 1954.

———. *El chico*. Destino, 1967.

———. *El perro del extraño rabo*. Editorial Tecnos, 1953.

———. *Este Mundo*. Taurus, 1955.

———. *Invierno*. Ediciones G.P, 1959.

———. *La cometa y el eco: Novela*. 2nd ed., Planeta, 1967.

———. *Las mariposas cantan: Comedia en tres actos*. Ediciones Alfil, 1953.

———. *María Elena, ingeniero de caminos*. La Novela Ideal, 1940.

———. *Pasaron por aquí*. Ediciones Destino, 1985.

———. *Taller*. Ediciones Destino, 1960.

———. *Verano*. Ediciones G.P, 1959.

Bourdieu, P. *La dominación masculina*. Anagrama, 2000.

Carrasco, Antonio M. *La novela colonial hispanoafricana*. Sial, 2000.

"Carreras para la mujer." *Y. Revista para la mujer*, no. 44, Sept. 1941, p. 19.

Castro Rodríguez, Mayca. "Fascismo, colonialismo y masculinidad: la regeneración homonacional a la luz de literatura producida en el espacio colonial de Guinea Ecuatorial." *V Encuentro Internacional Jóvenes Investigadores en Historia Contemporánea*, 6–8 Sept. 2017, Zaragoza.

Connell, Raewyn. *Masculinities*. London, Allen and Unwin, 1995.

"Consultorio." *Medina, Semanario Oficial de la Sección Femenina*, no. 34, 11 Nov. 1941, p. 21.

Corcuera, María Gabriela. "La mujer universitaria." *Y. Revista para la mujer*, no. 50, Mar. 1942, pp. 24–25.

"Destino de la mujer falangista." *Medina, Semanario Oficial de la Sección Femenina*, no. 1, 20 Mar. 1941, p. 3.

"Diez mandamientos de la universitaria." *Medina, Semanario Oficial de la Sección Femenina*, no. 216, 6 June 1945, p. 18.

Domingo, Carmen. *Coser y cantar. Las mujeres bajo la dictadura franquista*. Lumen, 2007.

Fernández Gutiérrez, José María. *La Novela del Sábado (1953–1955): Catálogo y contexto histórico literario*. CSIC, 2004.

Flecha García, Consuelo. *Las primeras universitarias en España (1872–1910)*. Narcea, 1996.

———. "Las mujeres en el sistema educativo español." *Las mujeres en la construcción del mundo contemporáneo*, edited by Teresa Marín Eced and María Mar del Pozo Andrés, Diputación de Cuenca, 2002, pp. 209–26.

Formica Corsi, Antonio. "Fuera escrúpulos." *La Independencia Médica*, no. 10, 1878, p. 172.

Formica, Mercedes. *Espejo roto. Y espejuelos*. Huerga y Fierro, 1998.

Fuero del Trabajo, 10 Mar. 1938. BOE 505.

Gajić, Tatjana. *Paradoxes of Stasis: Literature, Politics, and Thought in Francoist Spain*. U of Nebraska P, 2019.

Gallego Méndez, María Teresa. *Mujer, falange y franquismo*. Taurus, 1983.

Jiménez-Landi, Antonio. *La Institución Libre de Enseñanza y su ambiente: Los orígenes de la Institución*. Editorial Complutense, 1996.

Johnson, Roberta. *Gender and Nation in the Spanish Modernist Novel*. Vanderbilt UP, 2003.

"La educación de la mujer." *Medina, Semanario Oficial de la Sección Femenina*, no. 75, 23 Aug. 1942, p. 14.

"La mujer, alma del hogar, preparada por la Falange." *Y. Revista para la mujer*, no. 59, Dec. 1942, p. 37.

"La mujer y la preparación intelectual." *Haz*, no. 13, May 1939, pp. 40–41.

"Las terribles intelectuales." *Medina, Semanario Oficial de la Sección Femenina*, no. 134, 19 Oct. 1943, p. 15.

López, Francisca. *Mito y discurso en la novela femenina de posguerra en España*. Pliegos, 1995.

"Lo que las armas victoriosas traen, mujer." *Y. Revista para la mujer*, no. 15, Apr. 1939, pp. 12–13.

McClintock, Anne. *Imperial Leather: Race, Gender, and Sexuality in the Colonial Contest*. Routledge, 1995.

Magallón Portolés, Carmen. *Pioneras españolas en las ciencias. Las mujeres del Instituto Nacional de Física y Química*. CSIC, 1998.

Martín-Márquez, Susan. *Desorientaciones. El colonialismo español en África y la performance de identidad*. Bellaterra, 2011.

Medina-Doménech, Rosa. "Scientific Technologies of National Identity as Colonial Legacies: Extracting the Spanish Nation from Equatorial Guinea." *Social Studies of Science*, vol. 39, no. 1, 2009, pp. 81–112.

Morales, María Pilar. *Mujeres (Orientación femenina)*. Editora Nacional, 1944.

"Muchachas en la Universidad." *Medina, Semanario Oficial de la Sección Femenina*, no. 64, 7 Jun. 1942, pp. 18–19.

Muñoz-Páez, Adela. *Sabias. La cara oculta de la ciencia*. Penguin Random-House, 2017.

Núñez Puente, Sonia. *Reescribir la femineidad: La mujer y el discurso cultural en la España contemporánea*. Pliegos, 2008.

Pardo Bazán, Emilia, and Guadalupe Gómez-Ferrer Morant. *La mujer española y otros escritos*. Cátedra, 2018.

Radway, Janice A. *Reading the Romance. Women, Patriarchy and Popular Literature*. U of North Carolina P, 1984.

"Retornos a las aulas." *Y. Revista para la mujer*, no. 81, Oct. 1944, p. 34.

Richmond, Kathleen. *Las mujeres en el fascismo español. La Sección Femenina de La Falange*. Alianza, 2004.

Rodríguez-Puértolas, Julio. *Literatura fascista española*. Akal, 1986.

Sánchez López, Rosario. *Mujer española, una sombra de destino universal: Trayectoria histórica de la Sección Femenina de Falange*. Universidad de Murcia, 1990.

Scanlon, Geraldine. *La polémica feminista en la España contemporánea 1868–1974*. Akal, 1986.

Sussman, Herbert. *Masculine Identities: The History and Meanings of Manliness*. Praeger, 2012.

Vázquez Ramil, Raquel. *La Institución Libre de Enseñanza y la educación de la mujer en España: la Residencia de Señoritas*. Lugami Artes Gráficas, 2001.

Zemon Davis, Natalie. "Women on Top: Symbolic Sexual Inversion and Political Disorder in Early Modern Europe." *The Reversible World: Symbolic Inversion in Art and Society*, edited by Barbara A. Babcock, Cornell UP, 1978, pp. 147–90.

CHAPTER 10

Unorthodox Theories and Beings

Science, Technology, and Women in the Narratives of Rosa Montero

MARYANNE L. LEONE

Rosa Montero's publications might be read as a barometer of key issues in Spain since the nation's transition to democracy, and several of her novels engage directly with the central concerns of this collection, namely the involvement, marginalization, and exclusion of women in STEM fields. Spain's gender equality law (*Ley de Igualdad*, 2007) and various initiatives to augment equity in the workplace may improve women's opportunities in sciences, math, engineering, and technology, and yet invisible barriers continue to impede equivalent participation (Osca Lluch 7, 13). While the following numbers are just one measure, men produced 73 percent and women 22 percent of the total science and technology publications in Spain from 1999 to 2008, and women's production grew from 21 percent in 1999 to 31 percent in 2008 (Osca Lluch 30–31). Making visible gender asymmetry and women's contributions and challenges in STEM is a strategy that feminists have employed to protest inequity and create greater opportunities and recognition. This essay analyzes Montero's representation of female figures connected to science and technology in the narratives *Instrucciones para salvar el mundo* (2008; Instructions to save the world), *Lágrimas en la lluvia* (2011; *Tears in Rain*, 2012), *La ridícula idea de no volver a verte* (2013; The ridiculous idea of never seeing you again), and *El peso del corazón* (2015; *Weight of the Heart*, 2016).[1]

Unorthodox scientific theories and genres in these texts call attention to the non-normative place of women in STEM fields—their imposed absences, invisibilities, and relegation to the sidelines of scientific work— yet also their persistence despite substantial obstacles. Montero's portrayals of women in or associated with the sciences and technology also suggest an imperative to understand the individual self within a broader ecological co-dependence that encompasses human and non-human life. This study will begin with a brief overview of the four novels and analyze how they represent the exclusion and marginalization of women in STEM. Next, it will address how Montero offers unorthodox scientific theories and textual strategies to highlight women's ostracism and their challenges to norms in these fields. Finally, the study will analyze how Montero's works construct community for women in STEM. These works blend the humanities and sciences to bring critical perspective to discoveries and innovations in the past, present, and future, suggesting the need to weigh positive and negative consequences not only for humans but for all life forms.

Exclusion and Marginalization of Women in STEM

Instrucciones para salvar el mundo, *La ridícula idea de no volver a verte*, and *El peso del corazón* all include female scientist characters while *Lágrimas en la lluvia* and *El peso del corazón* feature a technohuman protagonist that embodies bioengineering. Most of the women associated with science in these works experience exclusion from scientific, professional, or social communities. These novels take place in or reference a range of time periods, places, and socio-political contexts—from the turn of the twentieth century in Paris (*La rídicula idea*) to the contemporary era in Spain (*Instrucciones* and *La ridícula idea*), and one hundred years into the future in a globalized world that extends beyond planet Earth (*Lágrimas* and *El peso*). When considered together, these novels suggest the enduring ostracism of women in STEM, along with their continued resolve and insistence of their place in these fields.

LA RIDÍCULA IDEA DE NO VOLVER A VERTE

In *La rídicula idea de no volver a verte,* Montero reflects on Marie Curie's professional achievements and personal life, and she considers the premature death of Curie's husband and research partner Pierre in relation to the loss of her husband and fellow journalist, Pablo Lizcano, to cancer in 2009. Prompted by Seix Barral editor Elena Ramírez's request that Montero write a prologue to the twenty-page diary that Curie composed and addressed to Pierre following his death, Montero contemplates not only Marie Curie but also societal gender expectations, women in the sciences, and writing to process mourning. Montero's emphasis on parallels between her own life and Curie's makes patent the transferability of this early twentieth-century woman's experiences to the twenty-first century and beyond. Through making this connection, the narrative suggests that foremothers support and inspire today's women, who like their predecessors, face gender discrimination and obstacles to studying and working in STEM.

The narrative emphasizes the formidable magnitude of Marie Curie's accomplishments in their own right and their amplification in the context of her gender and times. Curie formed part of a small group of scientists who predicted the instability of atoms, which challenged Newtonian theory and linked without precedent the fields of chemistry and physics (*La rídicula* 106). As Montero's narration states, Curie is the first woman to win a Nobel Prize, the first person to win two, first in physics in 1903 with her husband Pierre and then in chemistry in 1911 on her own, one of only three people ever to win two Nobels, and one of two to win in two different fields. Moreover, very few women have become Nobel Laureates in the sciences, with only six total in chemistry and three in physics, including Curie: "O sea que Madame Curie permanece imbatible" (11; In other words Marie Curie remains unbeatable).[2] The narrator emphasizes that Curie's notable scientific accomplishments are even more astounding in the pre-twentieth century European environment that restricted middle-class women to the home, employment as teachers or ladies in waiting, or "las tres ocupaciones tradicionales: monja, puta o viuda" (55; the three traditional professions: nun, prostitute, or widow). Before studying at the Sorbonne, Marie worked in Poland as a teacher; nonetheless, her exercise of this traditional female profession broke with socio-political norms, for she created a clandestine school to teach farmers Polish, a language the Russians had prohibited. Drawing on biographies and Curie's diary, the narrator surmises that Curie's childhood living under Russian occupation contributed to her revolutionary character and her atypical future.[3]

As a female scientist, Curie faced disdain from colleagues despite her achievements. Four male scientists who nominated Pierre Curie and lab partner Henri Becquerel for the Nobel for the discovery of polonium and radium omitted Marie. At Pierre's insistence, she was included, but she and Pierre received a monetary award for one person, as if she did not count, rather than the customary award for each winner. Only Pierre accepted the prize on stage, though he attributed full credit to Marie. Such lack of public recognition discourages women from pursuing work in so-called male fields and causes self-doubt about their abilities. After Pierre's death, when Marie first taught his classes, she directed a remark to Pierre in her diary about the need to prove that the classroom and research lab were her places too: "quizás sea también el deseo de demostrar al mundo y sobre todo a mí misma que aquella a quien tú amaste realmente valía algo" (149; perhaps it is also the desire to show the world and above all myself that the one whom you loved really was worth something). As the narration points out, Marie Curie taught for two more years before the Sorbonne would grant her a professor title. Her exclusion from the Nobel nomination as well as from recognition by Sorbonne colleagues both exemplify the obfuscation of women in STEM and other traditionally male fields. Truly a pioneer, Curie was the first woman to teach at the Sorbonne (149).

Montero's narrative connects Curie's self-doubt to women past and present who seek to enter disciplines dominated by men, have insufficient female models, and face social pressure to relinquish their goals: "Cuando todo el entorno y tu propia educación te están diciendo que no eres, que no sirves, que no correspondes a ese #Lugar, es difícil no sentirse una impostora" (149; When the whole environment and your own upbringing are telling you that you do not exist, that you are not useful, that you do not belong in that #Place, it is difficult to not feel like an imposter). *La ridícula idea* highlights that women must possess extraordinary determination to pursue work in STEM and other typically male domains and that Curie challenged social norms for much of her life.[4] Moreover, by examining Curie's exclusion along with her extraordinary professional achievements, Montero points out the double standard that women must outperform men to receive validation.

INSTRUCCIONES PARA SALVAR EL MUNDO

While *La ridícula idea* focuses on memoir and the challenges faced by highly successful female scientific figures like Curie, another of Montero's works,

Instrucciones para salvar el mundo, presents a fictional female scientist who is unable to continue her career in the field due to political oppression for non-normative sexuality.[5] The youngest full professor in Spain, Cerebro is imprisoned for nine months near the end of the Franco dictatorship under the *Ley de Peligrosidad y Rehabilitación Social* (law on dangerousness and social rehabilitation) for an affair with a female doctoral student. Her decision, when she is released in 1975, to never again seek an intimate relationship with a woman attests to the trauma of her social and professional expulsion (251). In the bar that Cerebro frequents nightly to anesthetize herself to those painful memories, she meets the novel's protagonist, Matías, a loner figure who is mourning the loss of his wife and best friend to cancer.

As a female scientist and gay woman during the Franco years, Cerebro embodies a double marginalization. Like Cerebro's outsider status, the places that Cerebro inhabits—the bar that functions as her second home and the inherited family house that serves as her official one—lie on the perimeters of Madrid close to the M-40. As dominant political and social norms have consumed Cerebro and transformed her into a ghost of her former self, urban development, with its office buildings, corporate campuses, stores, and nightclubs, engulfs her large stone "palacete" (187; mansion), now in ruins, in a once-wealthy residential neighborhood. The almost empty house, with its possessions sold to support her and her bar tab, personifies her ousting from academia, and the bar, appropriately named the Oasis, acts as refuge from sleepless nights in a house that mirrors her professional demise. Ironically, a scientific approach informs Cerebro's alcoholic self-obviation and mission to forget the public humiliation and termination of her scientific career: "había cumplido con exactitud y perseverancia de investigadora científica su férreo programa de embrutecimiento" (182; she had achieved with precision and perseverance of a scientific researcher her unwavering plan of desensitization). Her professorial scientific persona endures, however, despite the decades that have passed since she last taught.

LÁGRIMAS EN LA LLUVIA AND *EL PESO DEL CORAZÓN*

Bruna Husky, the protagonist of *Lágrimas en la lluvia* and its sequel, *El peso del corazón*, is a product of science rather than a scientist herself, a former warrior turned detective and one of a new species of beings bioengineered to carry out specific tasks, such as combat, mining, and mathematical calculations.[6] Montero's narration identifies Bruna Husky as female, using the subject pronoun "ella" (she). In the first novel, she saves her own species from an annihilation and demonization plot, and in the second, she discovers the illegal sale of radiation and secures medical care for a ten-year-old girl from Earth's most contaminated area who suffers from radiation-induced illness. In Montero's trilogy, women throughout the singular polity called *Los Estados Unidos de la Tierra* (The United States of the Earth) have the same legal rights as men; yet, gendered hierarchies continue in this imagined future. Even the genetically engineered, physically strong Husky has internalized some gendered roles, suggesting the inefficacy or disinterest on the part of the male-dominated sciences to address gender issues. In both novels, a stereotypical male figure, the police inspector Paul Lizard, with whom Bruna has a professionally competitive and personally intimate relationship, rescues her from likely death. Bruna critically observes his tendency to have the last word, yet his corpulent dominance sexually excites her (*Lágrimas* 149).[7] Nonetheless, Bruna's social ostracism on Earth stems largely from her android rather than her female identity: "Somos una especie subsidiaria y unos ciudadanos de tercera clase" (*Lágrimas* 61; "We're a secondary species and third-class citizens" [*Tears* 48]). Above Earth, however, gender inequality undergirds the floating territory Labari, a so-called perfect, human-only society: "Las mujeres no eran nadie, no eran nada. Menos que los reps en la Tierra" (*El peso* 188; "Women were nobodies. Their standing was even worse than that of reps on Earth" [*Weight* 156]). Montero's futurist protagonist, a doubly discriminated bioengineered female, points back to the marginalization of women in STEM today.

Unorthodox Research, Theories, and Texts

While scientific inquiry by its nature pushes boundaries of the known, the theories, discoveries, and scientists in these narratives occupy the margins of their fields or propose hypotheses that challenge widely accepted understandings of the natural world. Emphasizing ecological interconnectedness, these texts also stress the need to consider the benefits and the harm that scientific work may cause to all life forms.

LA RÍDICULA IDEA DE NO VOLVER A VERTE

If male scientists negated Marie Curie's contributions when she and Pierre worked in partnership, colleagues insisted that Marie would not accomplish any more significant work after Pierre's death, and they diminished her role in the discovery of radium. Yet the narration points out that while Pierre Curie and other peers were working on more newsworthy experiments, Marie Curie was developing a means to measure radium that would constitute a critical service to industry, medical research, and the general public (145–46).

Moreover, Curie insisted on the scientific rigor of her work. When British physicist and mathematician Lord Kelvin published a letter to *The Times* asserting that radium was not an element but rather a compound of helium, Marie worked with another scientist for three years to successfully extract pure radium metal from uraninite (also called Pechblende), which until then was known only in its salt form. In recounting Kelvin's challenge to Marie's reputation, Montero repeats a remark by one of Curie's female biographers that had Marie been a man, Kelvin would not have questioned her discovery.[8] Furthermore, he would have written to a scientific journal rather than a widely read general newspaper: "Qué manera de desdeñar a Marie; y de intentar rebajarla públicamente" (146; What a way to spurn Marie; and to try to publicly humiliate her). The dichotomies of male/female, noble/commoner, journal/newspaper in Kelvin's attempt to discredit Marie underscore the inferiority generally associated with women researchers.

In contrast to the male scientists' patriarchal dismissal of Marie's work, Montero's narration proposes that her perspective as a woman leads to unique contributions that men might not make. The text equates her isolation of radium to giving birth, thus suggesting the transferability to research of patience during gestation and dedication after parturition. Marie Curie put her work before her person, even before she began to research radium. As a student at the Sorbonne, she lived on little food, often without enough money to heat her apartment. If, as the text suggests, Marie viewed her research as a type of motherhood, abnegation was part of both roles.

Motherly sacrifice to her work, a surrogate child, obscured Curie's recognition of physiological harm from contact with radium. She and her daughter Irène only implemented safety measures in their lab in 1926, well after such precautions had become standard, and then they ignored them despite evidence of radium's danger: "incluso hacían cosas tan bárbaras como pasar radio y polonio de un recipiente a otro aspirando las sustancias con la boca por medio de una pipeta" (121; they even did things as frightful as passing radium and polonium from one recipient to another by breathing in the substances with their mouth through a pipette). A 1954 photo of Irène sucking on such a pipe and Montero's narration highlight the Curies' bodily sacrifice for scientific discovery: "Ese cuerpo traidor; pero, también ese pobre cuerpo maltratado y cometido a una radiactividad brutal durante tantos años. Al final, ¿quién termina siendo el rehén de quién?" (190; That betraying body; but, also that poor abused body dedicated to brutal radioactivity during so many years. In the end, who ends up being the hostage of whom?). Indeed, Marie, Irène, and Irène's husband, with whom Irène won a Nobel for discovering artificial radioactivity, all died of radium-induced illnesses (125–26).

Although the Curies were not alone in their fascination with radium, whose magic was widely touted, the narrative suggests that their willful blindness renders them at least partially responsible for the public's casual contact with this noxious element. The text refers not only to the health problems of Marie's lab employees, but also to the death of nine North American female factory workers (122–28). The "radium girls" pointed paintbrushes with their lips to draw the numbers and hands on watch faces and, for fun, painted their nails and teeth (Orci). Moreover, as the narrative explains, radium was added to a variety of cosmetics such as creams that purportedly delivered youthful skin or combatted cellulitis, endangering women's health in the name of beauty: "En esto de la belleza las mujeres siempre hemos hecho barbaridades" (103; When it comes to beauty, we women have always done foolhardy things). Ironically, Marie Curie prioritized intellect, yet her discovery fed an industry focused on women's physical appearance.

Congruent with Marie Curie's atypical achievements as a woman scientist of the first half of the twentieth century, *La ridícula idea* presents an unorthodox narrative form that blends biography, memoir, essay, scientific writing, and social media. Rather than fit a pre-determined convention, the text is in process, a few lines on the author's tablet: "Pero éste no es un libro sobre la muerte. En realidad, no sé bien qué es, o qué será" (10; But this is not a book about death. In reality, I do not know for sure what it is, or what it will be). In addition to blurring genres, Montero combines the disciplines of technology, biology, and writing through metaphors that employ the language of science. The sentences on her tablet are "células electrónicas aún indeterminadas" (still undetermined electronic cells) and books "crecen como zigotos, orgánicamente, célula a célula . . . cada vez más complejas, hasta llegar a convertirse en una criatura completa y a menudo inesperada" (10; grow like zygotes, organically, cell by cell . . . each time more complex, until becoming a complete and often unexpected creature). The narrating voice also merges philosophy and biology in the observation that loved ones' births or deaths bring us face to face with our existence, "bordeando esas fronteras biológicas" (9; bordering those biological frontiers). As a narrative that does not conform to a traditional genre, that cannot be easily defined, and that refuses to stay in an assigned place, *La ridícula idea* the text acts as a metaphor for women pioneers in the STEM fields.

The inclusion of hashtags throughout the text contributes to its unorthodoxy, voices a feminist perspective, and communicates solidarity in protest. At the same time, an index of these hashtags at the end of the book places the unconventional within convention and thus challenges the boundaries of textual practices and gender/genre. The hashtags call attention to major themes—restrictive expectations for women, the judged strangeness of nonconforming women, a metaphysical theory of coincidences, intimacy, and the therapeutic and political power of words and writing: #HacerLoQueSeDebe (#DoWhatOneShould), #CulpaDeLaMujer (#WomensGuilt), #LugarDeLaMujer (#WomensPlace), #HonrarAlPadre (#HonorTheFather), #HonrarALaMadre (#HonorTheMother), #HonrarALosPadres (#HonorParents), #Ambición (#Ambition), #Raro (#Strange), #Coincidencias (#Coincidences), #Intimidad (#Intimacy), #Palabra (#Word), and a few other similar hashtags (211).

These single words and brief phrases invite others to identify with the narrated story and post with the same hashtags to form common interest groups. Embedding the contemporary phenomenon of hashtags in the text suggests the pertinence of Marie Curie's gender-based exclusions, restrictions, and expectations for women of the twenty-first century. Common use of this social-media tool to garner support for protests and social change indicates that Montero's text seeks the same.[9] For example, naming Curie with the hashtag #Mutante recognizes marginalization as socially constructed and, in the context of this narration, presents her counter-conventional persona positively. The multi-genre text of *La rídicula idea*, with its hashtags, reflects the boundaries that women cross, complicate, challenge, and confound when participating in the sciences.

LÁGRIMAS EN LA LLUVIA AND *EL PESO DEL CORAZÓN*

For female scientists in these narratives, science and research form part of an identity that others view with suspicion. Curie had to battle the disregard of male peers, and Bruna Husky faces marginalization as a technohuman. Husky's critical views about the environment's compromised state also set her apart from a dominant discourse regarding technological progress. In the futuristic world of the year 2109, air contamination has worsened and is graver in the poorer southern hemisphere, the glaciers have melted, and the polar bear has become extinct.[10] While there are many specific examples of this contaminated world in *Lágrimas en la lluvia* and *El peso del corazón*, the polar bear's extinction exemplifies Bruna's perceptiveness of human activity's negative impact on Earth's ecological balance. Although in the novel the last polar bear had already died fifty years earlier, Madrid's president constructed a World Expo-like pavilion to house a bio-engineered replica, Melba, that replaces Madrid's symbol of the bear and the mulberry tree which today stands in the Plaza del Sol. The replica bear's placement in an imitation arctic environment highlights humans' manipulation of the natural world to suit their own ends. The narration describes in detail the final polar bear's excruciating death, noting also that a world war distracted the public from saving her. Visitors watch Melba's final moments on film as a moving belt transports them along the tank that houses her:

> Realmente parecía que uno estaba allí, viendo cómo se partía el último pedacito de hielo al que la osa pretendía aferrarse; cómo el animal nadaba cada vez más despacio, cómo resoplaba al hundirse bajo la superficie, cómo sacaba con un esfuerzo agónico su oscuro morro del agua y lanzaba un gemido escalofriante, un gruñido entre furioso y aterrado. Y cómo desaparecía al fin debajo de un mar gelatinoso y negro. (*Lágrimas* 187)

> You really felt as if you were there, watching the last small piece of ice that the bear was trying to hang on to breaking up; the animal swimming more and more slowly, snorting as it sank beneath the surface, then thrusting its dark snout out of the water with one last, agonizing effort and letting out a chilling wail, a furiously terrified growl. And then finally disappearing under a black, gelatinous sea. (*Tears* 159–60)

This passage evokes in the reader deep sadness and urgency to help this species now, before it is too late. The switch from the female "la osa" (the bear) to a generalized "el animal" (the animal) prompts the reader to transfer climate change's impact on one species to all beings. Bruna, who constantly recites the years, months, and days left in her ten-year lifespan, identifies with the fabricated animal, who also will die at an early age of a cancer that affects all replicas. Her knowledge that this genetically exact reproduction will be followed by "[u]na infinita cadena de Melbas en el tiempo" (*Lágrimas* 189; "an infinite chain of Melbas down through the ages" [*Tears* 162]) may suggest to Bruna not only her own mortality, but also that she is replaceable.

Bruna's empathy with this animal, whom she spots every time she visits, even when others cannot find the bioengineered bear, aligns with her sensitivity to the environment's deterioration and is coherent with her denouncement of the global energy company Texaco-Repsol's "parques-pulmón" (*El peso* 36; "lung parks" [*Weight* 23]) within Madrid's Retiro Park, consisting of artificial trees that create a supposed "espacio ecológico y puro" (*El peso* 36; "ecological, pure space" [*Weight* 23]): "después de haber esquilmado el planeta, ahora aparentaban ser los sumos sacerdotes de la ecología" (*El peso* 37; "after having overexploited the planet, the company now pretended to be the high priest of ecology" [*Weight* 23]). As a genetically engineered being, the Bruna Husky character's counterhegemonic critique of unsustainable resource exploitation and her message that bioengineering cannot substitute the natural environment is especially patent.[11]

INSTRUCCIONES PARA SALVAR EL MUNDO

In *Instrucciones*, the female scientist character emphasizes ecological interconnectedness in her references to several scientists' work: biologist Paul Kammerer; evolutionary biologist and science historian Stephen Jay Gould; cell biologist and biochemist Rupert Sheldrake; fictional physicist and mathematician Aaron Fieldman; and the scientist, inventor, and environmentalist John Lovelock. Cerebro identifies with marginal scientists who, like her, "triunfiaron y luego cayeron en el abismo . . . o aquellos que habían sido criticados y maltratados por sus pares" (251; triumphed and then fell in the abyss . . . or those who had been criticized and mistreated by their peers). The male gender of all of the researchers with whom Cerebro identifies suggests an implicit message that women in STEM historically have few female models. Despite this gender difference, Cerebro regains her identification as a scientist, and her self-esteem grows as she discusses the other scientists' theories with Matías.[12]

The concepts that Cerebro introduces suggest that life has positive purpose, justice occurs, and explanations for the inexplicable exist. Kammerer's Law of Seriality, published in 1919, counters the widely accepted Second Law of Thermodynamics, which states that energy in the physical world causes entropy, or disorder and uncertainty (Atkins 49–78). Based on his observation that coincidences occur in serials, Paul Kammerer asserted that the universe also tends toward order. For the character Matías, the notion that "las coincidencias tenían que tener un porqué" (coincidences had to have a reason) assuages the existential angst from his wife's death: "tal vez el mundo pudiera recuperar algún sentido" (69; perhaps the world could recuperate some meaning). Kammerer challenged an accepted law not only of physics but also of biology. This scientist became well known for an experiment that disputed the Darwinist position that genetic adaption occurs over many generations, instead supporting Lamarck's theory that organisms can pass on environmental adaptations to offspring. Kammerer was accused of falsifying the results yet claimed innocence in his suicide note.

Cerebro takes the side of the ostracized scientist Kammerer, remarking that Stephen Jay Gould had affirmed that the experiment probably did show this generational change. However, this Harvard professor was also controversial. As Richard York and Brett Clark explain, Gould was part of the organization Science for the People, which protested the Vietnam War, and he critiqued the prevailing notion of neutrality in scientific research, sought connections between science and the humanities, challenged biological determinism, and was critical of the atomic bomb (15–17).[13] Gould's positions align with Cerebro's questioning of scientific pursuits that compromise life, a view also found in *El peso del corazón*'s indictment of nuclear waste. In summary, Gould's life's work emphasized social equity and the need for the humanities and sciences to inform each other, themes that traverse the Montero narratives analyzed in this essay.

It may be coincidental that the film *A Glorious Accident: Understanding Our Place in the Cosmic Puzzle* (Wim Kayzer, dir.) places two of the scientists to whom Cerebro refers in *Instrucciones* in conversation about their understanding of man's existence in the universe: Gould and Sheldrake. The characterization of the six men in the film as "creative thinkers" who "push the boundaries of scientific theories and philosophical ideas" (Kayzer, DVD back cover) is congruent with Montero's focus on renegade scientists whose work addresses broader human experiences. Through the creation of her character Cerebro, however, Montero narratively injects a female, lesbian voice into the male-dominant conversation portrayed in the film that is representative of gender inequality in STEM. Cerebro's view is consistent with Sheldrake's questioning of hegemonic precepts, such as human-centered theories of one-directional progress in evolutionary biology that justify humankind's exploitation of nature (Kayzer 19:57–21:35). Sheldrake holistically notes that "everything is nested within something else" (9:50–9:52). Earth sits within the solar system, the solar system within a galaxy, this galaxy within the universe, and so on. In the opposite direction, ecosystems house organisms, composed of organs, made of tissues, consisting of cells, then atoms, and finally subatomic particles (9:52–11:10). Connection also undergirds his morphogenetic fields theory, which proposes that groups within a species transmit the memory of a certain practice to others in the species even when they have had no physical contact.[14] In short, Sheldrake's theories consider how matter and beings cooperatively develop.

Also focused on interconnection, Cerebro's conversations with Matías about Aaron Fieldman, a disenchanted physicist who worked on the atomic bomb with the Manhattan Project, emphasize that human actions have consequences for all other beings and the physical world. Unlike the other scientists whom Cerebro references, Fieldman seems to be a fictional character; nonetheless, as Serra-Renobales points out, his theories provide Matías and the reader a moral compass that prioritizes one's impact on others (*Instrucciones* 76).[15] Cerebro positions Fieldman as a singular critical voice, explaining that he died in a fight with soldiers at Los Alamos, or perhaps was killed, because he was going to inform the public about the atomic bomb. Cerebro's attribution of the bomb's use to scientific ambition rather than national security affiliates her with this fellow scientist who critiqued research quests with destructive ends: "Yo creo que simplemente querían ensayar sus bonitas bombas, a ver cuál de las dos funcionaba mejor" (225; I think that they simply wanted to try out their lovely bombs, to see which of the two worked better).[16] Additionally, Cerebro highlights a theory by Fieldman that human actions reverberate energetically to physically impact the world, which she likens to the physical principle of communicating vessels, or Pascal's law on fluid mechanics and pressure equilibrium ("Pascal's"). Every single action has consequences, either contributing to or detracting from systemic harmony; in other words, positive actions beget more positive actions and vice versa.

Lastly, Cerebro also introduces the work of John Lovelock, one of the earlier scientists to discuss climate change and who is also credited with leading the creation of the field of Earth Sciences.[17] Like the aforementioned scientists, Lovelock proposes hypotheses that challenge prevailing assumptions in science and society: "the whole key to my work, throughout my career, was that whenever they started saying something was the standard wisdom, I started saying 'it can't be'" (Gribbin and Gribbin 72). His controversial Gaia hypothesis asserts that Earth is a "single complex system" that self-regulates to support life, which scientists scorned when first published in the *New Scientist* in 1975, in part because of its popularity among hippies and the green revolution that Rachel Carson's fictional *Silent Spring* spurred (Highfield; Gribbin and Gribbin xx, 118, 144–50, 161).[18] Today, the Gaia principle has its supporters and detractors (Highfield). Lovelock's ecocentric stance that Earth will adjust, if not necessarily for human survival, and his pragmatic support of technologies that he believes will allow humans to survive longer, such as genetically modified foods and fracking, place him at odds with the majority of environmentalists (Gray). Cerebro refers to Lovelock when she positions her concern for climate change as counter normative as well: "'Ni una sola nube en la península, cielos completamente despejados y cuarenta grados de temperatura, de manera que ¡sigue el buen tiempo!' . . . Pues se van a enterar. Con el cambio climático, España será por fin un desierto perfecto" (*Instrucciones* 106; Not a single cloud on the peninsula, skies completely clear and forty degrees, so the good weather continues!' . . . Well they are going to realize. With climate change, Spain will be at last a perfect desert). Moreover, Cerebro connects Lovelock to Kammerer when she explains Lovelock's work on an instrument to determine if life exists on Mars based on the detection of entropy or order.[19] The scientific work of Lovelock, Kammerer, Fieldman, Gould, and Sheldrake help us understand what it means to be human in the context of an interconnected ecology. Cerebro characterizes these scientists' ideas as "hermosa[s]," "consoladora[s]," and "conmovedora[s]," "poesía" more than "ciencia" (68; beautiful, consoling, and moving, poetry more than science). The reassuring nature of their theories notwithstanding, Montero's narratives make clear that this continuity among male scientists often is not afforded to women, who many times work in isolation.

In all four works, unorthodox research, theories, scientists, and texts express that we must be sensitive to how scientific innovations may positively and negatively impact biodiversity and quality of life. The narrating voice of *La ridícula idea* explains that polonium, the first element Curie discovered, "últimamente se ha puesto de moda como una eficiente manera de asesinar" (11; recently it has become fashionable as an efficient method to murder). Radium, the second element she isolated, is toxic, yet its medical applications save lives, too.[20] In *Lágrimas* and *El peso*, engineered beings help multiple species survive; however, human inventions have also deteriorated Earth's capacity to sustain life. Cerebro, of *Instrucciones*, schools Matías on nefarious uses of science along with theories that highlight connection, order, and equilibrium in what seems like a chaotic, nonsensical world.

Creating Community for and by Women in STEM

Beginning in the 1960s and gaining momentum in the 1970s, feminist literary critics recognized the value of studying women's writing separately from men's in order to understand social, political, and market influences, and to highlight a tradition of female-authored literature (Moi 61–63). Elaine Showalter's *A Literature of Their Own* (1977) mapped continuity in the development of British literature by women from the mid-nineteenth to the late-twentieth centuries, and Sandra M. Gilbert and Susan Gubar's *The Madwoman in the Attic* (1979) studied the female artist and her literary strategies within male-dominant society (Moi 68–72). These and many other works provided theoretical approaches to affirm the literary and political significance of women's writing while also creating literary histories. In the sciences, as well, feminists have understood that to achieve gender equality, the often forgotten or marginalized contributions of women must be acknowledged (Osca Lluch 7). The Montero novels in this study suggest that establishing community for scientific women is critical to supporting them in their professional and personal lives and encouraging more women to enter STEM fields.

LA RIDÍCULA IDEA DE NO VOLVER A VERTE

Montero's *La ridícula idea* asserts the important role of foremothers in countering women's experiences of exclusion in STEM. Narrated congruence between Curie's difficulties and those of other women scientists constructs female solidarity out of isolation. The text briefly tells the stories of four women—Lise Meitner (1878–1968), Rosalind Franklin (1920–1958), Henrietta Swan Leavitt (1868–1921), and Jocelyn Bell Burnell (1943–)—who made significant contributions in, respectively, nuclear fission, DNA, a standard for measuring stellar brightness ("Henrietta Swan Leavitt"), and radio pulsars, or signals from stars ("Jocelyn Bell"). By including their stories, Montero's narrative vindicates women whom male mentors or colleagues attempted to render invisible. In the cases of Meitner, Franklin, and Bell, only their male colleagues received Nobel prizes. German chemist Otto Hahn refused to credit Meitner when he won in 1944 because of her gender and Jewish identity. Franklin, who died young of ovarian cancer, probably from exposure to X-rays, may not have known that two male colleagues based their work on her photo of DNA (*La rídicula idea* 12–13). While these biographies alone attest to patriarchal dominance in STEM, placed together they show systemic discrediting of women who have made significant contributions to the sciences.

By inserting four women scientists from different periods in a narrative about Marie Curie, in a span of four pages, *La rídicula idea* emphasizes a shared experience of erasure and forges a community of women in STEM across time and place. Montero further highlights the importance of community with a story about her own fan network and social-media connections. A female Facebook friend in Canada, unaware that Montero was writing about Curie, sent her a note about women's lack of acknowledgment in the sciences and named the four female scientists Montero mentions in her text, two of whom the author was not aware before the fan's message. In this way, *La rídicula idea* emphasizes a shared feminist project of recognizing female scientists' contributions as an end in itself and employs the history of key women in order to inspire contemporary girls' and women's pursuit of studies and work in these fields.

Montero also forges a feminist solidarity with one of Curie's biographers, Barbara Goldsmith, who authored *Obsessive Genius: The Inner World of Marie Curie*.[21] Goldsmith delves into Curie's scientific studies more than Montero does, yet both emphasize that Curie's female identity shaped how she experienced the hardships of poverty in Russian-occupied Poland, the death of a loved one, and the lack of support for her work. Intimating their shared perspective, Montero repeats the juxtaposition with which Goldsmith initiates her text. In the 1995 ceremony in which François Mitterrand bestowed on the Curies the national honor of burial at the Panthéon, the French president affirmed that Marie possessed "the exemplary struggle of a woman who decided to impose her abilities in a society where abilities, intellectual exploration, and public responsibility were reserved for men" (*La ridícula idea* 15), yet the words etched into the Panthéon's façade attest to women's exclusion from the public sphere: "To Great Men from a Grateful Country" (15). Montero, however, goes a step further than Goldsmith when she notes Mitterand's use of the past tense (as quoted by Goldsmith) in reference to gender inequities: "Estaban, dijo. Como si esas desigualdades ya hubieran sido superadas por completo en el mundo contemporáneo" (20; They were, he said. As if those inequalities already had been completely overcome in the contemporary world). By referencing Goldsmith's biography in her narrative, Montero makes patent a community of female writers who seek to honor women in STEM and understand their lives as scientists, women, and individuals.[22]

To further her presentation of women's shared experiences, the narrative emphasizes a quotidian rather than idealized super-woman portrait of Marie Curie. While the achievements of this impressive scientist might be difficult to emulate, her struggles in managing a family while studying and working make her more relatable. Marie speaks proudly in her diary about her intellectual parity and emotional partnership with Pierre, yet she also comments that housework and childcare were her responsibilities alone.[23] Although Marie does not critique her husband, Montero's remark highlights the inequity of the double work-shift for women: "Durante estos años, Pierre publicó bastantes más artículos científicos que Marie. No puedo decir que me extrañe" (97; During these years, Pierre published considerably more articles than Marie. I cannot say that I am surprised). The narration connects Curie's challenges as mother, wife, and researcher to Montero's working female friends (not necessarily scientists) who felt overwhelmed after having a baby, albeit with domestic help (95). Linking the humanities and the sciences, Montero broadens the circle of women whose careers suffer under patriarchy with a reference to Carmen Laforet's 1945 *Nada*. The narrator hypothesizes that had it not been for her tenacity, Curie's life would have resembled those of the novel's tormented old women with unrealized dreams, as well as that of Laforet, who was unable to maintain her creative force "en medio del machismo ramplón de la posguerra española" (52; in the midst of the crude machismo of the post-Spanish Civil War). By writing about parallel experiences from different time periods and professional fields, Montero amplifies the recognition that social expectations, which mandate women care for the family, often hinder women's professional success yet benefit men.

Montero returns often to the theme of "#Intimidad" (69; #Intimacy) in her portrayal of Curie, showcasing the commonness of her personal life in order to augment her potential as a model for many women. Typical of long-time couples, Marie understood banal details about Pierre's preferences (70). Montero herself relates to the "#Intimidad perdida" (lost #Intimacy) that she and Curie suffer as widows (69). Curie's devastation when her first love ended the relationship, desperation after Pierre's death, and desire for intimacy, friendship, and professional partnership with the man she loved after Pierre's passing (physicist Paul Langevin), as well as her torment when Langevin is with his wife, highlight her humanity. In light of these challenges, *La rídicula idea* narrates a Marie Curie who is part of a female "us," in possession of a strength that eclipses men. Referencing Langevin's fear of his wife's revenge for his affair, Curie remarks to a friend: "Tú y yo somos duras . . . él es débil" (156; You and I are tough . . . he is weak). The narrator broadens this observation: "Y aquí hay que hacer un punto y aparte para hablar de la #DebilidadDeLosHombres, una gran verdad que todas conocemos pero ninguna menciona. Quiero decir que el verdadero sexo débil es el masculino" (156; And here we must make a full stop to speak about the #WeaknessOfMen, a great truth that we all know but no one mentions. I mean that the true weak sex is the masculine one). In this instance, and in others, one notes echoes of second wave, difference feminism, which acknowledges gender distinctions even if socially constructed.[24] Montero's narrative forcefully creates a community of successful women who have faced gender-based obstacles in their work and intimate lives. In this way, rather than an unattainable ideal, the text proposes a Marie Curie with whom other women can identify for their own ambitious aspirations.

Although Montero's text provides a feminist perspective of Marie Curie's life, the narrating voice notes with disappointment that Marie never wrote of the difficulties she confronted because she was a woman, and the work notes that Curie did not view herself as a champion of women in the sciences (130). Whereas Marie reserved one notebook to record her work and kept another for personal reflections, thus signaling their separation (Goldsmith 141–42), Montero's narrative blends the two to suggest that Marie Curie can best be understood as a woman scientist, rather than as one or the other. That is, *La rídicula idea* narrates Curie's achievements in the context of enduring gender-differentiated expectations in order to present a feminist reading of this remarkable female scientist who challenged the boundaries of science and gender even as those boundaries circumscribed her.

INSTRUCCIONES PARA SALVAR EL MUNDO

Cerebro narratively creates a community of scientists from various disciplines as she shares their theories with Matías during conversations in the Oasis bar. Granted, these individuals are all male, yet through discussing their work, Cerebro places herself within a scientific network from which she had been excluded decades prior. In addition, she forges a bond with an individual who seems drastically different from her, a working-class taxi driver, which enables both to find meaning in living. At first, each character is guarded. Due to fear of humiliation from slurring her words in her inebriety, Cerebro refrains from telling Matías of a science-based idea that would have lifted his spirits: because atoms recycle, existing eternally, it is possible that his and his wife's atoms will unite one day in another body (183–84). Cerebro's silence on this occasion suggests the silencing of her scientific voice, of her as a minority figure, and of a theory out of the mainstream. By the narration's end, however, the attention that Matías gives Cerebro when she shares scientific concepts transforms this "vieja insomne y solitaria aferrada a una copa" (insomniac and solitary old woman taking refuge in drink) into "la profesora que había sido" (182; the professor that she had been). The use of the past tense seems to indicate the late recognition of her contributions, like that of other female scientists, as well as the enduring professional repercussions of expulsion from academia for Cerebro. Nevertheless, through validating Cerebro as a scientific thinker, Matías and this former professor develop a friendship and together create community for each other.[25]

In addition to the bond between characters within this novel, Montero discursively constructs intertextual connections. Cerebro thinks of Marie Curie when she contemplates that atoms of past lives form part of present ones: "se puede decir que todos los seres humanos que ha habido en la Tierra viven en mí, y que yo viviré en todos los que vendrán en el futuro" (184; you might say that all of the human beings that have been on Earth live in me, and that I will live in all of those that will come in the future).[26] This reference in *Instrucciones* to Curie, the subject of *La rídicula idea*, published five years later, illustrates the texts' insistence on the need for female scientists to inspire women of future generations. Textual intermingling continues when Montero the narrator discusses in *La rídicula idea* Sheldrake's ideas on coincidences, a notion that Cerebro had explicated in *Instrucciones*: "creo que los científicos como Rupert Sheldrake son muy dudosos, pero, con los años, tengo la creciente sensación de que hay una continuidad en la mente humana; de que, en efecto, existe un inconsciente colectivo que nos entreteje . . . Y las #Coincidencias forman parte de esa danza, de ese todo" (*La rídicula idea* 142; I think that scientists like Rupert Sheldrake are very dubious; however, as I get older, I have the growing feeling that there is continuity in the human mind, effectively, that an unconscious collective exists that intertwines us . . . And that #Coincidences form part of that dance, of that whole). Noting the uncanny similarities between Curie's reflections on Pierre in her year-long diary and her own thoughts on Pablo Lizcano's death during the previous year, Montero, in *La rídicula idea*, like her character Matías in *Instrucciones*, finds comfort in scientific theories that emphasize positive, deep associations among living beings (142). Montero creates an intertextual dialogue between these two narratives that mimics the connectivity proposed, in different ways, by Sheldrake, Gould, Kammerer, and Lovelock. The four novels examined here consistently draw on scientists and theories that suggest our actions, positive or negative, conjoin us with and have repercussions for other beings.

LÁGRIMAS EN LA LLUVIA AND *EL PESO DEL CORAZÓN*

Montero's Husky narrations extend the theories of interwoven lives seen in *Instrucciones* to encompass an intersectional approach to community and a care-based ethic.[27] In *El peso del corazón*, a female nuclear scientist named Mai Burún cares for children in a war zone and seeks to protect Earth's environment from further contamination in her work on an underground nuclear-waste storage facility. The password that Bruna must give to Mai to gain information about the site, "Tranquilidad" (*El peso* 328; "Tranquility" [*Weight* 275]), signals the peace that Mai offers orphaned children in her home and also alludes to her opposition to the greed that, she asserts, presents the greatest threat to peace and environmental sustainability (*El Peso* 332). Although Mai revealed the nuclear-waste cemetery's location in exchange for money, an action that enabled Labari to fuel its misogynist, racist society, she did not know of this end and took the money to provide for the rescued children (335). In addition to her actions, her physical appearance signals an affinity with nature; her unaltered body and grey hair are unusual in a future in which most people recur to plastic surgery and hair color. Mai's and Bruna's care for others opposes dominant prejudices, political maneuverings, and capitalist-driven motives.

The android Husky performs a regenerative role congruent with Donna Haraway's cyborg model of "oppositional consciousness" in a "post-gender world," in which affinities with others do not depend on race, class, or gender.[28] Husky proclaims concern with only her needs, yet she attends to others throughout both novels, in the process facilitating unconventional alliances. Bruna Husky herself defies traditional identity boundaries as a female with a gender-fluid sexuality like that of most technohumans and humans in this fictional future (122). Although Husky cannot biologically reproduce, she contributes to a caring network that broadens the notion of mothering to community with diverse species.[29] She provides shelter and food to two extraterrestrial beings, the Balabi Bartolo and the Omaá Maio, and convinces a friend to accept Maio, who is a flutist, into her orchestra. The positive impact of Bruna's kind action amplifies when Maio and Mirari become intimate partners. The protagonist also intends upon her death to give her engineered arm to Mirari, a violinist with a dysfunctional prosthetic one.[30] Further, Bruna takes an orphaned ten-year-old girl named Gabi from Earth's most polluted area, cares for her, gets her medical treatment, and finds her a permanent home with her friend Yiannis. Her detective work calms exacerbated prejudices against technohumans in the first narrative. In the sequel, she prevents the spread of global war and nuclear-waste contamination, and she terminates Labari's energy source, at least momentarily.[31] In summary, Husky, a bioproduct of scientific innovation, and Mai, a woman scientist, serve as restorative, nurturing forces.[32]

Toward the end of *El peso del corazón*, Bruna finishes a story that she has been telling Gabi throughout the novel. In her narration about the mythical three-headed Cerberus who guards Hades, the monster transforms into a gentle companion who represents friendship and interspecies collaboration. Bruna delivers the tale in segments, a form consistent with women's oral storytelling and serial publications. She explains to Gabi that she includes this frightful dog, "Porque los monstruos somos hermosos" (*El peso* 368; "Because we monsters are beautiful" [*Weight* 309]). Relating herself with the monstrous in an oral, serial narrative, Bruna voices solidarity with women and feared others, as well as acceptance of her own engineered being. Nonetheless, *Lágrimas en la lluvia* and *El peso del corazón* make clear that one technohuman hero, several caring humans, and a pair of extraterrestrial beings cannot alone address Earth's environmental contamination, political strife, and social inequities. Furthermore, technical innovations may prolong resources and help sustain life, but we must recognize the codependence of all species and sustain each other in a world with fewer and contaminated resources for life on planet Earth to continue.

In the novels examined here, individually and read together, Montero narrates women, women scientists, and a woman made by science who face marginalization in their professional fields and in general society. All of these women persist in spite of their exclusion. In *Staying with the Trouble: Making Kin in the Chthulucene*, Haraway argues that unexpected kinships must be forged that address environmental damage and contemporary economic and social inequalities. The characters in Montero's narratives exhibit a drive to succeed in their respective disciplines, to assist those in need, to seek alliances, and to deeply connect with others, despite challenging circumstances. Recognizing the solitude and ostracism women in the sciences often experience, Montero's texts discursively create support networks of, for, and by women in STEM. If the Bruna Husky science-fiction narratives address more directly Haraway's assertion that "science fact and speculative fabulation need each other, and both need speculative feminism" (3), all four of these works propose that women's scientific research—past, present, and future—is crucial if we are to address the pressing issues of our times. These texts also suggest that literature and, more broadly, the humanities bring critical perspective to the impact of scientific developments in order to realize a more just and sustainable world for women, marginalized persons, and for all that inhabit Earth now and in the future.

NOTES

1. Montero published the third novel in the Bruna Husky series, *Los tiempos del odio* (2018; Times of hatred), after I had completed this study. The introduction of a new character, Ángela, continues the themes of women in STEM and female solidarity discussed here. A mathematics and computing genius, she helps Husky understand a terrorist case and funds space travel so Husky can search for her kidnapped lover.
2. I have used English editions for the translations of *Lágrimas en la lluvia* and *El peso del corazón*. All other translations are mine.

3. As Goldsmith notes, in the Poland of the late nineteenth and early twentieth centuries, women were viewed as physically and mentally unfit for work. Peasant women worked in factories due to necessity, but for less than men's wages. And, while women worked in jobs formerly held by men during a Polish uprising against the Russians, the women returned to housework and childcare when the men took their jobs back after the rebellion failed (23).
4. Bill Bryson observes that Curie was not elected to the Academy of Sciences because the men who made such decisions disapproved of the widow's affair with married physicist Paul Langevin (111).
5. Pilar Valero-Costa and Ellen Mayock have centered their respective analyses of *Instrucciones* on the impact of globalization on Spanish society and on Spanish responses to immigration. Luis I. Prádanos considers the novel's network narrative structure and representation of capitalism's negative impact on the environment and on community ("La degradación ecológica"). Prádanos also cites the novel as an example of Spanish literature congruent with degrowth and slow growth ("Towards a Euro-Mediterranean").
6. Critics have found many themes to explore in Montero's Bruna Husky series. For example, both Maryanne Leone and Carmen Flys Junquera examine the first two works from an ecofeminist perspective while Pilar Martínez-Quiroga studies the transfeminism embodied in the protagonist and transmitted through the science-fiction genre. Prádanos identifies a critique of global capitalism ("Decrecimiento"). Juan C. Martín Galvan addresses the posthuman in *Lágrimas*, while Dale J. Pratt attends to the relationship of memory to personhood and identity. Todd Mack discusses a Vermeer painting in Husky's apartment in light of Levinas's understanding of alterity, and Irene Sanz focuses on the perceived other through the interactions between humans and animals in *Lágrimas* and *El peso*. Given the interest that the series has attracted thus far, it will not be surprising if more studies emerge.
7. Fátima Serra-Renobales argues that Lizard exemplifies a more recent tendency in Montero's narratives of positive cooperation between women and men (78).
8. In the last section of this essay, I discuss how Montero narratively joins biographer Barbara Goldsmith in defense of female scientists.
9. This connection with protesting male-centric norms became even more patent when the hashtags #NotOkay and #NastyWoman went viral during the 2016 United States presidential campaign in response to then-candidate Donald Trump's misogynist comments about women and his opponent Hillary Clinton. The #MeToo hashtag to call out sexual harassment and change the culture of tacit acceptance followed one year later.
10. For more on the unequal distribution of wealth and resources between the northern and southern hemispheres in *Lágrimas en la lluvia* and *El peso del corazón*, and on Bruna's relationship with the polar bear, see Leone.
11. Katarzyna Beilin and Sainath Suryanarayanan locate *Lágrimas en la lluvia* within a broader discussion of Spain's enthusiasm for genetically modified foods and a bioeconomy. They argue that the novel supports bioengineered responses to climate change, while also critiquing a lack of ethical consideration for the bioproducts of humans' scientific endeavors (253).

12. One has to wonder if Montero might have found female models for Cerebro whose scientific work would also support the narrated ethical stances on caring for others. Serra-Renobales argues that *Instrucciones para salvar el mundo* is representative of twenty-first century inclusive feminism, of masculine protagonists in solidarity with women, and of the general public's growing interest in science, especially chaos theory and Darwinism.
13. Bill Bryson explains that Gould's 1989 *Wonderful Life* was a commercial success, but scientists disagreed with some of Gould's conclusions and suggested that his eloquent writing often superseded scientific rigor. For example, Richard Dawkins and others were at odds with his assertion that the Cambrian period's evolutionary process was unique, and scientists had not believed for fifty years prior to Gould's supposed revelation that evolution climatically progresses toward man (330–31). In the acknowledgments section of *Instrucciones*, Montero notes her indebtedness to Bryson's *A Short History of Nearly Everything* for the idea that the atoms of Marie Curie and Cervantes reside in us today.
14. In the film, Gould doubts the validity of Sheldrake's morphogenetic theory. Kayzer explains that, for some, Sheldrake is "the new Darwin and Einstein combined" (16:46–16:51) while, for others, he is "someone who . . . has developed a brilliant theory which is hopelessly wrong" (17:06).
15. Searches for information on Aaron Fieldman in books on the Manhattan Project, academic databases, and Google did not turn up an actual person similar to the novel's character.
16. Only a handful of the 150,000 people who worked on the Manhattan Project knew of the overall goal of building an atomic bomb (Kelly 93). Scientists on the interim committee and advisory group to War Secretary Henry Stimson thought that American military lives would be saved if the bomb were dropped, though they acknowledged the scientific community's varying views (Kelly 290–91). A letter from 155 Manhattan Project scientists urged Stimson and President Truman to consider the moral consequences and communicate the conditions of surrender to Japan to avoid the bomb's use (Kelly 291–93).
17. Lovelock's extensive work includes crayons that write on wet glass, methods to protect soldiers from burns, inventions to search for life on Mars, the creation of a device to detect pollutants, observations that led to the discovery of the ozone layer, and more. He received the Wollaston Medal from the Geological Society in 2006, was named Commander of the Order of the British Empire, is a Fellow of the Royal Society (since 1974), and has received many awards in chromatography. Continued interest in his work prompted a speaking tour in the United States in 2006, at 86 years old (Gribbin and Gribbin xix, xxi, 73–76).
18. Lovelock's invention in the 1950s of an electron capture detector (ECD) to measure the presence of molecules, including pollutants, "revolutionized our understanding of the relationship between human activities and the environment" and still is used today (Gribbin and Gribbin 112). The scientist disagrees with environmentalists that use the ECD to ban substances even if levels are low and not necessarily harmful, and he argues that substances like DDT also saved lives (Gribbin and Gribbin 119).
19. Disequilibrium, where there is energy transfer from high (warmer) to low (cooler) systems, would indicate life was present. Lack of energy flow, where there is high equi-

librium, would indicate no life. When Congress's funding of the Voyager mission to Mars ceased in 1965, Lovelock's experiments ended, yet in that same year, scientists in France used Lovelock's method and determined that life did not exist on Mars (Gribbin and Gribbin 137–40).

20. Diane Preston explains in *Before the Fallout: From Marie Curie to Hiroshima* that the creation of the atomic bomb began with Curie's discovery of radium.
21. See *La ridícula idea*, pp. 209–10, for a list of sources Montero cites for her research on Curie.
22. Like Montero, Goldsmith also highlights Curie's many first-woman achievements (16–17).
23. Goldsmith critiques biographies that focus only on Curie's scientific work or gloss over hardships; she instead deconstructs the Curie legend that women can "do it all—and perfectly" (145). Montero praises Goldsmith for addressing the work-family pressure Curie experienced (*La ridícula idea* 95).
24. The text proposes that women and men approach relationships differently, women with idealized romanticism (61).
25. While Serra-Renobales focuses on Matías and the other male characters in this novel and I focus more on Cerebro, we both conclude that solidarity, acceptance, and inclusion underpin Montero's proposal of a better world (76). Though outside this essay's focus on women scientists, it is worth noting that because Matías assists a trafficked young woman from Senegal to escape entrapment, she gives birth to a son who will become an environmental scientist who studies oceanic methods to mediate global warming.
26. With this comment, Cerebro also suggests her affinity with Sheldrake, who asserts that "the past is potentially present everywhere, and that . . . we tune into or access aspects of past experience" (Kyzer 50:14–50:34).
27. For a more complete analysis of a care-based model of ecological sustainability in *Lágrimas en la lluvia* and *El peso del corazón*, see Leone.
28. Beilin also argues that Bruna Husky might be read in light of Haraway's cyborg figure. Haraway explains that Chela Sandoval introduces the term "oppositional consciousness" to describe a conscious, unifying political identity that does not totalize or assume natural categories ("A Cyborg Manifesto" 479–80, 490).
29. Beilin asserts that because androids cannot become biological mothers, they might better "resist cultural codes that lead us to destroy nonhuman life" (252). Leone also cites Beilin on this point in "Trans-species Collaborations."
30. Bruna blew off one of her arms with a plasma gun to free herself from a fallen tree during a chase. Promising her arm to Mirari transforms Bruna's fatalistic perspective into lightness, felt in "lo poco que pesaba un corazón feliz" (*El peso* 371; "how little a happy heart weighed" [*Weight* 312]; Leone 76).
31. Brief mentions of Labari in the series' third novel suggest that Bruna's actions do not finish this discriminatory polity.
32. We see the same with Cerebro, who has no children but supports Matías. Curie's struggles to care for her children and dedicate herself to her lab are discussed earlier in this piece.

WORKS CITED

Atkins, P. W. *Four Laws that Drive the Universe*. Oxford UP, 2007.

Beilin, Katarzyna Olga, with Sainath Suryanarayanan. "Debates on GMOs in Spain and Rosa Montero's *Lágrimas en la lluvia*." *In Search of an Alternative Biopolitics: Anti-Bullfighting, Animality, and the Environment in Contemporary Spain*, Ohio State UP, 2015, pp. 235–61.

Bryson, Bill. *A Short History of Nearly Everything*. Broadway, 2003.

Flys Junquera, Carmen. "Ecofeminist Replicants and Aliens: Future Elysiums through an Ethics of Care." *Women's Studies: An Interdisciplinary Journal*, vol. 47, no. 1–4, 2018, pp. 232–50.

Gilbert, Sandra, and Susan Gubar. *The Madwoman in the Attic: The Woman Writer and the Nineteenth-Century Literary Imagination*. Yale UP, 1979.

Goldsmith, Barbara. *Obsessive Genius: The Inner World of Marie Curie*. Norton, 2005.

Gray, John. "James Lovelock: A Man for All Seasons." *New Stateman*, 27 Mar. 2013, www.newstatesman.com/culture/culture/2013/03/james-lovelock-man-all-seasons. Accessed 10 Oct. 2016.

Gribbin, John, and Mary Gribbin. *John Lovelock: In Search of Gaia*. Princeton UP, 2009.

Haraway, Donna J. "A Cyborg Manifesto: Science, Technology and Socialist-Feminism in the Late Twentieth-Century." *Simians, Cyborgs and Women: The Reinvention of Nature*, Routledge, 1991, pp. 475–497.

———. *Staying with the Trouble: Making Kin in the Chthulucene*. Duke UP, 2016.

"Henrietta Swan Leavitt." *Encyclopaedia Britannica*, 1 July 2019, www.britannica.com/biography/Henrietta-Swan-Leavitt. Accessed 6 July 2019.

Highfield, Roger. "The Vanishing Face of Gaia: A Final Warning by James Lovelock, review." *The Telegraph*, 19 Mar. 2009, www.telegraph.co.uk/culture/books/bookreviews/5017620/The-Vanishing-Face-of-Gaia-A-Final-Warning-by-James-Lovelock-review.html. Accessed 24 Oct. 2016.

"Jocelyn Bell." *A Science Odyssey: People and Discoveries*. WGBH, 1998, www.pbs.org/wgbh/aso/databank/entries/babell.html. Accessed 29 Oct. 2016.

Kelly, Cynthia C. *The Manhattan Project: The Birth of the Atomic Bomb in the Words of Its Creators, Eyewitnesses, and Historians*. Blackdog and Levanthal, 2007.

Kayzer, Wim, director. *A Glorious Accident: Understanding our Place in the Cosmic Puzzle. Rupert Sheldrake*. Films Media Group, Films for the Humanities and Sciences, 2004.

Leone, Maryanne L. "Trans-species Collaborations in Response to Social, Economic, and Environmental Violence in Rosa Montero's *Lágrimas en la lluvia* and *El peso del corazón*." *Ecozon@: European Journal of Literature, Culture, and the Environment*, vol. 8, no. 1, 2017, pp. 61–78.

Mack, Todd K. "El arte de *Lágrimas en la lluvia*." *Alambique: Revista Académica de Ciencia Ficción y Fantasía/Jornal Académico de Ficção Científica e Fantasia*, vol. 5, no. 1, art. 7, 2017, dx.doi.org/10.5038/2167-6577.5.1.7. Accessed 2 July 2019.

Martín Galvan, Juan C. "El universo posthumano de *Lágrimas en la lluvia*: Memoria artificial, identidad, historia y ficción." *Alambique: Revista Académica de Ciencia Ficción y Fantasía/Jornal Acadêmico de Ficção Científica e Fantasia*, vol. 5, no. 1, art. 5, 2017, dx.doi.org/10.5038/2167-6577.5.1.5. Accessed 2 July 2019.

Martínez-Quiroga, Pilar. "La detective Bruna Husky de Rosa Montero: Feminismo, distopía y conciencia cyborg." *Hispania*, vol. 101, no. 2, 2018, pp. 306–17. *Project MUSE*, doi:10.1353/hpn.2018.0122. Accessed 30 June 2019.

Mayock, Ellen. "West Meets East in Rosa Montero's *Instrucciones para salvar el mundo*." *Cuaderno Internacional de Estudios Humanísticos y Literatura*, vol. 16, 2011, pp. 162–71.

Moi, Toril. *Teoría literaria feminista*. Cátedra, 1995.

Montero, Rosa. *Instrucciones para salvar el mundo*. 2nd ed., Santillana, Feb. 2011.

———. *Lágrimas en la lluvia*. Seix Barral Biblioteca Furtiva, Planeta, 2011.

———. *El peso del corazón*. Seix Barral Biblioteca Furtiva, Planeta, 2015.

———. *La ridícula idea de no volver a verte*. Planeta, 2013.

———. *Tears in Rain*. Translated by Lilit Žekulin Thwaites, Kindle ed., Amazon Crossing, 2012.

———. *Los tiempos del odio*. Seix Barral Biblioteca Furtiva, Planeta, 2018.

———. *Weight of the Heart*. Translated by Lilit Žekulin Thwaites, Amazon Crossing, 2016.

Orci, Taylor. "How We Realized Putting Radium in Everything Was Not the Answer." *The Atlantic*, 7 Mar. 2013, www.theatlantic.com/health/archive/2013/03/how-we-realized-putting-radium-in-everything-was-not-the-answer/273780. Accessed 11 Feb. 2018.

Osca Lluch, Julia. *La aportación de la mujer a la historia de la ciencia y de la técnica en España*. Instituto de Historia de la Medicina y de la Ciencia López Piñero, 2011.

"Pascal's Principle." *Britannica Academic*, Encyclopædia Britannica, 26 Apr. 2018, academic-eb-com.lib.assumption.edu/levels/collegiate/article/Pascals-principle/58620. Accessed 5 July 2019.

Prádanos, Luis I. "Decrecimiento o barbarie: Ecocrítica y capitalismo global en la novela futurista española reciente." *Ecozon@: European Journal of Literature, Culture and Environment*, vol. 3, no. 2, 2012, pp. 74–92.

———. "La degradación ecológica y social del espacio urbano en *Instrucciones para salvar el mundo* de Rosa Montero." *Letras femeninas*, vol. 39, no. 2, 2013, pp. 45–61.

———. "Towards a Euro-Mediterranean Socioenvironmental Perspective: The Case for a Spanish Ecocriticism." *Ecozon@: European Journal of Literature, Culture and Environment*, vol. 4, no. 2, 2013, pp. 30–48.

Pratt, Dale J. "Vidas virtuales, memorias postizas: Teorías de la identidad personal en *Lágrimas en la Lluvia*." *Alambique: Revista Académica de Ciencia Ficción y Fantasia/Jornal Acadêmico de Ficção Científica e Fantasia*, vol. 5, iss. 1, art. 6, 2017, dx.doi.org/10.5038/2167-6577.5.1.6. Accessed 2 July 2019.

Preston, Diane. *Before the Fallout: From Marie Curie to Hiroshima*. Walker Books, 2005.

Sanz, Irene. "Human and Nonhuman Intersections in Rosa Montero's Bruna Husky Novels." *Science Fiction Studies*, vol. 44, no. 2, 2017, pp. 323–30.

Serra-Renobales, Fátima. "La ciencia para salvar el mundo: Una propuesta de Rosa Montero." *La incógnita desvelada: Ensayos sobre la obra de Rosa Montero*, edited by Alicia Ramos-Mesonero, Peter Lang, 2012, pp. 71–81.

Showalter, Elaine. *A Literature of Their Own: British Women Novelists from Bronte to Lessing*. Princeton UP, 1977.

Valero-Costa, Pilar. "*Cosmofobia*, de Lucía Extebarría e *Instrucciones para salvar el mundo*, de Rosa Montero: Complejidad cultural en la España global." *La mujer en la literatura del mundo hispánico*, Instituto Literario y Cultural Hispánico, 2009, pp. 31–48.

York, Richard, and Brett Clark. *The Science and Humanism of Stephen Jay Gould*. Monthly Review, 2011.

PART III
ON GENDER
Using STEM to Critique Gendered Roles

CHAPTER 11

Biotech, Barceló, Bustelo

Reproduction, Motherhood, and Gendered Hierarchies in Spanish Science Fiction

MIRLA GONZÁLEZ

During the second half of the twentieth century, concurrent with the emergence of second-wave feminism, production of North American and British feminist science fiction (SF) increased, as exemplified by the publication of Ursula Le Guin's *The Left Hand of Darkness* (1969), Joanna Russ's *The Female Man* (1975), Marge Piercy's *Woman on the Edge of Time* (1976), Doris Lessing's *Canopus in Argos* series (1979–1983), and Octavia Butler's *Xenogenesis* trilogy/*Lilith's Brood* (1987–1989). This wave of female writers employed SF to question values put forth by patriarchal society and to advocate for feminist goals. Science fiction and fantasy have served as mechanisms of feminist thought, bridging theory and practice through the "textual exploration of theoretical and activist ideals for progressive and social change" (Helford 291). Though feminist SF boomed during the end of the twentieth century in the United States and parts of Europe, in Spain SF (let alone feminist SF) faced a series of social, political, and economic challenges that delayed its development (M. Barceló 483, Díez 9, Moreno 407, Sánchez Conejero 5).[1] Among these obstacles was Spain's predilection for realist manifestations, especially during the postwar years of the Spanish Civil War (1936–1939), in which novelists were more interested in social realism than speculative fiction. Spain's ties to Catholicism, particularly during Franco's dictatorship, also hindered the development of SF. As Díez and Moreno state, by highlighting scientific discovery, providing an alternative to religious teachings, and creating imagined worlds where authority could be questioned and the patriarchal model subverted, SF posed a threat to the regime and was thus rejected (72). Additionally, SF was a genre that both in quantity and quality was monopolized by Anglo-Saxon writers, limiting the market space for SF authors from other countries (Martínez de la Hidalga 13). All of these factors contributed to the characterization of Spanish SF as a marginal genre aimed at a niche audience. In *Ciencia ficción en español: una mitología moderna ante el cambio*, Yolanda Molina-Gavilán argues that to this day both Spanish and Spanish-American SF is seen as part of a "cultural ghetto" by critics and academics (23).

Despite these negative perceptions of Spain and Spanish SF, contemporary Spanish women writers have published innovative texts demonstrating how scientific discourse is part of the larger Spanish cultural context and the important contribution women's voices have made to this often-trivialized genre.[2] This study explores how utopias and dystopias have been used to invert traditional gender roles in Elia Barceló's *Consecuencias naturales* (1994; Natural consequences) and Gabriela Bustelo's *Planeta hembra* (2001; Female planet).[3] Barceló's and Bustelo's novels illustrate how scientific discourse has informed cultural production in Spain, particularly related to the issue of reproductive technologies. In their works, Barceló and Bustelo invert the traditional sex-gender paradigm, offering texts that deal with reproductive rights and biotechnologies while including various familial and societal models to highlight the power struggles that exist between different genders and sexes. This analysis focuses on how biotechnology in these novels serves to control and manipulate various populations, ultimately creating gendered hierarchies that place restrictions on an individual's rights over his/her body to the detriment of society.

While scientific discourse has at times served to relegate women as biologically inferior, contemporary authors are incorporating biotechnology, including reproductive technologies, in order to re-examine how gender roles are socially constructed.[4] Biotechnology involves employing living organisms or cell processes to make useful products; this essay will focus on biotechnology's medical applications by taking a close look at contraceptives, genetic engineering, and cloning. As highlighted in an August 2016 *Marca España* article, "Biotecnología en España: sector competitivo e innovador," Spain's biotechnology sector has become one of the most competitive in the world and is continuing to grow.[5] It is important to note, however, that even though biotechnologists and multinational companies founded in Spain have made significant inroads, the field has not been without its challenges, in part as a result of the country's Catholic tradition. The very nature of the discipline's objects of study and techniques (embryos, genetic manipulation, etc.) often places biotechnological progress in conflict with ethical and religious belief systems that can constrain research and development. The presence of biotechnology as a central motif in contemporary novels demonstrates the ongoing anxiety in Spain over scientific innovation and the evolving role of biotechnology in reproductive theory. Along with analyzing the use of biotechnology in SF novels to invert traditional gender roles, attention will also be given to how narrative techniques (such as the confluence of temporal and spatial dimensions, estrangement, and parody) underscore the unconventional social arrangements proposed in both texts. On a larger scale, this study aims to provide a Spanish complement to existing Anglo-American scholarship on feminist SF and to serve as a foundation for future comparative studies.

When coupled with a feminist lens, utopias and dystopias, both subgenres of SF, can portray futures that radically differ from the present state of affairs and thus raise questions about existing sex/gender dynamics. In *Metamorphoses of Science Fiction: On the Poetics and History of a Literary Genre*, Darko Suvin argues that SF is the literature of "cognitive estrangement" (4).[6] In other words, SF empirically validates that which otherwise appears as alienating to the reader. Utopia, he argues, is "the verbal construction of a particular quasi-human community where sociopolitical institutions, norms, and individual relationships are organized according to a more perfect principle than in the author's community, this construction being based on estrangement arising out of an alternative historical hypothesis" (49). Suvin's definition emphasizes the link between the utopian text and the historical context from which it emerges. Like Suvin, Fredric Jameson asserts in his work *Archaeologies of the Future, The Desire Called Utopia and Other Science Fictions* that utopias are grounded in socio-historical reality. While utopias may appear to be portrayals of a perfect future, they are in fact imagined spaces that allow readers to trace the problems and perils present in society so that they may reflect and perhaps change course. Located nowhere, utopias begin as blank fictitious spaces that are injected with institutions, value systems, and patterns of behavior and interaction. Eventually, these elements undergo a process of transformation, thereby becoming detached from their original context, until they are ultimately re-contextualized in a new world (Hutchinson 181). Via the reading of each text, readers are displaced into these worlds, which then allow them to examine the familiar through new configurations (182).

The presence of unfamiliar social arrangements and controversial scientific principles prompts one to question the underpinnings of society both in the fictional world and in the lived world. In Barceló's *Consecuencias naturales*, this estrangement is produced by the alien society, in which hierarchy is dependent solely on reproductive ability, while in Bustelo's *Planeta hembra* it is caused by the totalitarian matriarchal society's condemnation of heterosexuality and male homosexuality. The exaggerated sociological structures and oppressive forces present in both novels invite readers to critically examine the scientific principles, political systems, and social relations that govern those alternative futures, as well as consider how they compare to the dynamics present in each reader's society. While on the surface both novels offer satiric overviews of feminist fundamentalism, a deeper reading shows how this defamiliarization and estrangement allows readers to disengage from their own world and actively rethink how the present and future are always in a process of continual change: "rather than prepackaged knowledge delivered to a passive reader, defamiliarization functions as a discovery that occurs through the reader's imaginative participation in the world created by the text" (Patai 67). This defamiliarization facilitates criticism of how biotechnology affects gender hierarchies.

Moving readers outside familiar boundaries, particularly norms related to gender, is a recurring theme in Barceló's work. Although Barceló states in an interview with Cristina Sánchez-Conejero that she is not actively involved in the feminist movement, she affirms that society must demand mutual respect between both sexes (189). Barceló has addressed issues of sexuality in "La dama dragón" (1981) and "Piel" (1989), while deconstructing patriarchal paradigms "through the construction of alternative cognitive scenarios" in works such as *Sagrada* (1981) and *Consecuencias naturales* (Perriam 104). Despite writing in a male-dominated genre, Barceló has won numerous awards and is considered Spain's leading female SF writer, though she has also expanded into other literary genres, gaining wider readership. In an interview for *El País* with Juan Manuel Játiva, Barceló boldly claims that SF is the only genre that delivers new themes to audiences, as opposed to other genres, which she claims often merely produce new spins on old themes. Known as "la gran dama de la ciencia ficción," Barceló won the first Ignotus Award, created by the Spanish Association of Fantasy, Science Fiction, and Terror (AEFCFT) in 1991, a prize that acknowledges works of science fiction in categories including novel, novella, anthology, illustration, website, and short story. Subsequently, in 1993, Barceló became the first female to win the International Prize of the Polytechnic University of Catalonia (UPC) for her novel *El mundo de Yarek* (Yarek's world). Also created in 1991, the UPC award is the most important prize in Spanish SF (Romero 445). Awards such as Ignotus and UPC have helped to promote the visibility of the genre, specifically the contributions of Spanish authors.

Consecuencias naturales, one of Barceló's earliest and perhaps most polemical novels, provides a commentary on machismo and a portrayal of how reproductive theory is changing traditional gender roles. *Consecuencias naturales* recounts the journey of Nico Andrade, a male protagonist, as he works through the various stages of a typical female reproductive process: from copulation to gestation, parturition, and finally acceptance of a maternal role. Andrade's pregnancy raises many of the same ethical and legal questions that have surfaced regularly in contemporary Spanish society, such as issues surrounding the use of contraceptives, abortion, and rape, but here those issues are presented in the context of the male body. As Sánchez-Conejero points out in *Novela y cine de ciencia ficción española contemporánea: Una reflexión sobre la humanidad*, through Andrade, Barceló has created a character that fully embodies the "Iberian macho." Andrade is one of several hundred crewmembers that live aboard the *Victoria*, the space station located furthest from planet Earth. Aboard the station, Andrade has become notorious for being a Don Juan; when he learns that members of Xhroll, a humanoid alien species, are in need of mechanical help and about to enter their base, his first question is: "¿Hay alguna mujer? Hace siglos que no vemos carne fresca" (12; Is there a woman? It has been centuries since we have seen fresh meat). Engrossed in the idea of achieving fame, Andrade proceeds with his plan to become the first human to have sexual relations with an alien species. Before engaging in coitus, the Xhroll warns Andrade that she is not using any type of contraceptive. Rather than assume responsibility and take precautionary measures, Andrade decides to fool the Xhroll by pretending to ingest birth control by swallowing an aspirin. Unfortunately for Andrade, it is not the Xhroll but he himself who becomes impregnated, thus marking the beginning of his journey toward "motherhood." The text thus plays with gendered norms, misogynistic assumptions, and the fluidity of gender roles in this fictive future space.

Andrade's character and actions raise several issues regarding sex and gender that have been widely debated in Spain, including the role that contraceptives have played in controlling the female body, the function of abortion in unwanted pregnancies, and the primacy of fetal rights over maternal ones. To help elucidate how these issues play out in *Consecuencias naturales*, it is helpful to understand how definitions of sex have changed over time and the increasing role that biotechnology plays in manipulating and controlling the body. According to Laqueur, for centuries it was believed that "women were essentially men in whom a lack of vital heat—of perfection—had resulted in the retention, inside, of structures that in the male are visible without" (4). The male body was a symbol of perfection while the female body was seen as a weaker variation of male anatomy in which the vagina was viewed as an interior penis, the uterus as scrotum, and the ovaries as testicles. By the end of the eighteenth century, the two-sex model had gained ground, as female organs acquired their own names and were no longer considered inversions of male organs. Laqueur argues that this shift from a one-sex model to a two-sex model was due in large part to politics. With the rise of capitalism and the bourgeoisie in the nineteenth century, it became necessary to differentiate between males and females in order to preserve hierarchies and class distinctions (194).

This transition from the one-sex model of the sovereign society to the two-sex model of the disciplinary society was accompanied by increased surveillance and control of the human body by state apparatuses. In *Discipline and Punish: The Birth of the Prison*, Foucault argues that institutions such as schools, hospitals, and jails function to reform and normalize individuals into docile and utilitarian bodies. According to Paul B. Preciado, after WWII (or in the case of Spain, after the Spanish Civil War), sex and sexuality were once again at the center of political and economic activity. During the latter part of the twentieth century, there was a major paradigm shift as biotechnology and the visual era redefined sexual identity.[7] Preciado coins this third regime of subjectivization "the pharmacopornographic society" (77).[8] These new technologies of control include "biomolecular, digital, broadband, data-transmission [. . .] that could be injected, inhaled—'incorporated' [...] the technologies become part of the body: they dissolve into it" (77–78). These new forms of technology function at the molecular level and regulate the body at the cellular level. In other words, while in the disciplinary society of the nineteenth century the body was controlled externally, in the pharmacopornographic society repression originates from within one's own body. Made up of artificial hormones, the contraceptive pill was the first biochemical technique to facilitate the separation of heterosexual intercourse and reproduction.[9] In order for birth control to be effective, however, it requires that women be very disciplined about taking their daily dose. In short, Preciado argues that the contraceptive pill is an edible panopticon "where the surveillance tower has been replaced by the eyes of the (not always) docile user of the pill who regulates her own administration without the need for external supervision" and where "the prison cell has become the body of the consumer" (205).

In *Consecuencias naturales*, by refusing to use an actual contraceptive, Andrade fulfills his sexual desires at what he believes to be the expense of the Xhroll. His disregard for any precautionary method demonstrates how undisciplined he is and how little he esteems the life of the Xhroll with whom he is about to have sex. To Andrade, women are dispensable commodities created to satiate men's primordial desires; at one point in the story he even remarks, "Era muy agradable tener compañía en la cama cuando uno estaba despierto, pero una vez zanjada la cuestión primordial, lo correcto era que la chica tuviera el buen gusto de marcharse a su propia cama y lo dejara descansar a sus anchas" (110; It was nice to have company in bed while one was awake, but once the primordial matter had been resolved, the proper thing was for the girl to have the decency to return to her own bed and allow one to rest as he pleased). Tired of Andrade's archaic attitude and behavior toward women, Charlie, one of Andrade's superiors, adds a comedic touch to his misogyny when she states that Andrade is a type of *Homo erectus* living in the twenty-third century. By comparing Andrade to *Homo erectus*, who arose 1.8 million years ago and is associated with the first major innovation in stone tool technology, Charlie highlights Andrade's fossilized treatment of women. Moreover, the juxtaposition of time periods and the play of words with erectus and "erect" serve as subtle commentary that, although advances have been made in the scientific and technological arena, sociological progress has been much slower. There continue to be sectors of Spanish society where one finds traces of machismo and the reckless wielding of the phallus. In fact, in her interview with Sánchez-Conejero, Barceló admits that *Consecuencias naturales* caused public controversy since some "lectores se sintieron ridiculizados y agredidos a través del personaje de Nico Andrade" (190; readers felt ridiculed and attacked through the character of Nico Andrade), despite the novel's ironic undertones.

Like machismo, a social structure yielding patterns of negative consequences for women, the contemporary pharmaceutical industry is similarly an economic powerhouse that can also wield significant social effects. Indeed, with the rise in prescription drug use and ubiquitous medication, differences between biotechnology and the human body's natural mechanisms are becoming indistinguishable. In other words, according to Preciado, the pharmacopornographic regime has created a lucrative performative feedback mechanism through which artificial chemical substances and molecules are conflated with the body's physiological demands and desires: "The success of contemporary technoscientific industry consists in transforming our depression into Prozac, our masculinity into testosterone, our erection into Viagra, our fertility/sterility into the Pill . . . without knowing which comes first: our depression or Prozac, Viagra or an erection, testosterone or masculinity, the Pill or maternity" (34–35). These artificial substances have become trademarks of the human body.

In the SF world of *Consecuencias naturales* women are sterile until they actively take the drugs that will return their fertility, thus eliminating the need for either sex to use contraceptives. In a sense, this gives women greater agency over their own bodies because the decision to become fertile is now solely in their power. However, it has also freed the male sex of all liability. Men in Andrade's world are so removed from any sense of reproductive responsibility that Andrade muses on male contraceptive devices as part of ancient history:

> Prácticamente todas las mujeres humanas eran voluntariamente estériles hasta que decidían invertir la situación y tomaban los fármacos que les devolvían la fertilidad. Antiguamente existían unas fundas de goma que se ajustaban al pene y evitaban la entrada del semen masculino en el útero de la mujer pero estaba seguro de que en toda la *Victoria* no había uno solo de esos inventos antediluvianos, aquello era una estación espacial, no un museo. (27–28)

Basically all of the human women were voluntarily sterile until they decided to reverse the situation and take drugs that returned their fertility. Previously, there existed some rubber casings that adjusted to the penis and prevented the entry of masculine semen into the uterus of the woman, but I was sure that in all of *Victoria* there wasn't a single one of those antediluvian inventions, that was a space station, not a museum.

After centuries of women being in control of if/when fertilization will occur, Andrade is shocked to learn that in the society of the Xhrolls, contraceptives are still used and that it is his responsibility to take precautionary measures. Ultimately, Andrade has no qualms about taking the aspirin as a placebo to psychologically appease the Xhroll because he believes that even in the unlikely event of a pregnancy, it is the female that will carry the burden. Indeed, this same risk-benefit analysis represented in the novel by Andrade is one of the central issues that contemporary science has had to grapple with in creating a pill for men. In *The Male Pill: A Biography of a Technology in the Making,* Nelly Oudshoorn explains that pressure to create new male contraceptives did not originate in the scientific community but rather from the governments of India and China who were seeking population control and from feminists in the Western industrialized world who wanted both sexes to share equally in the responsibilities and health risks associated with contraceptives (7). Despite the technical feasibility of creating a male contraceptive akin to the pill, throughout the twentieth century research was primarily aimed at controlling women's reproductive functions, not men's. The assumption is that there is a risk-benefit analysis that applies to women but not to men, and thus it appears unlikely that there will be a male contraceptive anytime soon, as a recent NPR story argued after a clinical trial was suspended due to side-effects for men: "when women use a contraceptive, they're balancing the risks of the drug against the risks of getting pregnant. And pregnancy itself carries risks. But these are healthy men—they're not going to suffer any risks if they get somebody else pregnant" ("Male Birth Control"). In *Consecuencias naturales*, Andrade's assumption that the Xhroll are governed by the same physiological mechanisms as the human species leads him to the conclusion that unprotected sex is better than no sex, but to Andrade's horror, it is he who becomes impregnated and not the Xhroll. To mitigate the shock of a male pregnancy and to ease the reader into the events about to transpire, there is a change in setting; the second part of the novel takes place on planet Xhroll.

By having Andrade go through the process of gestation in unfamiliar territory, a degree of critical distance is created, allowing readers to reflect on their own experiences. Indeed, as details of the social hierarchy on planet Xhroll are disclosed, a parallel is established between the Xhroll's overbearing treatment of those who are pregnant and our own society's treatment of expectant mothers. On planet Xhroll, the alien species is better equipped to assist with what Andrade describes as a hostile relationship between the fetus and host, thereby portraying the fetus-mother bond as a parasitic relationship. Given the lack of a uterus in which to gestate, the fetus creates a network using Andrade's organs in order to sustain itself. Andrade feels trapped in other ways as well, describing the maternity ward as a maximum-security prison. Xhroll's small population and a low fertility rate have fostered a social structure that revolves around reproductive ability. Individuals are either "abba" (those who can give birth), "ariarkhj" (those who can impregnate), or "xhrea" (those who are incapable of reproducing and who oversee life on Xhroll). During his stay, Andrade's status becomes that of an abba. In many Semitic languages, "abba" means father, which in the novel serves to further emphasize the increased paternal role in the process of maternity. Though the abbas are greatly revered and respected given their role in keeping the species from extinction, they are also considered public patrimony and lack civil rights. Since they carry life, they are part of the common good that must be protected at all costs. Andrade comments, "Por milésima vez en los últimos meses se sentía ridículo y humillado, un mero objeto sin voluntad ni capacidad de decisión a quien se le hacían pruebas y se alimentaba y ejercitaba convenientemente, un ser de segunda categoría a quien no se le daban explicaciones directas, un mero pedazo de carne en crecimiento" (81; For the thousandth time he felt ridiculed and humiliated, a mere object without will and without the ability to make decisions on whom tests were performed and who was fed and exercised conveniently, a second-class being who wasn't given direct explanations, a mere piece of growing meat). Andrade's role as an abba parodies the state of inactivity and compliance that is often imposed and expected of women during the various stages of fetal development.

The novel highlights a variety of controversial social issues in the Xhrollian society that are also feminist issues of concern in Spain, including rape, unwanted pregnancy, and abortion. The passive role assigned to the abbas is due in part to the fact that Xhroll is a collectivist society, which puts the interest of the group above that of the individual. Given this mentality, in the event that Andrade gives birth to a viable being, Xhroll is willing to rape all of the men on *Victoria* if it will ensure the survival of the species. This is also the reason the Xhroll are unwilling to allow Andrade to have an abortion even though Andrade describes the fetus as a monster and a tumor and is horrified when he feels it begin to kick. Andrade is being forced to endure an unwanted pregnancy in which the fetus has more rights than he does, bringing to light the issue of reproductive rights.

In Spain, abortion has been a very controversial topic, given the country's conservative history and its deep ties to Catholicism. During most of Franco's dictatorship, the State and the Catholic Church were inseparable. Under the Francoist regime, women were relegated to the private sphere since the family "connected vertically with the state rather than horizontally with society" (Graham 184). The family was considered the backbone of society, and a woman's job was to be a perfect mother and wife, ascribing to the ideals of the Virgin Mary and Isabel la Católica. The Franco regime cited church doctrine, argued for a policy of pronatalism given the many lives lost during the Civil War, and criminalized contraception and abortion. However, as Graham notes, socio-economic class mediated Spanish women's experience, and many poor working-class women had to use "whatever birth control methods they could gain access to—and abortion, as before—to exert some degree of control in order to protect their own health, and stave off economic disaster" (192). After Franco's death, Spain underwent significant political and social changes during its transition to democracy. With the Partido Socialista Obrero Español (PSOE; Spanish Socialist's Workers Party) in power, the 1985 Organic Law 9/1985 legalized abortion in three cases.[10] Then in 2010 with PSOE once again in power, Organic Law 2/2010 provided an even more liberal abortion policy.[11] In 2011, Spain's Partido Popular (PP; Popular Party) regained control of the government and sought to tighten restrictions on abortion. Given the PSOE's and PP's opposing views, thousands of both pro-life and pro-choice supporters have taken to the streets in the last decades to voice their opinions. It is a continuous give-and-take legal battle between the rights of the mother and the rights of the fetus. Abortion in Spain continues to be an ongoing controversy, and this struggle is reflected in Barceló's novel. By having a male protagonist serve as an advocate for the rights of the "mother," one is prompted to re-examine the gender disparities of Spain's medical, legal, and political fields.

In *Consecuencias naturales*, Andrade's struggle to accept the monstrous being growing inside of him is a reminder of the physical and psychological toll a pregnancy can take on a woman, especially in the instance of an unwanted pregnancy. As long as Andrade's status remains that of an abba, he has no voice and no rights over his own body because the life of the fetus is considered more precious than his own. Throughout the novel, Charlie's (Andrade's superior) comedic and reflective digressions help place Andrade's complaints about pregnancy in historical perspective:

> Se había progresado mucho en cuanto a igualdad de derechos y oportunidades entre los sexos. . . . Lo que a nadie podía habérsele ocurrido era que un macho humano pudiera encontrarse en la situación que durante miles de años había sido tristemente normal para todas las mujeres del planeta: tener que llevar a término un embarazo no deseado sin contar ni siquiera con el consuelo de sentirse apoyada por una pareja . . . los machos habían podido durante milenios eludir su responsabilidad mientras que las hembras, aunque, de haber podido, probablemente también lo habrían hecho, se habían visto encerradas en los límites de su propio cuerpo que llevaba adelante la tarea de la reproducción sin contar con su permiso. (64–65)

> Progress had been made in equal rights and opportunities between sexes. . . . What no one could have imagined is that a male human would find himself in the situation that, sadly, for thousands of years had been normal for all the women on the planet: to have to bring to term an unwanted pregnancy without even having the support of a partner . . . for millennia, males had been able to elude their responsibility, while women, although, if they had been able to, would probably also have done so, had found themselves confined to the limits of their own body, which carried out reproduction without their permission.

Andrade is finally held accountable for his actions and treatment of women. Having followed Andrade's arduous journey through maternity, from fertilization to the postnatal period, the novel comes full circle in the third and final part as Andrade returns to his own space station with the newborn. The novel's last lines emphasize the fluidity of sex and gender. Charlie comments, "Seguro que alguien habría empezado a darle el biberón a Lenny, que se había convertido en la mascota oficial de la *Victoria*, pero ella quería hacerlo personalmente. Al fin y al cabo era su hija. O su hijo. Y ella era su madre. O su padre. O su madre. O su padre. O . . ."(185; Surely, someone would have given the bottle to Lenny, who had become the official mascot of the *Victoria*, but she wanted to personally. In the end it was her daughter. And she was its mother. Or its father. Or its mother. Or its father. Or . . .).

By using a SF framework, having the story unfold in a different spatial and temporal dimension, and inverting the traditional reproductive roles assigned to males and females, Barceló invites readers to distance themselves from their own cultural norms in order to critically examine sociocultural realities. Throughout the novel, Barceló meticulously draws the reader's attention to the contentious dynamic between sex, gender, and motherhood. At one point, Charlie wonders whether a man automatically becomes a woman when he takes on the role of a female during reproduction. Also, by relying on appearances, Andrade mistakes the Xhroll for a female because of her breasts, when in fact the Xhroll with whom he has intercourse is male. Through estrangement, parody, and exaggeration, Barceló shows that although society has evolved in terms of scientific advancement and allowing women greater legal control over their bodies, there is still work to do in terms of people's perceptions of motherhood and the complicity between sexual difference and social power inequities.

Indeed, redefining motherhood has been fundamental to feminist movements, and within literature this impulse has led to a complex and ambivalent relationship between honoring motherhood and critiquing it. Historically, patriarchal society has assigned steadfast gender binaries to one's role in reproduction and child rearing. Thus, at times, novelists and feminists have critiqued motherhood as a source of women's oppression, while at others they have honored motherhood and a woman's commitment to the family. The conception of motherhood has transformed from a lifetime identity or synonymous relationship between womanhood and motherhood "to a role—an identity that could be taken on, thrown off, or combined with other identities" (Allen 220). Motherhood is a social construct, easily molded by political ideologies. As society becomes more accepting of alternative familial configurations, motherhood is being reconceptualized to encompass the wide spectrum of sexual identities.

While the discussion of *Consecuencias naturales* has focused on contraceptives and maternity, the analysis of Gabriela Bustelo's *Planeta hembra* will focus on how genetic engineering and cloning are revolutionizing reproductive intervention and paving the way for societal structures that offer an alternative to traditional patriarchal society.[12] Although Bustelo's second novel, *Planeta hembra*, is categorized as science fiction, elements of

Spanish Gen X narrative are pervasive in its many references to advertising, pop culture, cinema, and television. Carmen de Urioste's work traces the characteristics that unite the narratives produced by female Spanish writers in the 1990s, noting that "las narradoras de los noventa escriben textos de características posmodernas—inclinación a la subjetividad, fragmentarismo, discursividad, búsqueda de una nueva identidad—, en los cuales se indaga de manera sistemática una escapatoria al logocentrismo de la tradición patriarcal" (203; the narrators of the 1990s write texts with postmodern characteristics—propensity to subjectivity, fragmentation, discursivity, the search for a new identity—which seek in a systematic manner an escape from the logocentrism of patriarchal tradition).[13] Although *Planeta hembra* was published in 2001, it too contains many of these characteristics as it seeks to redraw sexual hierarchies. *Planeta hembra* is the story of how the political "Party XX," constituted entirely of women, has taken control of planet Earth via censorship and the manipulation of history. As Dillon, one of the female protagonists, notes, Party XX destroyed all books "para acabar con toda la Historia de otros tiempos, cuando los Hombres eran los que mandaban" (42; to put an end to the History of other times, when men were the ones that ruled). The opposing all-male party led by Graf is "Party XY." While parties XX and XY are considered homosexual parties, the third party, "Comando H" is a heterosexual party, which is seeking to overthrow Party XX. As Marta E. Altisent comments, the society present in *Planeta hembra* reminds us of Aldous Huxley's *Brave New World* in that power and success don't depend on weapons and aggression but rather on artificial intelligence and eugenics (90).[14]

While reproductive theories from classical times into the nineteenth century have typically attributed to women a marginal role in the reproductive process, Bustelo's novel overtly criticizes the past and portrays a future in which, thanks to advances in science and technology, women no longer need men for reproductive purposes. As Juan Carlos Martín notes, Party XX has employed artificial reproduction to defy evolutionary and biological processes that for centuries placed women in a subordinate position to men. *Planeta hembra* is a dystopia in which the sex-role reversal serves as a form of defamiliarization, in the sense that in the novel's matriarchal society, science has been used to elevate females rather than males to primary positions of power. However, unlike some sex-role reversal stories, which argue that society will be better when women are in power, *Planeta hembra* shows that women's total freedom and power, at the expense of the enslavement of men, leads to the destruction of civilization. The leaders of Party XX indoctrinate their members to accept their position in a utopian subculture within what is otherwise an oppressive and dystopian society. As a result, although initially women seem to enjoy more rights due to advances in biotechnology, the dystopian outcome demonstrates that Bustelo's work ultimately presents a darker vision of reproductive technology than the one present in Barceló's more optimistic novel.

The importance of artificial reproduction in sustaining the matriarchal lineage in *Planeta hembra* is evidenced by the persistent number of scientific explanations about the reproductive processes that Party XX has refined:

> Mediante ingeniería genética se realizan en el ADN las variaciones necesarias para obtener individuos con características diferentes. Se mantienen los rasgos biológicos de cada etnia, pero el número de individuos perteneciente a cada una de ellas está preestablecido. [. . .] Los detalles de este proceso de selección genética se mantienen en absoluto secreto. Dado el alto coste que supone la ingeniería del ADN cabe pensar que las modificaciones que producen la individualización y mejora genética sólo se llevan a cabo en el caso de las Hembras reales. El resto de los seres humanos, Hembras No Reales y Hombres, serían una repetición *ad eternum* de personas que ya han existido. (128)

> Through genetic engineering, the necessary DNA variations are carried out to produce individuals with different characteristics. The biological traits of each ethnicity are maintained, but the number of individuals belonging to each one is pre-established. [. . .] The details of this process of genetic manipulation are top secret. Due to the high cost associated with DNA engineering, modifications that produce individuality and better genetics are only carried out in the royal Women. The rest of the human beings, Not Royal Women and Men, would be an *ad eternum* repetition of people that have previously existed.

Given that people are replicas of beings that existed in the past, the cloning program has been tweaked to ensure that no two identical individuals will coexist at the same time. Additionally, only the embryos destined to be future leaders have the privilege to grow inside the uterus of a surrogate mother. While all of the other embryos develop in machines, these surrogate mothers have passed numerous screenings and have pledged to surrender their children to Party XX so that the perfect lineage of female leaders will continue. In the novel, the leaders of Party XX have appropriated various scientific methods as well as scientific discourse in order to justify this new world order in which women are no longer subordinates. In order to stay in power, Party XX has perfected artificial reproduction by controlling cloning and gene therapy. The group has also been able to create a population in which over 50 percent of the individuals are females, thus ensuring electoral success every four years. Furthermore, as Graf learns in one of the forbidden books, *El siglo veinte: La muerte del hombre*, (Twentieth century: The death of man), "una vez obtenidos los embriones, se gestan en máquinas perfectas, que hacen las veces de útero. De esta forma, no es necesario el embarazo ni se precisan espermatozoides. En resumen, el Hombre no sólo ha dejado de ser imprescindible, sino que podría llegar a desaparecer. Bastaría con que las Hembras decidieran dejar de fabricarlo" (119; Once the embryos are obtained, they are gestated in perfect machines, in place of the uterus. This way, neither pregnancy nor sperm are needed. In summary, the Man is no longer indispensable, and could disappear. All it would take is for Women to decide to stop creating him). The male seed has become obsolete in this society.

By depicting a society that relies heavily on science, that favors homosexual preferences, and that gives women total control, Bustelo's *Planeta hembra* offers an alternative to the traditional patriarchal system that has been present in Spain for centuries and was fiercely maintained by Franco's regime. The Francoist regime sought to build a nation based on traditional gender roles, and Franco employed the power of the Catholic Church in an attempt to control the moral trajectory of the nation. As Gema Pérez-Sánchez notes, homosexuality posed a threat to the heterosexual family, which was the backbone of the regime and consequently had to be criminalized (24). Homosexuals were considered dangerous subjects who had to be punished and reeducated. Consequently, several laws were passed including La Ley de Peligrosidad y Rehabitalización (the law of social danger and rehabilitation) in 1970 (29). After Franco's death, a new queer politics emerged which facilitated conversations about identities and sexualities that differed from the heterosexual norm. In 2005, when the PSOE was in power, after decades of fighting for the same rights granted to heterosexual couples, same-sex marriage was legalized.

In the new society that Party XX has created, several elements reverse the former patriarchal order. First and most importantly, homosexuality is considered the new norm, and heterosexuality is seen as repulsive. For example, when Báez, who ultimately becomes the leader of Party XX, is trying to imagine what life is like in Comando H, she states, "por más que intentaba entenderlo, no lograba verle el atractivo al coito heterosexual. Era el ayer. Una rémora. Un lastre superado. Un acto más asociado a los simios que a los humanos" (13; no matter how much she tried to understand it, she couldn't see the appeal of heterosexual sex. It was the yesteryear. A hindrance. A burden surpassed. An act associated more with primates than with humans). In fact, every year, all members of Party XX must undergo a routine examination whereby any indications of heterosexuality are surgically corrected. This technological imposition of sexual preference shows how genetic engineering can be used to discriminate against an entire group of people, creating even greater disproportionate social and political power. Evolutionary engineering has the potential to create warring social classes. Báez's narrow point of view toward sexuality is attributed to the widespread censorship imposed by the leaders of Party XX. Several pages in the novel show lists of words and their synonyms that have been censored. For example:

FILE 06.ZHE, LINE 00190 HOMOSEXUAL N0106B CORRUPTO (corrupt)
FILE 06.ZHE, LINE 00191 HOMOSEXUAL N0107B DEGENERADO (degenerate)
FILE 06.ZHE, LINE 00192 HOMOSEXUAL N0108B DESVIADO (deviant)
FILE 06.ZHE, LINE 00193 HOMOSEXUAL N0109B INVERTIDO (queer)
....
FILE 06.ZHE, LINE 00441 HEMBRA (FEMALE) K0103B CORTESANA (courtesan)
FILE 06.ZHE, LINE 00442 HEMBRA K0104B CRIADA (maid)
FILE 06.ZHE, LINE 00443 HEMBRA K0105B EVA (Eve). (166–67)

There is also a list of new equivalencies generated by Party XX to mislead the population into condemning heterosexuality:

FILE 06.ZHE, LINE 00458 HETEROSEXUAL W0103B ANORMAL (abnormal)
FILE 06.ZHE, LINE 00460 HETEROSEXUAL W0104B ASESINO (assassin)
FILE 06.ZHE, LINE 00461 HETEROSEXUAL W0105B DELINCUENTE (criminal)
FILE 06.ZHE, LINE 00463 HETEROSEXUAL W0106B DEMENTE (demented). (168)

The manipulation of language and history is one technique Party XX uses effectively to divide society according to sex and gender, normalizing homosexuality while pinning factions against one another. For example, in an attempt to eradicate all traces of what Party XX perceives as a masculine-centric artistic and literary world, the leaders order the burning of books belonging to authors such as Cervantes, García Márquez, Goethe, and Faulkner, disguising the act as a recycling campaign. With no primary records showing an authentic version of history, the population blindly believes the information found online, a medium controlled entirely by Party XX.

Another subversion of patriarchal society is that Party XX has both eliminated the traditional family unit, opting instead for unipersonal housing, and has also abolished the concept of marriage as a partnership. In this sense, the novel points to contemporary historical analyses of marriage. As Báez notes in the novel, the notion of egalitarian coupling tends to be a historical fallacy, since one partner generally has more power than the other (145). One partner would make more money than the other, or one partner would be more in love than the other, and it would always be an unbalanced relationship with a winner and a loser. In the text, another major difference between matriarchal society and life under previous patriarchal societies is that maternity has been abolished and replaced with artificial reproduction because maternity was considered a form of slavery disguised as a miracle of life (159). Throughout the work, both egalitarian coupling and maternity become revealed as cultural constructs, systems that view the heterosexual family unit as the ideal to which one should aspire. As Jameson argues, the family plays an important role in utopias. The family may take on different meanings, but it will always be present in utopia. It may be present in the form of a nuclear family, a single-parent household, a reproductive arrangement, or any number of alternatives (206–07). Twentieth-century utopias see the traditional family as a failure and seek to provide alternatives that are better than the present. Criticism of the family includes:

> (1.) It is foolish to think that men (and later women) will be satisfied with one or only one major sexual outlet. (2.) It is a terrible way to breed. (3.) It is a worse way to raise children. (4.) It binds women to specified and limiting roles. (5.) There is too much tension between the sensuality and passion of sexual relations and the reason and order needed in child-rearing. (6.) The patriarchal family is dystopian and must be replaced. (Tower Sargent 116)

In addition to granting greater homosexual freedom and relying on artificial reproduction, *Planeta hembra* seeks to address many of the issues regarding the traditional family unit by implementing communal child rearing. As the novel states:

> En cualquier caso, los niños, tanto Hembras como Hombres, se educaban en escuelas de las que prácticamente no salían hasta la edad considerada adulta, los quince años. No era frecuente ver niños por las calles. Cuando uno de ellos se perdía o escapaba era de inmediato recogido por los equipos de seguridad y devuelto a su escuela correspondiente. En las últimas décadas del infame siglo veinte, los niños se volvieron extremadamente peligrosos debido a las malas influencias externas. El Partido XX había decidido cortar este asunto de raíz. La educación era un asunto de máxima prioridad política, y estaba férreamente controlado por las Hembras líderes. (39)

> In any case, children, Females and Males, were educated in schools until they reached adulthood at age fifteen. It wasn't common to see children on the streets. When one of them got lost or escaped they were immediately picked up by security and returned to the corresponding school. In the last decades of the infamous twentieth century, children became extremely dangerous due to bad external influences. Party XX had decided to kill this issue at its root. Education was of maximum political priority and was strictly controlled by the Female leaders.

From Party XX's point of view, they have created an ideal utopian society where females have finally liberated themselves from centuries of male domination and have normalized lesbian identities. Unfortunately, by oppressing the male sex, Party XX has taken its power to the extreme and simply reversed the oppression.

In *Planeta hembra*, Bustelo creates a parody of feminist fundamentalism and thus invites the reader to question the ideals backing extreme feminist movements, as well as the masculinist movements they inadvertently replicate. At the end of the novel, war breaks out between all of the parties when news spreads that Báez, who like all women has been programmed to only take pleasure from other women, has had intercourse with the leader of Party XY. To give the reader a detailed image of what is considered a heinous act by a member of Party XX, the encounter is described in filmic terms: "en primer plano, la vagina depilada de Báez, abierta bajo el enorme pene de Graf, a punto de penetrarla como una ávida bomba de petróleo" (211; in the foreground, Báez's waxed vagina, open below Graf's enormous penis, about to penetrate her like an eager oil pump). This image is transmitted to all corners of the earth, and in minutes the heterosexual act leads to a war that destroys the entire planet. Despite being in Party XX, which physiologically and psychologically inculcates lesbianism from the moment of conception, Báez develops and acts on an attraction for the opposite sex, showing that sex and gender are not identities that a society can simply impose on its members, even with the most advanced scientific techniques; moreover, the novel's scientific premises are often offset by ethical digressions, primarily on the subject of eugenics.

In *Planeta hembra,* eugenics, which aims to improve the human population by controlling desirable heritable characteristics, plays a significant role in Party XX's attempts to erase history and create a society founded on science. Until Party XX came into power, society's survival depended on heterosexual intercourse for procreation. However, new reproductive technologies have made society more open to homosexual relations since alternative methods of conceiving have become readily available. Even so, Party XX's manipulation of genes, generation after generation, raises many concerns among the other factions, which is not surprising given that, since its inception, eugenics has raised many ethical questions. As Juárez González states, "aunque aparentemente digan luchar por la mejora de la raza, estas doctrinas biologizantes [la eugenesia] llevan en su aplicación al control férreo del individuo y conducen a la desvalorización del mismo, mediante el establecimiento y aceptación de escalas de valores que permiten aprobar a unos y descalificar a otros" (119; Even if seemingly they're said to fight for the betterment of the race, in their application these biological doctrines (eugenics) strictly control the individual and lead to the devaluing of oneself, through the establishment and acceptance of value scales that esteem some and discredit others). Eugenics is just one example of how scientific discourse has been appropriated by social and political movements to maintain power and subdue those individuals it deems a threat.

In both novels, via mechanisms such as contraceptives, artificial reproduction, cloning, and genetic engineering, Barceló and Bustelo deconstruct the human body at the cellular level to show how science plays a role in the creation of potential social arrangements, how new technologies question traditional definitions of "family," and how laws and reproductive models have evolved to give women greater freedom and choice. Today, methods of reproduction without heterosexual intercourse include artificial insemination by donor (AID), in vitro fertilization (IVF), and surrogate embryo transfer (SET). New reproductive technologies have forced society to re-examine social arrangements and have led to the development of new laws on abortion, maternal-fetal relations, and surrogacy (to name a few). These novel arrangements have in turn raised a new series of controversies. For example, "some critics have objected to commercial surrogacy on the ground that it improperly treats children and women's reproductive capacities as commodities," while others argue that surrogacy empowers women (Anderson 233).

The Catholic Church has also raised many questions and is likely to play a major role in how slowly or quickly progress is made in biotechnological research. In 2007 in a speech to the participants of the Pontifical Academy for Life, Pope Benedict XVI stated that attacks on the right to life throughout the world had assumed new forms:

> The pressures to legalize abortion are increasing in Latin American countries and in developing countries, also with recourse to the liberalization of new forms of chemical abortion under the pretext of safeguarding reproductive health: policies for demographic control are on the rise, notwithstanding that they are already recognized as dangerous also on the economic and social plane. At the same time, the interest in more refined biotechnological research is growing in the more developed countries in order to establish subtle and extensive eugenic methods, even to obsessive research for the "perfect child," with the spread of artificial procreation and various forms of diagnosis tending to ensure good selection.... All of this comes about while, on another front, efforts are multiplying to legalize cohabitation as an alternative to matrimony and closed to natural procreation. (Benedict XVI)

In his speech, Pope Benedict touches on many of the scientific subjects that feminist utopian projects are concerned with, including the legalization of abortion, artificial procreation, and same-sex marriage. Indeed, the fact that contemporary feminist authors and global church leaders are all concerned by these issues demonstrates the ongoing social dilemmas facing many societies and countries, not just Spain. While the novels studied here highlight the power of science, they also highlight the fact that science alone cannot bring about change; society's attitudes toward sex and gender, as well as institutional structures, must also change, and literature can play an important role in that process.

The SF novels analyzed in this chapter provide not only perspectives on the implications of biotechnology but also insights into the legal and ethical dilemmas that often accompany scientific advancement. While it is naïve to think that SF can or should serve as a moral guide on the various applications of science, the genre does offer readers greater insight and engagement with other points of view and possible alternatives. When presented with dynamic characters and future-oriented settings, the reader can reflect on possible consequences of biotechnological advancement in a way that scientists cannot, because it allows for the presentation of issues that cannot be analyzed or quantified in a laboratory but that nonetheless have a major impact on society. Spanish SF will continue to play an important role in that ongoing global discussion as societies grapple with their own patriarchal and hetero-normative pasts and look toward more inclusive and empowering futures. Novelists like Barceló and Bustelo offer a reminder that women writers in Spain are on the vanguard of these important global conversations.

NOTES

1. See Lola Robles's article "Escritoras españolas de ciencia ficción" for an analysis of the additional factors that limited SF production by women.
2. In 2018, a two-volume anthology titled *Antología de escritoras españolas de ciencia ficción: Vol. 1, Distópicas* and *Vol. 2, Poshumanas*, edited by Lola Robles and Teresa López Pellisa, was published to combat the absence of female writers from SF anthologies.
3. Though this essay focuses on two female SF authors, it is important to note that in Spain, male SF writers significantly outnumber female SF writers; for an example of a novel with a similar thematic, written around the same time period but by a male author, see Gabriel Bermúdez Castillo's *El hombre estrella* (1988).
4. In the 1980s, concurrent with second-wave feminism and feminist science fiction, feminist science and technology studies coalesced as a field out of the need to critically examine how scientific inquiry is affected by culture and consequently how science has been a source of gender-based forms of oppression. See Evelyn Fox Keller's *Reflections on Gender and Science* and Sandra Harding's *Whose Science? Whose Knowledge? Thinking from Women's Lives*.
5. Biotechnology has been the flagship of science in Spain in large part due to the legacy of Dr. Severo Ochoa, winner of the 1959 Nobel Prize in Physiology or Medicine with Arthur Kornberg (Ariza). Held biannually, BioSpain is one of the biggest biotechnology business fairs in the world. For a detailed analysis of recent biotechnology produc-

tion in Spain and how it compares to world trends, see the 2017 report compiled by FECYT (Fundación Española para la Ciencia y la Tecnología).
6. See Vanessa Knights for an analysis of how Barceló uses cognitive estrangement in her works to facilitate the reader's ability to critically examine contemporary sociocultural structures.
7. The visual era refers to the proliferation and infiltration of television, cinema, Internet, video games, etc. into daily life.
8. The first two regimes Preciado refers to are Foucault's sovereign and disciplinary societies.
9. Representative of its disciplinary roots, the first large clinical contraceptive pill trials were performed between 1956–1957 on female psychiatric patients at Worcester State Hospital and on male prison inmates in Oregon (Preciado 175).
10. The three instances were as follows: if the pregnancy posed a serious risk, either for the mental or physical health of the woman, if the fetus had malformations, or if the fetus was conceived in the context of rape.
11. Abortion was allowed during the first fourteen weeks, and in cases of fetal deformities abortion was allowed up to the twenty-second week.
12. Prior studies focus on the characteristics that Bustelo and her two novels (*Veo veo* and *Planeta hembra*) share with other writers and works of Generation X, how Bustelo uses linguistics to subvert the Spanish language's masculine norm, or how Bustelo incorporates cyberfeminism into her novels. As Nina Molinaro notes in her essay, "Watching, Wanting and the Gen X Soundtrack of Gabriela Bustelo's *Veo veo*," though Gabriela Bustelo is not as well-known as Lucía Etxebarria, she too offers a neorealist and feminizing vision of contemporary Spain via the female protagonist's journey through Madrid in her first novel. For readings of Bustelo's second novel see Elizabeth Russell for an analysis of how *Planeta hembra* depicts the city as a psychogeographical space for the construction of identity and Susan Marie Divine's dissertation for an analysis of how capitalism impedes socially transformative ideas.
13. Marina Villalba Álvarez also provides a study depicting where Bustelo's novels fall among works produced by female authors in the 1990s.
14. See Altisent's essay for a summary of how *Planeta hembra*'s themes and utopian and dystopian techniques relate to Anglo-Saxon works and feminist writings.

WORKS CITED

Allen, Ann Taylor. *Feminism and Motherhood in Western Europe, 1890–1970: The Maternal Dilemma*. Palgrave Macmillan, 2005.

Altisent, Marta E. "El mundo antitético de *Planeta hembra* de Gabriela Bustelo." *Pluralidad narrativa*, Biblioteca Nueva, 2005, pp. 89–106.

Anderson, Elizabeth. "Is Women's Labor a Commodity?" *Contemporary Issues in Bioethics*, Wadsworth Publishing Company, 1994, pp. 233–43.

Ariza, Luis Miguel. "Los revolucionarios de la biotecnología." *El País*, 16 Oct. 2016.

Barceló, Elia. *Consecuencias naturales*. Miraguano Ediciones, 1994.

———. *El mundo de Yarek*. Lengua de Trapo, 2005.

———. "La dama dragón." *Sagrada*, Sportula, 2019, pp. 197–230.

———. "Piel." *Sagrada*, Sportula, 2019, pp. 179–342.

———. *Sagrada*. Ediciones B, 1989.

Barceló, Miquel. *Ciencia ficción: Nueva guía de lectura*. Ediciones B, 2015.

Benedict XVI. "Address of His Holiness Benedict XVI to the Participants in the General Assembly of the Pontifical Academy for Life." 24 Feb. 2007, Clementine Hall. w2.vatican.va/content/benedict-xvi/en/speeches/2007/february/documents/hf_ben-xvi_spe_20070224_academy-life.html.

Bermúdez, Castillo G. *El hombre estrella*. Ultramar, 1988.

"Biotecnología en España: sector competitivo e innovador." *Marca España*, 25 Aug. 2016.

Bustelo, Gabriela. *Planeta hembra*. RBA, 2001.

———. *Veo veo*. Anagrama, 1995.

Butler, Octavia. *Lilith's Brood*. Grand Central Publishing, 2000.

Díez, Julián. *Antología de la ciencia ficción española*. Ediciones Minotauro, 2002.

Díez, Julián, and Fernando Ángel Moreno, editors. Introduction. *Historia y antología de la ciencia ficción española*, by Díez and Ángel Moreno, Cátedra, 2014, pp. 9–117.

Divine, Susan Marie. *Utopias of Thought, Dystopias of Space: Science Fiction in Contemporary Peninsular Narrative*. PhD Dissertation, U of Arizona, 2008. UMI 3352592, 2009.

Foucault, Michel. *Discipline and Punish: The Birth of the Prison*. Vintage Books, 1995.

Fox Keller, Evelyn. *Reflections on Gender and Science*. Yale UP, 1985.

Fundación Española para la Ciencia y la Tecnología (FECYT). "Producción científica española en biotecnología. 2005–2014." Ministerio de Economía, Industria y Competitividad, 2017.

Graham, Helen. "Gender and State: Women in the 1940s." *Spanish Cultural Studies: An Introduction: The Struggle for Modernity*, edited by Helen Graham and Jo Labanyi, Oxford UP, 1995, pp. 182–95.

Harding, Sandra. *Whose Science? Whose Knowledge? Thinking from Women's Lives*, Cornell UP, 1991.

Helford, Elyce. "Feminism." *The Greenwood Encyclopedia of Science Fiction and Fantasy: Themes, Works, and Wonders*, edited by Gary Westfahl, Vol. 1, Greenwood Publishing, 2005, pp. 289–91.

Hutchinson, Steven. "Mapping Utopias." *Modern Philology*, vol. 85, no. 2, 1987, pp. 170–85.

Jameson, Fredric. *Archaeologies of the Future: The Desire Called Utopia and Other Science Fictions*. Verso, 2005.

Játiva, Juan Manuel. "A la gente le gustaría tener poderes especiales y cortar cabezas." *El País*, 25 Aug. 2013, elpais.com/ccaa/2013/08/25/valencia/1377427068_268133.html.

Juárez González, Francisca. "La eugenesia en España, entre la ciencia y la doctrina sociopolítica." *Asclepio*, vol. 51 no. 2, 1999, pp. 117–31.

Knights, Vanessa. "Transformative Identities in the Science Fiction of Elia Barceló: A Literature of Cognitive Estrangement." *Reading the Popular in Contemporary Spanish Texts*, edited by Shelley Godsland and Nickianne Moddy, U of Delaware P, 2004, pp. 74–99.

Laqueur, Thomas Walter. *Making Sex: Body and Gender from the Greeks to Freud*. Harvard UP, 1990.

Le Guin, Ursula. *The Left Hand of Darkness*. Walker, 1969.

Lessing, Doris. *Canopus in Argos: Archives*. Vintage, 1992.

"Male Birth Control Study Killed after Men Report Side Effects." *All Things Considered*, National Public Radio, 3 Nov. 2016, www.npr.org/templates/transcript/transcript.php?storyId=500549503.

Martín, Juan Carlos. "Tecno-femeninas y tecno-feministas: una aportación crítica en torno a la configuración del sujeto artificial femenino en la ciencia ficción española." *Omateca*, vol. 21, 2015, pp. 114–24.

Martínez de la Hidalga, Fernando. Introduction. *La ciencia ficción española*, by Martínez de la Hidalga, Ediciones Robel, S.L., 2002, pp. 13–22.

Molina-Gavilán, Yolanda. *Ciencia ficción en español: una mitología moderna ante el cambio*. Edwin Mellen P, 2002.

Molinaro, Nina. "Watching, Wanting and the Gen X Soundtrack of Gabriela Bustelo's *Veo veo*." *Generation X Rocks: Contemporary Peninsular Fiction, Film, and Rock Culture*, edited by Christine Henseler and Randolph D. Pope, Vanderbilt UP, 2007, pp. 203–15.

Moreno, Fernando Ángel. *Teoría de la literatura de ciencia ficción: Poética y retórica de lo prospectivo*. Portal Editions, 2010.

Oudshoorn, Nelly. *The Male Pill: A Biography of a Technology in the Making*. Duke UP, 2003.

Patai, Daphne. "Defamiliarization and Sex-Role Reversal." *Extrapolation*, vol. 23, no. 1, 1982, pp. 56–68.

Pérez-Sánchez, Gema. "Franco's Spain and the Self-Loathing Homosexual Model." *Queer Transitions in Contemporary Spanish Culture*. State U of NY P, 2007.

Perriam, Chris, editor. *A New History of Spanish Writing: 1939 to the 1990s*. Oxford UP, 2000.

Piercy, Marge. *Woman on the Edge of Time*. Knopf, 1976.

Preciado, Paul B. *Testo Junkie: Sex, Drugs, and Biopolitics in the Pharmacopornographic Era*. Feminist Press at the City U of NY, 2013.

Robles, Lola. "Escritoras españolas de ciencia ficción." *Mujeres novelistas jóvenes narradoras de los noventa*, edited by Alicia Redondo Goicoechea, Narcea, 2003, pp. 179–90.

Robles, Lola, and Teresa López Pellisa, editors. *Antología de escritoras españolas de ciencia ficción*. Vol 1, *Distópicas*, Libros de la Ballena, 2018.

———. *Antología de escritoras españolas de ciencia ficción*. Vol. 2, *Poshumanas*, Libros de la Ballena, 2018.

Romero, Pedro Jorge. "Voces propias en la CF española: notas sobre algunos autores y obras entre los ochenta y el 2002." *La ciencia ficción española*, edited by Fernando Martínez de la Hidalga, Ediciones Robel, 2002, pp. 441–46.

Russ, Joanna. *The Female Man*, Bantam, 1975.

Russell, Elizabeth. "Self and the City: Spanish Women Writing Utopian Dreams and Nightmares." *Spaces of Utopia: An Electronic Journal*, vol. 2, 2006, pp. 59–71.

Sánchez-Conejero, Cristina. *Novela y cine de ciencia ficción española contemporánea: Una reflexión sobre la humanidad*. Edwin Mellen P, 2009.

Suvin, Darko. *Metamorphoses of Science Fiction: On the Poetics and History of a Literary Genre*. Yale UP, 1979.

Tower Sargent, Lyman. "Utopia and the Family: A Note on the Family in Political Thought." *Dissent and Affirmation: Essays in Honor of Mulford Q. Sibley*. Bowling Green U Popular P, 1983.

Urioste, Carmen de. "Mujer y narrativa: escritoras/escrituras al final del milenio." *La mujer en la España actual: ¿Evolución o involución?*, edited by Jacqueline Cruz and Barbara Zecchi, Icaria, 2004.

Villalba Álvarez, Marina. "Dos narradoras de nuestra época: Gabriela Bustelo y Marta Sanz." *Mujeres novelistas jóvenes narradoras de los noventa*, edited by Alicia Redondo Goicoechea, Narcea, 2003, pp. 123–30.

CHAPTER 12

Challenging Boundaries of Time, Science, and Gender

Einstein's Theory of Relativity in Marina Mayoral's "Admirados colegas"

VICTORIA L. KETZ

Time travel has held a special fascination in the human psyche since its conception, and that yearning has given rise to varied cultural productions that highlight the desire to translocate through time while simultaneously offering important social insights. Such is the case in Marina Mayoral's short story "Admirados colegas" (Admired colleagues), included in the collection *Querida amiga* (1995; Dear friend). Time travel in this work serves to critique prescribed gender norms and the historical exclusion of women from STEM fields within Spanish culture. In the short story, two female scientists from twentieth-century Spain venture back in time, and their interactions after being temporally displaced reveal important perceptions about women's roles and choices at various periods in Spanish history. In the work, academic scholars (such as the women's physics professor) and canonical literary figures (such as Lope de Vega) both reinforce and challenge gender norms, stereotypes, and limitations for the two female scholars and scientists.[1] Mayoral's story connects the history of women's participation in and exclusion from STEM fields with Spanish literary and cultural history, offering important insights about Spanish women's access to and exclusion from scientific and other academic pursuits. This study of "Admirados colegas" focuses on how Mayoral employs scientific theories and intertextuality, particularly related to theoretical physics, space-time, and Spanish literary history, in order to highlight a frustrated feminist agenda and to critique gendered norms.

The chapter will first examine how the narrative and STEM elements of "Admirados colegas" are framed within the text in order to accentuate women's exclusion from scientific fields. The story hypothesizes several implications of Einstein's theory of special relativity when two young women scientists and their male academic advisor are displaced into parallel temporal universes. Second, the chapter will analyze Mayoral's rebuke of gendered biases, particularly stereotypes positing that women have inferior intellectual abilities, especially related to math and science. The story itself follows the format of the scientific method by formulating a hypothesis, logging impartial observations, testing the hypothesis, and drawing a conclusion. Through temporal displacement, Mayoral is able to interweave scientific knowledge with social commentary on the constraints that have restricted women's roles in Spanish society. Finally, the study will explore Mayoral's vindication of female scientific agency through the protagonist's (Rosa's) authoritative knowledge, capacity, and position.

The subtle undermining of prescribed gender roles has been a constant theme in Marina Mayoral's (1942–) writing since she found her literary voice in 1979 with *Cándida otra vez*. To date she has written nineteen novels, eight short story collections, four tomes of essays, and numerous academic articles in Spanish and Galician. This prolific contemporary Spanish author's extensive literary production is multifaceted, and as Glenn notes, it "weaves together disparate voices, styles, and registers; utilizes narrative strategies borrowed from the detective novel, feuilleton, Gothic, and romance; and mixes high culture with pop" (88). Mayoral is a professor of literature at the Complutense University in Madrid, and she is thus well versed in literary criticism and theory, which she regularly infuses into her texts. As a global citizen, she also reflects on contemporary issues in her articles in *La voz de Galicia*, where she comments on a variety of topics, including feminist matters and scientific issues.[2]

In Mayoral's collection of short stories *Querida amiga*, the second of the seven narratives, "Admirados colegas," attracts readers' attention due to the presence of science-fiction elements as well as its departure from the more realistic tone and contemporary settings employed in the other short stories in the collection.[3] Yet all share a focus on the reworking of literary history, and they also adopt an epistolary form that seems to confide in the reader a sense of exposing long-kept secrets. "Admirados colegas" centers on Rosa, a female researcher in the last stage of her career, who is in the process of imparting her final research findings to colleagues at her retirement gathering. Her address, besides signaling the culmination of her work, also exposes a long-held secret, offering information that alters the knowledge gathered to date on the topic of her lifelong investigative project. In her discourse, Rosa discloses that during her youth, she and her best friend, Marta, who both studied physics, were working on a nuclear energy project under the supervision of a male faculty member named Professor Arozamena. While conducting an experiment, the two young women were projected into the past, where the encounters they had in seventeenth-century Spain, including meeting Lope de Vega and the Duke of Sessa, deeply affected their scientific and personal futures. However, each of the women reacted very differently to their temporal displacement and historical encounters. Marta responded to the experience by falling in love with Lope and deciding to remain in the seventeenth century, assuming the identity of Marta de Nevares, Lope's mysterious historical muse.[4] In contrast, Rosa chose to return to the twentieth century, but she subsequently abandoned her aspirations of obtaining a physics degree. She instead changed her field of study to Golden Age literature, motivated by a desire to understand what she and her friend had experienced and how encounters with Spanish history and scientific exploration had influenced them both.

With this plotline, Mayoral invites a deeper scrutiny of women's intellectual roles in Spanish society, both historically and in contemporary scientific and literary contexts. By selecting academia as the setting, and STEM as the focus, Mayoral is able to point to the gender inequality that has traditionally existed in the Spanish academy. Rosa and Marta study physics at the Complutense University in Madrid, one of the oldest universities in Europe. Founded in 1293 as *El Estudio de Escuelas Generales de Alcalá de Henares*, it was not until 1785 that the institution first granted a female student a doctoral degree, and even then, the degree was in Humanism and Letters.[5] Female participation in the university setting, particularly in the sciences, was and remains frustratingly unequal (Biosca and Sánchez). Wyer et al.'s essay collection *Women, Science, and Technology* addresses this issue from a global perspective, outlining how for centuries, women's intellectual pursuits and scientific ambitions have been stymied and censured.[6] Mayoral's short story presents similar themes within the overlapping context of literary, artistic, and scientific pursuits, or STE(A)M, offering the dilemma of young female scientists struggling with societal limits and gendered expectations, and finding within the artistic realm opportunities to resist the ostracism that many women in scientific fields have long faced while pursuing advanced studies.

In Mayoral's story, Rosa serves as the connection between science and literature, and her characterization brings to light the conflict of the methodologies associated with the respective fields, "one dedicated to the accurate observation of a pre-existing reality; the other to the premise that art is an illusion under the complete control of its creator" (Hayles 128). When Rosa eventually chooses to study literature over pursuing a degree in science, her decision recalls the dialectic tension sometimes posited between the sciences and the humanities.[7] Science has often been perceived as a field governed by rational thought and prescribed procedures, and historically it has been dominated by men. The humanities, on the other hand, are often viewed as a less rigorous intellectual endeavor based on subjective sensibilities, and they have historically been somewhat more open to women. This false dichotomy, at times, has led to labeling women as emotive and thereby less suited to participation in scientific pursuits. Such attitudes are apparent among the male characters in Mayoral's text, such as when Rosa recalls Marta's father's reaction to their intended field of study: "Marta y yo estudiamos juntas desde niñas. Las dos preferíamos las Ciencias, con cierto disgusto del padre de Marta" (35; Marta and I have studied together since we were girls. The two of us preferred Sciences, to the disappointment of Marta's father).[8] Rosa and Marta both recognize the authority and weight given to scientific fields, and they are acutely aware of the way patriarchal society rejects their desire to participate.

Mayoral's short story highlights how women's intellectual contributions have often been undermined, particularly in the scientific realm. As Martin argues in "Science and Women's Bodies," scientific knowledge in Western thought has been viewed as the highest form of mental reasoning, and it was only expected from men (69). In Mayoral's text, from the very onset of the story, the protagonist is discredited for her critical work. As she begins her address to her colleagues, Rosa notes, "mi prestigio académico ha ido decreciendo de forma paulatina: he pasado de 'hallazgos geniales' a 'brillantes intuiciones no documentadas,' hasta venir a dar en fantasías, novelerías y chocheces" (33; my academic prestige has been decreasing in a gradual manner: I have passed from 'genius findings' to 'brilliant undocumented intuitions,' to arrive at fantasy, novelties, and elderly musings). Indeed, she states that her last book received a poor assessment in *Hispanic Review*, the prestigious literary journal, from someone she describes as a "ratoncillo de biblioteca, criado, como quien dice, con las migajas que caían de mis manteles" (34; a library mouse, raised, as one would say, from the crumbs that fell from my tablecloths).

The bitterness of the protagonist's reflection on her decreasing stature is mitigated, however, by other elements of the narration, which subvert these negative evaluations by presenting a logically elucidated retirement speech that serves to prove Rosa's keen analytical insights and the social struggles she has faced to help those insights be recognized.[9] Though she is described as "chocha," or old and feeble-minded, by her younger colleagues, Rosa's narrative is constructed following rational language and using the steps of scientific process such that a hypothesis is formulated, a method is constructed, data is collected, and a theory is either proved or disproved. By adhering to the format of scientific inquiry, the story illustrates Rosa's ability to process thoughts with intellectual rigor, and it highlights her scientific formation via what Fox Keller ironically calls "thinking like a man" (134). Hence, the first segment of the short story, which focuses on the speech Rosa is delivering about her research, allows Rosa to establish her hypothesis. Next, she transcribes the overheard dialogue evaluating her work, which functions as data collection. The third section of the story allows her to interpret the data by connecting it to the story she relates about her younger years as a scientific researcher. The last two paragraphs of the story reference the content of the initial section of the story, functioning as the conclusion in which her hypothesis is proven via the revelation of the events that occurred during Rosa's teleportation. Mayoral not only frames the story within the steps of a scientific method, but she also bases the content of her story on a set of complex scientific theories that are, like all theories, hypothetical conjectures, and which in the past were often believed to approximate fantastical fiction. By grounding the short story with scientific theories that the reader willingly accepts as plausible, even though the lack of data makes the theories currently not provable, Mayoral challenges the negative cultural evaluation of Rosa's (and many other women's) work, arguing that what might appear to be "fantasy" can become reality, and that innovative and experimental thinking deserve recognition beyond gendered, literary, and scientific boundaries.

Mayoral's story emphasizes the scientific approach that the two female protagonists employ in their effort to make sense of and react to events, particularly their displacement in time and space. The two female protagonists conceptualize events through the lens of scientific observation. Initially, after the experimental accident, the young women fear that they are dead, but they begin to perform tests to ascertain that, in fact, they are integral living entities. Although they can see and hear each other, they dread being mere projections of an image that has visual and auditory qualities. To confirm their corporeal integrity, they engage in a series of experiments to determine if their bodily functions continue intact (38). When they encounter a group of boys and need information, they approach the issue as they would a scientific experiment via data collection from animal subjects. Thus, they decide to offer the boys a treat in order to earn their trust and gain information: "Solo tenía dos terrones de azúcar de los que dábamos a las cobayas" (39; I only had two sugar cubes of the kind we gave to guinea pigs). Recalling their scientific training, the young female scientists rely on methods of interacting with subjects as if they were in a laboratory setting, using the sugar cubes as a method of positive reinforcement. Even the women's questions about their own existence are tempered through this scientific lens. For example, Rosa's analysis of whether she is actually still alive exemplifies her methodical and objective approach to interpreting data. She determines that she has not died by identifying "[a]quel conjunto de material orgánica viva y consciente que era yo" (41; that set of live and aware organic material that was me). Her realization that she is constituted of subatomic particles occupying a particular space evokes Heisenberg's uncertainty principle, where one can only calculate to a certain limit a body's position and behavior. When the young women reach the most vulnerable point of their existence, believing their lives hang in the balance, they focus on "la misteriosa fuerza que mueve el sol y las otras estrellas" (45; the mysterious force that moves the sun and the other stars). Persistent references to the female characters' scientific acumen and perspectives demonstrate Mayoral's deliberate emphasis on the women's training and how their aptitude helps them manage the challenges of temporal and spatial displacement as well as the divisions created by gendered norms.

Time travel is employed in Mayoral's work to drive the plot, but it also highlights the complexities of quantum physics as well as the way gendered norms have determined access and reception of new and innovative ideas. Many scientists and authors have pondered hypotheses that could theoretically permit traveling to the past, but few of the scientists who have historically engaged the question of time travel have been women.[10] The most famous postulation is perhaps the Einstein Rosen Bridge, which posits a hypothetical topological feature that serves as a connection between two different points in the space-time continuum. Albert Einstein and Nathan Rosen conceptualized a tunnel with two ends at separate points in space-time. These wormholes could be traversed if exotic matter with negative energy density was stabilized, which would thus form a shortcut through time and space. To escape the wormhole would require the ability to travel at superluminial speeds in order to overcome the gravitational pull and exit at a different place in the space-time continuum. Other theories of time travel, including black holes, cosmic strings, and the Tipler Cylinder, rely on similarly complex and imaginative theories, mostly proposed by male scientists, and all unproven.[11] Thus, Rosa and Marta's studies in the field serve to highlight them as "chicas raras," or odd girls, who refuse to adhere to societal norms relating to both scientific experimentation and gender.

In the short story, Rosa and Marta demonstrate an aptitude and understanding of the complexities of quantum physics, and they are integrally involved in exploring time travel, rebuking claims of women's inferiority and offering a level of gendered parity related to the topic. Rosa describes the scientific work in Professor Arozamena's lab as dedicated to "la descomposición y reconstrucción de estructuras moleculares complejas" (35; the decomposition and reconstruction of complex molecular structures). Their time travel occurs shortly after Rosa registers higher-than-average readings on their machinery, suggesting that they were able to accelerate particle vibration at a rate approximating the speed of light. When what might be described as the Einstein Rosen bridge opens, "El profesor, que estaba frente a la pantalla, se convirtió de pronto en una masa de luz: fulguró un instante, se borraron sus contornos e inmediatamente se desvaneció en el aire" (36; The professor, who was in front of the screen, suddenly became a mass of light: he flashed for an instant, his outline erased, and immediately he vanished into thin air). At that moment, a wormhole expands, and the two student research assistants are engulfed in the tunnel. Since the three characters working in the lab were spatially distanced, they enter the wormhole at different points, and thus their exit points are altered as well. Rosa and Marta later learn that Dr. Arozamena, who

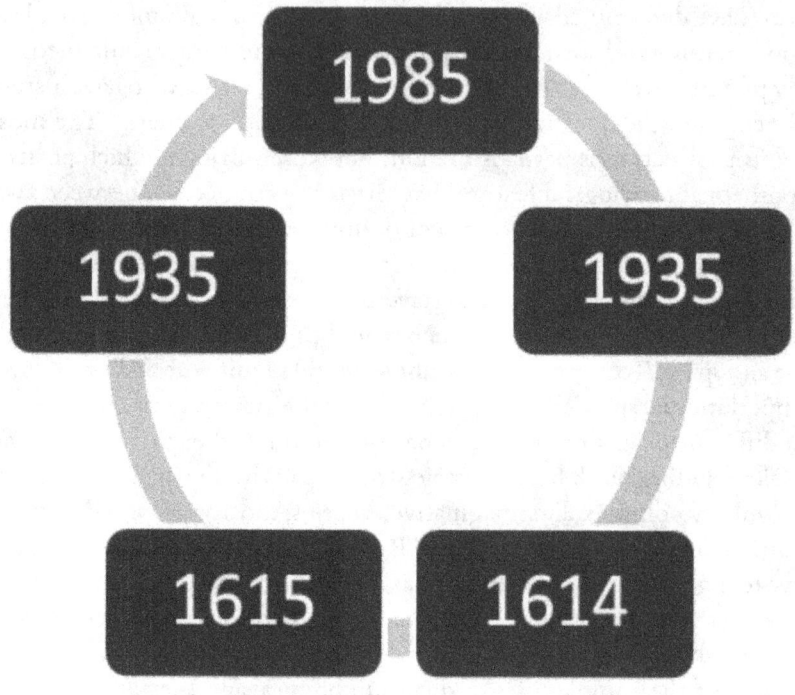

FIGURE 12.1. Temporal frames in "Admirados colegas."

disappeared before them into the tunnel, emerged a century before and assumed the identity of don Enrique de Villena (49). They themselves emerge from the wormhole in Dehesa de la Villa, a locality near the university, and though they were not displaced spatially, temporally they land in 1614.[12] Thanks to their knowledge of the Einstein Rosen Bridge, Rosa and Marta comprehend that cyclically oscillating energies might make it possible for the portal to reopen so that they can return to their point of origin, which is the unmoving or stable side of the wormhole. Marta articulates this possibility: "si el circuito de proyección seguía abierto y sin cambios, se repetiría cíclicamente hasta recuperar todos los elementos y en una de las vueltas podría recogernos" (49; if the projection circuit remained open and not changed, it would cyclically repeat itself until it recuperated all of the elements, and in one of the turns it could pick us up). The danger of time travel is present in their minds as they realize that "disintegration" at the portal is just as likely as returning home.

The time travel in the story occurs in 1935, at a time when Einstein's theories were being debated globally in nearly every scientific circle.[13] His work on the special theory of relativity appeared in 1905 in "On the Electromagnetics of Moving Bodies" with the explanation of objects moving uniformly in relation with each other. Ten years later the general theory of relativity appeared, which took into consideration the effects of acceleration and gravity on objects in motion. Einstein's theories revolutionized the concept of time and space since they disrupted the Newtonian understanding of a three-dimensional world, which reduced time to a linear succession, while Einstein posited time as a fourth dimension (Bay-Peterson 150). In the context of Mayoral's narration, there are three temporal frames referenced: 1985, 1935, and 1614, which are configured to form a circular conceptualization of time, as can be seen in Figure 12.1.

The beginning and ending of the short story take place in 1985, when Rosa is giving her retirement speech. During the speech, she flashes back to an occurrence in 1935 that shaped her life, and this timeframe is later revisited toward the end of the narrative. The central and longest portion of the story occurs in 1614, when Rosa and Marta meet Lope de Vega and the Duke of Sessa. This regression in time of three centuries is achieved through the theory of relativity. The length of time spent in each of the years also exhibits different ranges. Mirroring the time dilation, which occurs in time travel, the characters spend only an hour in 1985, a day in 1935, and a year in 1614. The compression of time is evident upon Rosa's return. She notices the metallic watches that all of them were wearing: "Los tres marcaban la hora y el día en que todo había comenzado: las doce de la mañana del veintitrés de junio de 1935" (54; The three showed the time and day that it all had begun: twelve noon on June 23, 1935). The fact that time has not changed when Rosa returns is explained by Einstein's theory of special relativity, which posits that time stops for an object when it moves at the speed of light, resulting in time dilation or the phenomenon known as the "twin paradox."[14] Yet, even though three centuries pass during the narration, Rosa and Marta do not see much of a difference in the behaviors and expectations of society. By playing with the temporal frame and creating the feeling of continuity or fusion between the different centuries, Mayoral is able to address the treatment that women have received over time.

Mayoral melds the three distinct time periods in the narrative and presents time as a continuity in the text, through a masterful handling of the linguistic elements in the short story. Constructed through linguistic parallels, repeated elements in the story allow the reader to easily contrast contemporary times to the seventeenth century. Mayoral carefully builds the story by using repeated vocabulary and evoking similar concepts in the multiple time periods. The usage of the same lexicon or variants of the words in each of these three time periods assists in making time seem constant and reinforces the notion of the fluidity of time. In the narration, the repetitive mention of people (such as Lope de Vega), as well as words (such as laboratory) and related ideas (such as academic knowledge), across the various periods produces the effect of fusing time, as can be seen in Figure 12.2. This meticulous manipulation of lexical aspects in the different temporal frames extends the original experience across time, creating a continuum while granting control to the female characters.

This gendered agency is continued when Mayoral utilizes linguistic cues to introduce quantum physics into the text. Although observation is central in every field of inquiry, this is especially true with quantum mechanics, as it is based on the positioning of the observer in relationship to the observed. In Mayoral's text, Rosa is placed in the position of the observer, that is, in the place of control or dominance. Rosa records her presence in the seventeenth century as one who carefully perceives the environment. Thus, her chronicle highlights sensory verbs such as "seeing" (36, 37, 46, 48), "hearing" (37, 39), "touching" (37, 38), and "feeling" (37, 43). When Rosa learns that her colleague Marta wants to remain in the seventeenth century with Lope de Vega, she manages the emotional import of the scene by focusing on recording it with as many of the physical indicators as possible: "El cielo está muy azul, el aire templado, el agua tranquila. Olía a mar, y solo se oía el rumor de las olas" (51; The sky is very blue, the air temperate, the water calm. It smelled like the sea, and the only sound was the rumor of the waves). The shift in verb tense here echoes the teleportation in the plot as the narrative voice moves between present and past. Mayoral emphasizes, in both diction and form, Rosa's role as scientific observer by repeatedly assigning to her verbs of cognition and reflection throughout the text. These include verbs such as "darse cuenta" (35, 37, 41, 54; realize), "mirar" (36, 37, 39, 49, 54; look), "pensar" (37, 40, 48, 50, 51, 53; think), "parecer" (37, 50; seem), "reconocer" (41, 43; recognize), "saber" (42, 45, 47; know), "constarme" (44; know), "entender" (49; understand), and "enterarse" (53; find out).

Rosa's analytic state serves as a reversal of the traditionally ascribed roles of womanhood. As Marilyn Strathern notes, "the constant rediscovery that women are the Other in men's accounts reminds women that they

FIGURE 12.2 Linguistic parallels linked to time travel

1985	1935	1614–1615	1935	1985
Medio siglo (33)	—	—	—	50 años (54)
Bibliotecas (33, 34)	Bibliófilo (35)	Biblioteca Nacional (43)	—	—
Ponencia (33), Artículos (34)	—	Poema (45), Papel escrito (46), soneto (47), cartas de amor	—	Escribiendo estos papeles (55)
Academia (34)	Ciencias (35)	Facultad (39)	—	Investigación (55)
Loca (34)	—	Loca (52)	—	Locura (55)
—	Laboratorio (35)	Laboratorio (36, 38)	Pantalla del laboratorio (54)	—
—	Brujería (35)	Herejes (42) magia (43, 45, 53) brujería (45, 46)	—	—
—	Proyección (36, 37)	Proyección (38, 48, 52, 53)	Proyección (54)	—
Lope (34)	—	Lope (42-48)	—	Lope (54)
Arozamena (34, 35)	Arozamena (35, 36)	Arozamena (47, 49)	Arozamena (54)	—
—	Masa de luz (36)	Luz del día (44)	—	—

must see men as the Other in relation to themselves. Creating a space for women becomes creating a space for the self, and experience becomes the instrument for knowing the self" (288). The objects of Rosa's observation vary, but often include men who are associated with a series of active verbs. Lope "actuó" (44; acted), "dio paso atrás" (45; took a step back), "avanzó" (46; advanced), "[se] apart[ó]" (46; moved away), and "romp[ió] en carcajadas" (47; broke into laughter), while the Duke of Sessa learned to "bailar al charlestón, a cantar tangos, a jugar al futbol y al balonvolea, a nadir estilo crawl" (48; dance the Charleston, sing tangos, play soccer and volleyball, swim freestyle) and "hacer footing" (51; jog). In opposition to the traditional gender divide of women as passive observers and men as actors, the reframing of these acts through the lens of quantum mechanics yields a different reading. The female, as the observer, becomes the scientific expert who views the subjects, males in this case, and records them enjoying frivolous freedoms. As Hubbard notes, scientists are "fact makers" and as such are endowed with the power to assert theories as laws (268).

The application of the theory of relativity in this short story is transformative as it allows for social commentary. By exploring different aspects of relativity and quantum theory, the story "offers a glimpse of a new reality, and in various ways they [the characters] are exploring the implications of that new perspective" (Kinch 289). Important to this is the transformational aspect of relativity, through which interconnectivity gives rise to a collective. As Hayles explains:

> Relativity then contains two fundamental and related implications that were to be absorbed into the field concept: first, that the world is an interconnected whole, so that the dichotomies of space and time, matter and energy, gravity and inertia, become nothing more than different aspects of the same phenomena; and second, that there is no such thing as observing this interactive whole from a frame of reference removed from it. Relativity implies that we cannot observe the universe from an Olympian perspective. Necessarily and irrevocably, we are within it, part of the cosmic web. (49)

Extrapolated from this interconnectivity implied in the theory of relativity is the Quantum Entanglement theory. This physical phenomenon occurs when particles that are in close proximity interact jointly and cannot be described independently. By evoking the Quantum Entanglement Theory, Mayoral implies that the actions of one woman affect all other women. As the short story progresses, women assume power over knowledge and challenge the traditional roles assigned to them. This allows Mayoral to remark on the roles ascribed to women, the societal views held about female scientists, and a possible way to further the feminist agenda.

By granting the female characters the position of power in the narrative, the story more sharply contrasts with the limited roles available to women. While the young women inhabit the seventeenth century, they become locked away in a palace, "sin nada que hacer sino esperar a nuestros protectores y divertirlos cuando llegaban" (48; without anything to do but wait for our protectors and entertain them when they arrived). During this period, Rosa and Marta relinquish their intellectual duties and assume a secondary role to amuse their protectors. Unfortunately, this is not the only peril to which their gender exposes them: "Si lográbamos escapar al cólera, a la peste, a la hoguera y hasta a una vulgar pulmonía, quedaba la cuestión de que éramos mujeres, es decir, aparte de los peligros del parto las fiebres puerperales y los maridos calderonianos, el sencillo hecho de que nos separaban tres siglos de los primeros brotes del feminismo" (50; If we were able to escape cholera, the plague, the stake, and even a common pneumonia, the problem remained that we were women, that is to say, that besides the dangers of childbirth, the puerile fevers, and the Calderonian husbands, the simple fact that we were separated by three centuries from the first buds of feminism). Rosa notes that, beyond the biological risks of diseases, women in the seventeenth century also struggled to overcome the sociological hazards of their sex. She exasperatedly states to Marta that "En esta época las mujeres son consideradas seres inferiores. ¡Hasta hace muy poco tiempo se discutía si teníamos alma!" (50; In this period women were considered inferior beings. Until very recently they discussed if we had a soul!). Readers observe, along with Marta and Rosa, how women have been systematically disadvantaged for centuries, and even though three hundred years have passed, the narrative emphasizes that not enough has yet changed for women. Stagnation in the progress of gender equity is made patent via the treatment of time in the text.

Due to the way that time is conceived in the work, the reader is able to take note that the treatment of women has not changed significantly. To elucidate this, it is important to study the relationship established between the sexes. When the women arrive in the seventeenth century, they are in a quandary as to whom they should turn for assistance. They consider several possibilities: "los Austrias, a partir de Felipe II, eran incompetentes y degenerados; Cervantes, manco, pobre e incomprendido; Quevedo, cojo y misógeno; Góngora, un cura malhumorado que escribía complicadísimo; Calderón, otro cura que escribía dramas de honor donde siempre mataban a las mujeres" (42; The Austrians, from Phillip II, were incompetent and degenerates; Cervantes, one-armed, poor, and misunderstood; Quevedo, lame and misogynist; Góngora, an ill-humored priest that wrote complicated verses; Calderón, another priest that wrote honor plays where they always killed women). Even though the male characters of Lope de Vega and the Duke of Sessa are portrayed as progressive, they cannot conceive of women beyond their traditional roles.[15] For example, Lope, a man who is willing to flaunt social mores by ignoring his vows of chastity, is unwilling to conceptualize the women as independent beings. Although Lope de Vega was considered advanced for his time due to the way he conducted his private affairs, it did not translate into his literary creations. The concepts in his works reinforce the patriarchal views held by the society of his time.[16] While the men enthusiastically embrace many of the practices of the future that the women teach them, they nonetheless view gender norms much more rigidly (48). When Marta decides to remain, Lope and the Duke of Sessa "[l]e consiguieron un apellido, Nevares, una familia, y un marido de pacotilla, dispuesto a servir de tapadera a unos amores ilícitos, o, mejor dicho, sacrílegos" (52; obtained for her a last name, Nevares, a family, and an inferior husband, who was willing to serve as a beard for an illicit love affair, or better said, sacrilegious). Ironically, Marta is purveyed with all of the elements of respectability so that she can continue in her love affair with Lope. Her status is normalized as she leads "una existencia legal" (51; a legal existence). For Marta to pursue her love, she must assume the role of a traditional wife. Tellingly, the woman who chooses to adopt a traditional gender role remains in the past, while the one who wishes to pursue research returns to the future. Mayoral makes patent that women have options, whether it be to follow traditional pre-established roles or to forge ahead with new ones.

In addition to imposing traditional roles on the women, the males in the story discredit the young female scientists. When Lope doubts that the women are from the future, their words do not persuade him, and they must use his own verses to validate their claim. Both Marta and Rosa gain credibility when they recite the first stanza of Lope's "Poema del alma" from his *Rimas sacras*, which he is concurrently composing (46).[17] The selection of this lyrical work for an intertextual reference is not casual. This baroque poem, written during Lope's mature years, reflects on his religious calling and morality. The quatrain they recite focuses on the poetic voice that reveals Lope's indifference to Jesus's calling over the years because of his disbelief in a higher power.[18] In this case, the verses echo the doubt that Lope has regarding the females, who astutely challenge him to accept their explanation of being from the future by paralleling it to his incredulity about Jesus's desire to save a sinner. Lope's inability to grant females validity is not isolated to his own time or to the Church's views on women's place in the social and religious hierarchy. Indeed, Rosa finds that doubts about her work and ideas follow her from the past into the future.

Another way Mayoral emphasizes women's limited roles and the devalued state of womanhood in social hierarchy is by evoking religious symbolism, which serves to mirror patriarchal beliefs. The notion of teleportation to another time period is referenced in the text as "projection." A projection is a technique that reproduces a spatial object upon another surface such as a screen. This word comes from the Latin term *proicere*, which means, "to throw forth" or in this case to launch into another dimension. In essence, Rosa and Marta are expelled from the wormhole into a new world. The fact that two female scientists, who seek knowledge, are expelled into another world evokes the biblical story of Adam and Eve. Due to Eve's curiosity, (that is, her thirst for knowledge), humans were punished by being banished from the Garden of Eden. Here the young women are expelled into an unfamiliar and harsh world where they have either to assume a role designated for women by society or leave that domain. Marta accepts a traditional role, while Rosa chooses to return to the twentieth century. This choice relegates Marta to obscurity, as little is known about Lope's lover, while it allows Rosa to blend distinct fields and potentiate the interaction of women, literature, and the sciences.

The male disapproval, prevalent in the text, demonstrates how female scientists have tended to be viewed as an oddity or worse. Women who lie outside the bounds of accepted norms are often portrayed as heretical. In the text, Rosa and Marta are accused of being witches for their knowledge: "¿Sois bruja?" (46; Are you a witch?). Just as the men here question these young women's scientific identities, so women throughout history have often been labeled negatively when engaging in scientific pursuits. For example, Wyer notes that women who practice science are often perceived as cultural aberrations:

> Though symbolic gender reveals empirical information about the everyday lives of women and men only indirectly and partially, its historical durability speaks to its power to represent an enduring cultural fit between conceptions of men and those of science—a fit that discourages, if not precludes, the full participation of women in science. The popular-culture images of women who are active agents with power over others tend to be villains or psychotics. (80)

Part of scientific innovation and inquiry involves breaking with accepted norms and introducing new ways to envision reality, but such hypotheses are often initially labeled fantasies. The first mention of fantasy in the story is pronounced early in Rosa's discourse when she speaks of how her research work is seen by the critics (33). She notes that the physics she and Marta practiced was seen as magical, since they dedicated themselves to studying "algo que entonces parecía cosa de brujería y que ahora es habitual en cualquier centro de energía" (35; something that then seemed like witchcraft and today is common in any atomic energy center). Rosa's speech highlights the antithetical relationships used to label women's experimentation, such that witchcraft is conceptualized as something performed without knowledge, while science is based on verified insight. Rosa acknowledges that the world has not always been accepting of progressive scientific ideas, resulting in a past where "[se] quemaba herejes y científicos sin distinguir unos de otros" (42; they burned heretics and scientists without distinguishing one from the other).

When they first meet Lope, he believes that the Duke of Sessa has arranged for the women to represent "comedies de magia" (magic comedies) or to participate in "una broma, pensaron que era cosa de brujería" (45; a joke, they thought it was witchcraft). When they insist that they are from the twentieth century, Lope accuses the women of being "witches" (46). Witches have traditionally been associated with unbridled female sexual power and were feared because they had the ability to defy laws of nature. Bovenschen has noted that programmed persecution of heretics and witches allowed for the Church to restore its power over society (96). It is interesting to note that when Professor Arozamena teleports, his knowledge is viewed differently. He is seen as a magician and gains the king's and the court's respect as he "practicaba magia: podía congelar el aire en forma esférica, adivinar el porvenir y hace invisibles los objetos. . . . Consiguió la protección del rey Juan II, vivió feliz y respetado, y murió en su cama de muerte natural" (49–50; practiced magic: he could freeze air in a spherical shape, prognosticate the future and make objects invisible. . . . He gained King John II's trust; he lived happily and respected and died in his bed from natural causes). Here a gender distinction is made clear as the male scientist is hailed and rewarded for his knowledge, while the female scientists are discouraged from continuing to learn and are hidden away from society.

In addition to women being represented as demonic forces, Mayoral's story highlights the way women in science have at times been portrayed as socially unstable or crazy. As Hilary Rose has noted, when females who are "intensely creative scientists" have tried to convey their theories to the scientific community without empirical evidence, they have been met with resistance. She uses the example of when the geneticist Joshua Lederberg "observed that 'the woman is mad or a genius,' he was only articulating publicly what many geneticists more privately thought" (61). At the beginning of the story, during Rosa's academic speech, she mentions that she can hear Marta's voice saying, "De ti también dirán que estás loca" (35; They will also say that you are crazy) when she confesses that her colleagues view her as demented. This notion is echoed when the young women arrive in the new spatial-temporal frame, laughing "como dos locas" (38; like two crazy women), but this time, their laughter surfaces out of their own scientific agency as well as happiness at finding themselves intact after traveling through the wormhole. While in the seventeenth century, Marta pronounces the same words that were invoked at the beginning of the story, "De ti también dirán que estás loca . . . si consigues regresar" (52; They will say that you are also crazy . . . if you manage to get back). She does so in an attempt to retain Rosa by her side in the seventeenth century instead of returning to the twentieth century to be labeled as insane. The idea of mental instability is also used as a way to cover Marta's odd behavior in the 1600s, since she does not fit into the mold of established and accepted norms: "su locura y su ceguera fueron cortinas de humo inventadas por Lope y el duque para justificar el apartamiento social de Marta y su comportamiento, extravagante en una mujer de aquella época" (55; her insanity and blindness were smoke screens invented by Lope and the Duke in order to justify Marta's social reclusion and her behavior, extravagant for a woman of that epoch). Marta could never have continued to pursue her intellectual inquiries in the seventeenth century, since her knowledge posed a threat to her existence. In the story's dénouement, Mayoral revisits the idea that the knowledge Rosa gained through her scientific investigation led to her being considered insane. Even though three centuries have passed, society's perception of women researchers has not changed enough to offer Rosa a reprieve from being labeled crazy or being excluded based on limited imaginations.

This leads Rosa to remain quiet about her knowledge, only daring to break the silence by using her retirement speech to attempt to alter society's conclusions. In her speech, Rosa uses her individual memories to try to influence the collective memory of history (Bay-Peterson 144). The fact that her work has been attacked highlights how female researchers have often been labeled as less credible in academic circles. Rosa has spent fifty years of her life attempting to prove what she knows to be true but not gaining any acceptance for her research. At the end of her presentation, she says, "Mis palabras de hoy son mi última aportación al tema" (55; My words today are my last contribution to the topic), and she resigns herself to anonymity. The barriers that Rosa experiences recall examples of scientific discoveries made by women that required male presentation to gain acceptance. For example, the scientific community only validated Marie Curie's work when her husband Pierre pressed for her recognition, and Rosalind Franklin's discovery of DNA's structure was acknowledged only recently, while male colleagues Crick, Watson, and Wilkins were awarded a Nobel Prize in 1962 for this work. This bias against acknowledging the contributions of female scientists is known as "the Matilda effect" and has plagued many women, among them Nettie Stevens, who deciphered the XY sex determination; Mary Whiton Calkins, who worked with brain stimuli recall; Marthe Gautier, who discovered the chromosomal abnormality that causes Down Syndrome; Marian Diamond, who posited brain plasticity; and Jocelyn Bell Burnell, who discovered radio pulsars. Mayoral's story uses the devaluation that Rosa has faced as a researcher to underscore the manner in which women in science have often been sidelined throughout history.[19]

The resignation that the protagonist feels at the end of her speech does not imply that women should accept the limitations placed on them by patriarchal systems. Instead, Mayoral vindicates women's expertise and perspectives by granting Rosa a critical role as scientific observer. The text emphasizes the relativity of time and space and thus highlights how both are dependent on the location and state of motion of the observer. Therefore, the act of observation becomes an act of control and not one of submission in the story (Coale 438). Rosa, by being granted the role of observer, becomes the creator of knowledge and is able to share her wisdom with others (Mulvey 438). By using linguistic elements to create continuity, Rosa establishes the primacy of her narrative over time. By interjecting the past into the present in hopes of redeeming her narrative, she allows the past to critique cultural structures and allows readers to determine the significance of her theories (Hayles 121). Rosa, by recreating the past, rewrites history to make sure that *herstory* is heard.

In "Admirados colegas," Mayoral employs scientific theories not only to make her tale of time travel plausible but also to critique gendered norms and exclusions related to scientific women. Science, in addition to facilitating the development of the plot, serves to reveal established perceptions of women's roles in society. The temporal displacement of the two women scientists highlights how women throughout history have been relegated to traditional roles or silenced and ignored in order to maintain the status quo. Indeed, the story underscores how women's thirst for knowledge is often repressed as a sign of mental instability or wickedness. By juxtaposing two time periods, the seventeenth century and the twentieth century, Mayoral is able to highlight how women intellectuals and scientists have been excluded and discounted over time.

Yet, at the same time, Mayoral supports women who thirst for knowledge in society. She vindicates them by positioning them as observers or referents to challenge accepted perceptions. In addition, by utilizing the scientific process in the text, she opens up a realm of new possibilities. Quantum theory serves to demonstrate the way scientific experimentation pushes humanity to break away from prescribed norms of thought, including ideas that denigrate women scientists and researchers. Kinch believes that "quantum physics teaches us about the subjectivity inherent in nature; characters who appreciate this lesson and apply it to their own lives can liberate themselves from the patterns. . . . Therefore, quantum physics, if we internalize its deep meaning, will lead to our creating a newer, more peaceful world" (305). The exploration of quantum theory in Mayoral's work accomplishes just that. It undermines traditionally held worldviews allowing for a new vision and perspective on female roles in both science and literature.

NOTES

1. As part of the plot, the female protagonist uncovers previously unknown details about Lope de Vega and his enigmatic lover and muse, Marta de Nevares, who appears in the famous playwright's works as the enchanting Amarilis.
2. In *La voz de Galicia*, Mayoral has written about feminist concerns including gender violence, female sex toys, and the #metoo movement. Her scientific preoccupations have led Mayoral to write about climate change, vaccinations, and the extinction of species.
3. The others concern: a frustrated first love, a petition for advice from an author suffering from writer's block, a woman asking for forgiveness from her friend for meddling in her sentimental relationship, a student requesting to present a thesis on Fernán Caballero's true story, a man revealing to the doctor the reason why he took his bus and ran it into a hotel, and a final story of the husband of the woman from the first story revealing that she misrepresented her infatuation.
4. According to the *Real Academia de la Historia*, few details about this enigmatic figure remain. It is believed that Marta de Nevares met Lope de Vega in 1616 at a poetry recital when she was twenty-five years old. They lived together and had a daughter in 1617 and a stillborn son in 1619. Until her death in 1632, they lived under constant scrutiny since she was married to another and he was a priest ("Marta de Nevares Santoyo").

5. In 1785, María Isidra de Guzmán y de la Cerda received a PhD in Humanism and Letters and was granted an appointment to teach at the university after her degree conferral. For more see Smit.
6. This is particularly evident in the insights provided in the essays by Fox Keller, Sands, Eisenhart, and Holland (in Wyer et al.).
7. Connections between literary creation and scientific experimentation recall that both fields posit hypothetical ideas and both rely on human interaction and observation to determine the value and implications of their creative experiments. In the case of Mayoral's text, the work integrates science-fiction themes, particularly those related to theoretical physics, with literary and historical meta-analysis. Numerous scholars have explored the integration of scientific theories in contemporary fiction, particularly works that employ the genre of science fiction, and critics like Susan Strehle and N. Katherine Hayles have even applied quantum theory more broadly to the general analysis of literature. Strehle has posited a "quantum universe" based on the shared characteristics (statistical, discontinuous, energetic, relative, subjective, and uncertain) that she finds coexisting in science and literature. Hayles has analyzed contemporary works of fiction through the application of the "field concept" to show that there exists a "cultural matrix" that is grounded in contemporary thought evident in both science and literature (15). This line of inquiry has led critics such as Argyros to argue that "literature is a microcosmic representation of the natural hierarchy out of which it emerged. It is after all, composed of fundamental particles (words, phonemes, graphemes, traces, marks or whatever one chooses to postulate as the microstructure of literature) whose subsequent organization yields a stratification of increasing complexity and self-reference" (36). Mayoral's work fits clearly into the above theoretical descriptions, depending heavily on scientific theories and using them to question important aspects of cultural experience, in this case, gendered norms.
8. All translations are mine.
9. The reference to *Hispanic Review* critiques not only the gendered bastions of traditional literary criticism but also the way criticism operates in the contemporary marketplace mired in theoretical approaches as opposed to literary and lived experiences. Additionally, it calls to mind distinctions between Peninsular and North American critical circles.
10. For more on the theories for this phenomenon see J. Richard Gott's *Time Travel in Einstein's Universe: The Physical Possibilities of Travel Through Time*, Martin Gardner's *The New Ambidextrous Universe: Symmetry and Asymmetry from Mirror Reflections to Superstrings*, or James Gleick's *Time Travel: A History*.
11. Another possibility for reverse time travel is through a black hole, which is a region of space-time that holds a strong gravitational effect absorbing objects outside into its interior, in the form of a vacuum. The conceptualization of strong unescapable gravitational fields found in a black hole originated in the eighteenth century with the work of John Michell and Pierre-Simon Laplace. In 1916, Karl Schwarzschild utilized general relativity to postulate the existence of a black hole, but it was not until 1958 that David Finkelstein was able to calculate the inescapability of such a field. Modern theorists such as James Bardeen, Jacob Bekenstein, Branden Carter, and Stephen Hawking

have continued to pursue studies on this phenomenon. There is also the theory of cosmic strings, one-dimensional topological defects formed during the earliest moments of the universe's evolution, serving as ways to time travel, which was sponsored by the work of Richard Feynman, Thomas Kibble, and Edward Witten. These strings, which are like phase transitions, remain out in space and are believed to link the eleven theorized dimensions. Insertion into one would allow an object to be displaced into another dimension. Finally, the invention of a time machine that would be able to accelerate to superluminal speed could permit time travel. Stephen Hawking mathematically proved the operational feasibility of a machine like the Tipler Cylinder. This final option has been the one most frequently embraced by fiction and was used in movies including *The Time Machine*, *Star Trek*, and *Avatar*. For more information please see Gleik pp.189–244 and Gott pp. 76–130.

12. Equally as important is the date of the teleportation, which occurs on June 23. This date coincides with the eve of St. John's feast, which has always had a relation with the supernatural. Typical of this celebration are the burning pyres, or *hogueras*, which have a purifying effect, six months after Christ's birth. Prior to the designation of this as a Christian holiday, it was known as the Celtic festival of *Alban Heriun*, which celebrated summer solstice. The Celts believed it to be a magical time when people were able to see their futures.

13. Albert Einstein had a connection to Spanish intelligentsia that is not widely known. The Complutense University of Madrid was the first institution to award Einstein the degree of Doctor of Science Honoris Causa on February 28, 1923. At the time of this visit, Einstein's theories were supported by all significant Spanish scientists, thanks to Levi-Civita's dissemination of them: "This unanimity of support was rare, if not unique, in European scientific circles" (Glick and Sánchez Ron 357–58). When Einstein entered exile from Germany, he intended to spend time as a visiting professor at various institutions, including ones in Leiden, Oxford, Paris, Caltech, Princeton, and Madrid. During the early 1930s, Spain was viewed as a safe haven, attracting many members of the European intelligentsia due to the rise of fascism in other continental countries. In April of 1933, Fernando de los Ríos, Minister of Education and Arts, offered Einstein a leadership role at the soon-to-be-named Instituto Einstein at the University School of Science. He declined the position due to the rapidly changing political climate in Europe and opted to accept a similar offer from Princeton University instead.

14. The twin paradox posits that if one twin travels in outer space, he will return younger than the twin who remained on earth since his bodily functions have slowed down during his journey due to the velocity at which he has traveled. For more, see Reynolds (164–65).

15. The election of these characters to be the ones the women choose to reconnect with in the past is not arbitrary. The figure of Lope de Vega has always held great respect and interest among the Spanish population. As a case in point, in the 2015 science-fiction television show *Ministerio del tiempo* (The ministry of time), written by Javier and Pablo Olivares for RTVE's Channel 1, the character of Lope de Vega has a reoccurring role. This immensely popular series centers on the exploits of the fictional

Ministry of Time's newest patrol, formed to correct anomalies so that current history is not altered, and one of the show's key features is the use of complex historical perspectives to highlight how cultural norms have shifted over time. In season 1, episode 2, the ministry's patrol is in charge of making sure that Lope embarks in the Spanish Armada. In season 2, episode 3, he is a character that is key to helping make sure that Cervantes's publication woes do not interfere with El Quijote's 1605 publication date. In season 3, episode 5, Lope de Vega appears again as the team has to ensure that the Treaty of London (1604) is signed.

16. In 2019, Yolanda Pallín's drama *Querellas de Lope y las mujeres*, performed at the University of Navarra, explores the complex relationship that this playwright had with the female figures he created for stage.
17. Coincidentally, Lope was in the midst of composing the very poem the girls recite when the young scientists arrive. He believes them to be witches because he thinks that they are reading his mind.
18. These lines are reminiscent of St. Augustine's Confessions VIII 12 in the fact that he does not heed the calling.
19. Notably, in 2020 three women received Nobel prizes, including the first all-female team. Andrea Ghez, along with two male colleagues, won the 2020 Nobel in Physics, and Emmanuelle Charpentier and Jennifer A. Doudna received the first all-female Nobel in 2020 (Chemistry) for their work developing the protocols of CRISPR technology for gene manipulation. It is important to note, however, that (as the argument above alludes) even the amazing accomplishments of Charpentier and Doudna have not come without controversy and debate, particularly as relates to their male colleagues (Begley).

WORKS CITED

Argyros, Alexander. "Deconstruction, Quantum Uncertainty, and the Place of Literature." *Modern Language Studies*, vol. 20, no. 4, 1990, pp. 33–39.

Augustine, of Hippo, Saint. *The Confessions of Saint Augustine*, edited by Michael P. Foley, translated by F. J. Sheehy, Hackett Publishing Company, 2006.

Bay-Peterson, Ole. "T. S. Eliot and Einstein: The Fourth Dimension in the *Four Quartets*." *English Studies*, vol. 2, 1985, pp. 143–55.

Begley, Sharon. "Three CRISPR Scientists Win Prestigious Award, Fanning Controversy over Credit." *Scientific American* May 31, 2018. www.scientificamerican.com/article/three-crispr-scientists-win-prestigious-award-fanning-controversy-over-credit.

Biosca, Patricia, and J. M. Sánchez. "Las mujeres lideran la tecnología en España pero aún queda mucho trabajo para lograr la igualdad." *ABC*, 8 March 2019, www.abc.es/tecnologia/redes/abci-mujeres-lideran-tecnologia-espana-201803072229_noticia.html.

Bovenschen, Silvia. "The Contemporary Witch, the Historical Witch, and the Witch Myth: The Witch, Subject of Appropriation of Nature and Object of Domination of Nature." *New German Critique*, vol. 15, 1978, pp. 82–119.

Coale, Samuel Chase. "Psychic Visions and Quantum Physics: Oates's Big Bang and the Limits of Language." *Studies in the Novel*, vol. 38, no. 4, Winter 2006, pp. 427–39.

Fox Keller, Evelyn. "Gender and Science: An Update." *Women, Science, and Technology: A Reader in Feminist Science Studies*, edited by Mary Wyer, Mary Barbercheck, Donna Geisman, Hatice Orun Ozturk, and Marta Wayans, Routledge, 2001, pp. 132–42.

Gardner, Martin. *The New Ambidextrous Universe: Symmetry and Asymmetry from Mirror Reflections to Superstrings*. Dover Editions, 2005.

Gleick, James. *Time Travel: A History*. Vintage Books, 2016.

Glenn, Kathleen. "Reading Postmodernism: The Fiction of Cristina Fernández Cubas, Paloma Díaz-Mas and Marina Mayoral." *South Central Review*, vol. 18, no.1–2, Spring-Summer 2001, pp. 78–93.

Glick, Thomas F., and José M. Sánchez Ron. "Science Frustrated: The 'Einstein Institute' in Madrid." *Minerva*, vol. 44, no. 4, Dec, 2006, pp. 355–78.

Gott, J. Richard. *Time Travel in Einstein's Universe: The Physical Possibilities of Travel Through Time*. Houghton Mifflin, 2002.

Hayles, N. Katherine. *The Cosmic Web*. Cornell UP, 1984.

Hubbard, Ruth. "Science, Power, and Gender: How DNA Became the Book of Life." *Women, Science, and Technology: A Reader in Feminist Science Studies*, edited by Mary Wyer, Mary Barbercheck, Donna Geisman, Hatice Orun Ozturk, and Marta Wayne, Routledge, 2001, pp. 265–71.

Jacobus, Mary, Evelyn Fox Keller, and Sally Shuttleworth, editors. *Body/Politics: Women and the Discourses of Science*. Routledge, 1990.

Kinch, Sean. "Quantum Mechanics as Critical Model: Reading Nicholas Mosley's Hopeful Monsters." *Critique*, vol. 47, no.3, Spring 2006, pp. 289–308.

"Marta de Nevares Santoyo." Real Academia de Historia, 2018. dbe.rah.es/biografias/6963/marta-de-nevares-santoyo. Accessed 15 September 2020.

Martin, Emily. "Science and Women's Bodies: Forms of Anthropological Knowledge." *Body/Politics: Women and the Discourses of Science*, edited by Mary Jacobus, Evelyn Fox Keller, and Sally Shuttleworth, Routledge, 1990, pp. 69–82.

Mayoral, Marina. *Querida amiga*. Alfaguarra, 2002.

Mulvey, Laura. "Visual Pleasure and Narrative Cinema." *Feminism: An Anthology of Literary Theory and Criticism*, edited by Robyn R. Warhol and Diana Price Herndl, Rutgers UP, 1991, pp. 432–42.

Olivares, Javier, and Pablo, writers. *The Ministry of Time*. Season 1, episode 2, "Time of Glory." Directed by Abigail Schaaff, featuring Rodolfo Sancho, Aura Garrido, Nacho Fresneda, and Victor Clavijo. Aired 2 March 2015. RTVE, 2015.

⸻. *The Ministry of Time*. Season 2, episode 3 "Time of *hidalgos*." Directed by Abigail Schaaff, featuring Hugo Silva, Aura Garrido, Nacho Fresneda, and Victor Clavijo. Aired 29 February 2016. RTVE, 2016.

⸻. *The Ministry of Time*. Season 3, episode 5, "Time of Splendor." Directed by Oskar Santos, featuring Hugo Silva, Aura Garrido, Nacho Fresneda, and Victor Clavijo. Aired 29 June 2017. RTVE, 2017.

Pallin, Yolanda. *Querellas de Lope y las mujeres*, directed by Ernesto Arias, Museo Universidad de Navarra, 25 October 2019.

Reynolds, Teri. "Spacetime and Imagetext." *Germanic Review,* vol. 73, 2001, pp. 161–74.

Rose, Hilary. "Nine Decades, Nine Women, Ten Nobel Prizes: Gender Politics at the Apex of Science" *Women, Science, and Technology: A Reader in Feminist Science Studies,* edited by Mary Wyer, Mary Barbercheck, Donna Geisman, Hatice Orun Ozturk, and Marta Wayne, Routledge, 2001, pp. 53–67

Smit, Theresa Ann. *The Emerging Female Citizen: Gender and Enlightenment in Spain*. UC Press, 2006.

Strathern, Marilyn. "An Awkward Relationship: The Case of Feminism and Anthropology." *Signs,* vol. 12, no. 2, 1987, pp. 276–92.

Strehle, Susan. *Fiction in the Quantum Universe*. U of North Carolina P, 1992.

Wyer, Mary, Mary Barbercheck, Donna Geisman, Hatice Orun Ozturk, and Marta Wayne. *Women, Science, and Technology: A Reader in Feminist Science Studies*. Routledge, 2001.

CHAPTER 13

Technological Portrayals

Framing Fernandinas in the Colonial Context through Photography and Press during the Spanish Second Republic

INÉS PLASENCIA

During the Spanish colonization of contemporary Equatorial Guinea, early technologies such as photography reflected and participated in the construction of the colonial notion of gender, ethnicity, and race, which implied the physical but also visual separation of the roles assigned to men and women, both white and black. This study analyzes how *Fernandinos*, the African elite of Fernando Poo, are represented via photography, charting the construction of gendered and racial norms via technological invention. In particular, *Fernandinas* (the female African elite) were introduced in the visual accounts of the colony at the intersection of colonialism, new technologies, gender, and race. Photography was employed not only as an early technology for describing, documenting, and illustrating historical processes but also as a way of consolidating hierarchies and evidencing power relations and tensions. This essay problematizes how colonialism dealt with gender and "racial difference" in the context of an elite African social class that challenged the hegemonic colonial visual culture and how early technologies in Spanish colonialism narrated, manipulated, and therefore constructed the social order.

Colonizing Difference

Colonialism in Equatorial Guinea, as in many parts of the world, operated under binary categories such as colonized and colonizers, black and white, women and men; however, it was also deeply characterized by the hierarchical separation of colonized agents themselves and by the definition of "Indigenous" in racialized and gendered terms as a strategy to reinforce colonial domination. In this context, photography served as a technology of colonization and was used as a privileged medium by European powers to colonize African territories, not only physically or politically, but also epistemologically. Indeed, in the colonial context photography became a "technology of power," as Michel Foucault would describe it—an instrument for the exercise of domination via institutions, symbols, and archives (131). It functioned as a technology of power not only in the moment the photograph was taken but also through the long life of its dissemination, when an image became part of an ideological and cultural visual frame and its potential meanings were achieved or manipulated. Indeed, photography was a visual technology that produced images, which is a notion substantially different and separate from the photographic referent, and thus a tool with its own life. As Vilém Flusser argues in *Towards a Philosophy of Photography*, photography is an apparatus with its own rules, and Flusser deliberately employs the expression "apparatus" in order to highlight the Latin origin of the term, from the word *apparare*: to prepare (21). Photography, understood as an apparatus of preparation, can be manipulated across broad time periods and contexts. It thus created within the colonial context (via its presence in a particular encounter) a certain ambience and, thereby, a unique agency in the construction of extremely hierarchical political, social, and gendered relations. The introduction of this technology enabled a new way of marking people's place in their social structure but also a new way of perceiving and interpreting reality.

The former Spanish Guinea is characterized by its pre-colonial and colonial diversity, and the region is not an exception in the use and importance of images as part of the colonization process. However, photographs from the colony have remained relatively unstudied, both in Equatorial

Guinea and Spain, until recently. Photographs from the colonial period must be analyzed not only as a way to make the colonial past visible but also as a way to understand the complexity of colonial social and gender relations by locating the technological mechanisms at the center of their construction. At the same time, many other technologies and issues must be considered to deepen the visual analysis of the colony, since both technical and iconographic conditions were part of what one might call the "photographic encounter," referring to the social dimension of photography as a technological act that creates new relations and offers its own unique implications.

In order to understand the significance of colonial-era photography from Spanish Guinea, it is helpful to grasp the historical context of Fernando Poo and the Fernandinos, or the African elite, in the region. Fernando Poo and Corisco, an island and part of what would become postcolonial Equatorial Guinea after independence in 1968, came under Spanish rule beginning in 1777, but it is not until 1858 that one finds an active Spanish Colonial administration in the territory. Before that year, the Spanish presence and colonizing efforts were marked by intermittent activity and a lack of national investment, with limited and unsuccessful expeditions, such as one by Miguel Martínez y Sanz. This absence on the part of the Spanish government enabled the settlement of Fernando Poo by the British government and the creation of the city of Port Clarence in 1827. The English port city was founded in the context of British pursuit of slave traders after slavery's abolition in 1807, and Dolores García Cantús points to the irony that while the Spanish had colonized the island of Fernando Poo in order to support the slave trade, the British built the first city and port as part of efforts to abolish the institution: "por ironías de la historia, Fernando, buscada por los españoles como centro abastecedor de esclavos, se convirtió durante la ocupación británica en un centro de emancipación de los mismos" (160; ironically, Fernando Poo, founded by Spain as a slave supply center, became during the British occupation a center for their emancipation). Port Clarence became the first urban settlement on the island of Fernando Poo, inhabited until then by the Bubis, a Bantu group and diverse community that had lived on the island they called Oche in the North and Oricho in the South since at least the fifth century (Sundiata 14). For the British, Fernando Poo was a strategic point from which to control commerce at the Niger Delta and, at the same time, to compete with France for the assault on West Africa. Although Spain maintained sovereignty and "rights" over the island, British presence and influence is especially important for the focus of this essay.

Fernando Poo became the first urban space in the region settled by Britons and also by freed African slaves and people from Sierra Leone, and these groups would become the African Creole elite of Fernando Poo: the *Fernandinos* or *krio*. This community, strongly marked by British culture in language, fashion, and traditions, would be key to the economic development of the island due to the cultivation and commerce of palm oil and to the later production of cocoa. The Fernandinos were also at the intersection between African traditions and a discourse about modernity that came from Europe and included new technologies such as photography as well as an ensemble of visual codes. Fernandinos were the group that most used the new photographic technology among the African people living in Spanish Guinea during the nineteenth and the first decades of the twentieth century.

In 1843, some years after the British abandonment of Fernando Poo (caused by the impossibility of buying the territories from Spain), Spanish colonization saw a resurgence. Juan José Lerena y Barry explored some of the territories, signed a treaty with the leader of Corisco, and visited Rio Muni, which was located in the continental area of the colony. He quickly changed the name of the capital from Port Clarence to Santa Isabel as a tribute to a Queen who admired both colonialism and photography. But he found very limited signs of Spanish presence or influence: he noted only two Spanish people among a substantial population comprising Bubis, British, Fernandinos, and Baptist missionaries. Conflicts and tensions between the different actors were commonplace, but what was most intriguing was that Port Clarence's particular history offered the Creole society a unique cultural space and economic role that developed not under a dominant colonial power but rather in a relative power vacuum, first due to the lack of a Spanish presence and subsequently influenced by a British settlement developing without property rights (Sundiata 111–118).

This unique backdrop of an elite African social class and Creole society must be considered in order to understand the development of photography in Fernando Poo and its implications for gendered representation. Researchers such as Jürg Schneider and Erin Haney, who focus on the history of photography in West Africa, have shown that the dynamism of the medium in the region was due not so much to Europeans, as some might assume, but rather to Africans, particularly in their roles as photographers but also as clients. In general, clients in Fernando Poo came from varied groups, including Spanish officials and colonists linked to the colonial administration, other Europeans and Fernadinos with commercial interests on the island, and British people linked to Baptist missions. Claretian missionaries, from Spain, were another key agent in the production of photography after their arrival and settlement on the island in 1883. Photography became a crucial tool of colonization: it was used by missionaries to document their practices and by colonists to record Catholic weddings, incursions into the mainland, and expeditions. Photographic images were even used to establish new relations with unknown residents of surrounding communities. The diversity of the former Spanish Guinea and of the photographic encounter in Fernando Poo is key to the historiographical tension evident in the images. Yet in the colonial era, incredible diversity was regularly reduced to power binaries, separating colonizers and colonized agents in terms of race, ethnicity, and gender. This was made possible through visual distinctions, both physical and via other external signs such as clothing, but it also included separation by spaces of life and labor, where origin, gender, and ethnic group were decisive for driving radical segregation.

While debates about the construction of African ethnicity are not the focus of this study, it is important to note that many photographic images produced in the colonies seemed obsessed with creating distinctions between European and autochthonous populations. Such distinctions became a primary focus for a large number of photographs, as evidenced via their titles, including generic and pejorative labels such as "mujeres bubis" (Bubi women), "Hausas" (referring to the largest ethnic group in sub-Saharan Africa), and "Congo boy."[2] Binarism was clearly dominant: acculturated or "civilized" versus "tribal"; men versus women; European versus African; plantation and home; sewing and carpentry (professions of mission-educated Africans); private versus public. Archives, magazines, and photographic collections are full of images that evince this troubling simplification.

In light of the complex uses and power structures surrounding photography in the colonies, it is important to acknowledge that while some individuals, including Fernandinas, were able to determine their relationship with the new technology, others were portrayed without their consent or initiative. Indeed, many Africans were photographed by Europeans in ways meant to incarnate Western fantasies about Africa and reinforce notions of the local population's cultural inferiority so as to justify colonization. However, there were also African subjects who were able to take some profit from this encounter, as was the case with leaders like the Bubi King Moka, who employed photography in an effort to contain Spanish colonial violence against the Bubis (Plasencia 204). It is also significant that within the clearly dualistic photographic material, particularly as it was divided into private versus public spheres, both Africans and Europeans were distributed in diverse ways across divided social and physical spaces. Indeed, in Fernando Poo, only one African community's images remained relatively private until the decade of the 1930s: those belonging to the Fernandinos.

A Place "In-Between"

The presence of Fernandinos in Spanish national archives during the nineteenth century via photographic media is surprisingly rare, and this absence can be explained in part by a colonial discourse sustained via the construction of a selective narrative that did not wish to emphasize the possibility of an African elite. However, it is helpful to recognize the role played by other cultural forces of the time, particularly Britain's dominance in the region, and also that photographs were commonly held in private collections and thus part of a cultural network distinct from standard media.[3] While Spanish archives appear to eschew the existence of Fernandinos, British archives hold multiple photographs of the African elite, such as the John Holt archive in the Liverpool Record Office. The Fernandina photographs present in the Holt archive take the form of *cartes de visite*, photographic objects that were usually offered as gifts to establish or reinforce a social relation.[4] It is clear from the content of the archive that John Holt, one of the wealthiest businessmen of Fernando Poo, was socially well connected within the region. However, the presence of four portraits of Fernandinas in Holt's archive raises questions not only about the limited presence of such photographs in the Spanish archives, but also why all four Fernandino portraits in Holt's collection would be representations of women (see figs. 13.1–13.4).

Considering the status that the use of *cartes de visite* implied during the period, it is likely that the Fernandinas themselves were the agents of dissemination of these expensive and culturally valuable photographic forms, highlighting the economic status of the African elite in Fernando

FIGURE 13.1. *Miss R. L. Williams of Fernando Poo, daughter [unreadable text] Barleycorn*. Shadrack St. John (ca. 1880). © Liverpool Record Office, Central Library, 380 HOL I, 11/3/11–13.

FIGURE 13.2. *Miss A. W.* Shadrack St. John (ca. 1880). © Liverpool Record Office, Central Library, 380 HOL I, 11/3/11–13.

FIGURE 13.3. *G. B. Williams of Sierra Leone* (ca. 1880). © Liverpool Record Office, Central Library, 380 HOL I, 11/3.

FIGURE 13.4. *Mrs. Betsy Top.* Francis W. Joaque (ca. 1885). © Liverpool Record Office, Central Library, 380 HOL I, 11/39.

Poo. The visuals in the photographs themselves further mark Fernandino financial success: note for example the luxurious dresses that two of the women are wearing, highlighting their economic class and social status. The images help to clarify the extent to which Fernandinos were incorporated as part of the elite society of Fernando Poo. Holt himself was personally linked to Fernandinos, since his business provided services for Fernandinas, including one who was the wife of the British officer Lynslager in a mixed-race marriage that was not uncommon on the island. Considering the prevalence of inter-racial marriage and socialization, it is not surprising that Holt held photographs of the African elite in his archive. However, one still might question why Holt only kept portraits of women. One possible explanation might be that the photographic selection was related to Holt's social and family links, and particularly the social contacts of his wife. Regardless, the most intriguing element of the Holt archive for this study is its contrasting content and purpose vis-a-vis the Spanish archives. The presence of multiple Fernandina images in Holt's archive highlights the absence of similar *cartes de visite* related to the African elite in the Spanish archives. Indeed, when photographs of Fernandinos do appear in Spain, the images reveal that they were created and disseminated for a quite distinct purpose: not as social capital in a surprisingly egalitarian racial context, but rather as part of the public press serving to forward Spanish colonial ambitions. In order to understand the contrasting context and purpose of the photographic representation of Fernandinos in Spain, it is essential to analyze the few images of Fernandinas present in the Spanish national archives.

Two particular images of Fernandinas in the Spanish archives warrant attention in the context of gendered representation and the use of a photographic genre known as typing. The first one is a postcard of a formally dressed Fernandina published in the 1920s in the bi-weekly *La Guinea Española* (published from 1903–1969) and edited by current and former colonists. This full-length portrait of a Fernandina in formal dress was likely not originally intended for publication.[5] The woman stands facing the camera, but with her gaze to the viewer's left, her hand on a chair, and tall foliage in the background. The Fernandina appears regal, but the expression on her face and the wandering gaze also make her look uncertain. If one analyzes the archive as the result of what Elizabeth Edwards and Janice Hart describe as "affective decisions" in *Photographs Objects Histories: On the Materiality of Images*, the inclusion of this image of the African

FIGURE 13.5. *Album, no title* (ca. 1930–1940). Ministerio de Educación, Cultura y Deporte. Archivo General de la Administración, AGA-15, Fondo 4, caja 81/06822, GUINEA 553, exp. 41.

elite within the Spanish archive is exceptional (49). Images of emancipated African women, even when being categorized by "type," did not appear frequently within colonial society.

The second image, from the 1930s, also portrays a woman standing alone, but this time the image is a close-up, and the Fernandina is gazing directly at the camera (see fig. 13.5). This second portrait is part of an album linked to the Spanish colonial administration, a sort of catalogue containing economic and ethnographic descriptions of Spanish Guinea. The page of the album where this photograph appears contains four photographs of women representing a range of groups: two of the women pictured on that same page are labeled "Indigenous"; the last one is described as "mujer pamue" (pamue woman), the name used for the Fang ethnic group in colonial times. Two of the images, that of the Fernandina and also a group picture, describe the women as "bellezas" (beauties). The album's labeling seems to both sexualize the women portrayed and also to erase their cultural and class differences and identities, linking and collapsing the complex social structures behind the images based on binary gender categories.

The titles of both photographs, *Mujer fernandina* and *Belleza fernandina* (Female Fernandino and Fernandina beauty), highlight the way these images functioned within the photographic genre called "types." Photographic types consisted of the metaphoric representation of a whole community through a set of features that were considered essential or common. Many photographers used portraits in this mode to categorize and solidify ideas related to social class, racial hierarchy, and gender roles, and the use of photographic typing constituted a very popular genre in the nineteenth and early twentieth centuries.[6] Indeed, in the two photographs sited above, we note that the women are typed not just via the titles, but also by their clothing, stance, and surroundings. Yet while photographic typing within the Spanish archive was common, the use of Fernandinos as the focus was not, just as European middle and upper classes were not often the focus of photographic typing either. But when Fernandino photographs do appear in the Spanish archives, the images are focused on "type" and nearly always also on women.

The relationship between photography and gender in colonial contexts has been the object of important studies that reveal the ways the emerging technology served to subjugate native populations, with feminized representations employed as a key modality of control.[7] Black women were often depicted in highly sexualized ways, using the visual to degrade, at times even offering quasi-pornographic representations. Women in Spanish Guinea did not escape this history of gendered and sexual abuse, and the process of deconstruction that colonialism implied affected women and gender relations particularly.[8] This intersection of gender and race within colonial contexts seemed to open the possibilities for the public distribution of images that otherwise would have been deemed private, even for the economically elite Africans of Fernando Poo. The use of gendered images of the African elite is even more striking if one notes that at the time, many Fernandinas were actually more economically powerful and socially well situated than their white Spanish female counterparts. They were part of a privileged economic class, even as they were linked to an oppressive colonial regime in other aspects of their lives, particularly sexually and via affective relations.

Indeed, perhaps one of the reasons that representations of Fernandinos so often focused on women and gendered typing has to do with the way a focus on gender allowed Spanish colonists to ignore or avoid the key visual implication of Fernandino photographs: that the Fernandinos were economically elite, capable not only of running their own affairs but also of thriving in a region where Spaniards had struggled to survive let alone thrive, both economically and governmentally. Yet this focus on gender and the general scarceness of African elite images also hints at the intriguing social power of the Fernandinos and connects with their own agency in the use of photography, as both producers and clients. Fernandinos did not have an economic need to be observed by European eyes, and Europeans could not easily control or disseminate a set of images that were private. While colonial control and dissemination were common when photographs were focused on African migrant workers or people from the rural areas, Fernandinos tended to exercise much more agency in both the production of photographs and their distribution. Indeed, when considering African agency and self-representation, there is evidence that Fernandinos, such as Walter Dougan, functioned as photographers and produced family portraits (see fig. 13.6).[9]

In the familial image included here and taken by Dougan, viewers can see three children sitting facing the camera, embracing each other in a formal sitting style on a bench, dressed in English attire and adorned with watches and lace. The backdrop reveals the photographer's intentions and also his professional practice. The children are captured close-up, with the background deliberately unfocused, highlighting the serious expressions

FIGURE 13.6. *Walter Dougan's Children.* Walter Dougan (ca. 1890–1900). © Trinidad Morgades / Laida Memba.

and elaborate clothing of Dougan's offspring. The image, taken by a Fernandino father, clearly marks the high-class status of these African children and their family, and it does not focus on gender. Indeed, it is not immediately clear which children are of which gender, although closer inspection of the clothing offers clues, with the two girls on the left in subtly frilled white and the young boy on the right in a slightly gray buttoned jacket and billowing white tie.

Indeed, Dougan's family portrait demonstrates how the photographic activity undertaken by Fernandinos challenged the norms of race, gender, and class common to colonial rule and often assumed within colonial photographic history, even in the present. While the children in the photograph were part of a colonized native population, they also experienced life as an elite economic class; Fernandino families held so much economic power that at a time when Spain was struggling economically, many Fernandinos were wealthier than the Spanish colonists (Sundiata). Therefore, it is not hard to understand why photographs of this African elite were not often included in Spanish journals and magazines, since they dismantled stereotypes, including racial and gender norms, of the time. For the Spanish, images of Fernandinos did not easily support an imagined national capacity to colonize and acculturate Africans; instead, Fernandino photographs recalled not just the British influence in Fernando Poo but, perhaps worse, the purchasing power of a Black community surpassing that of Spanish colonists via African economic and cultural agency.

In the context of visual culture, Fernandinos as a type were thus difficult to categorize, neither clearly on one side of the colonial binary nor the other, instead embodying an "in-between" symbolic space. Indeed, that would elicit, some years later, the creation of the figure of the *emancipado* (freed former slave) in an attempt to contain them within social colonial hierarchies.[10] The postcolonial scholar Homi Bhabha describes this in-between state in his ground-breaking analysis of colonial hybridity, *The Location of Culture*, defining concepts that can be useful to challenge the binary categorization of colonial subjects:

> The move away from the singularities of "class'"or "gender" as primary conceptual and organizational categories, has resulted in an awareness of the subject positions—of race, gender, generation, institutional location, geopolitical locale, sexual orientation—that inhabit any claim to identity in the modern world. What is theoretically innovative, and politically crucial, is the need to think beyond narratives of originary and initial subjectivities and to focus on those moments or processes that are produced in the articulation of cultural differences. These "in-between" spaces provide the terrain for elaborating strategies of selfhood—singular or communal— that initiate new signs of identity, and innovative sites of collaboration, and contestation, in the act of defining the idea of society itself.
>
> It is in the emergence of the interstices—the overlap and displacement of domains of difference—that the intersubjective and collective experiences of *nationness*, community interest, or cultural value are negotiated. How are subjects formed "in-between," or in excess of, the sum of the "parts" of difference (usually intoned as race/class/gender, etc.)? How do strategies of representation or empowerment come to be formulated in the competing claims of communities where, despite shared histories of deprivation and discrimination, the exchange of values, meanings and priorities may not always be collaborative and dialogical, but may be profoundly antagonistic, conflictual and even incommensurable? (Bhabha 1–2)

FIGURE 13.7. *Mariana Dougan*. Unknown photographer (maybe Walter Dougan) (ca. 1890–1900). © Trinidad Morgades / Laida Memba.

Bhabha is speaking broadly about the colonial experience, but his insights clearly apply to the complex "in-betweenness" of the African elite of Fernando Poo. The usual strategies of representation used by colonists could not operate in the same way with Fernandinos, particularly when Fernandinos themselves were behind the camera lens.

The image of Mariana Dougan (see fig. 13.7), demonstrates the complex inbetweeness of Fernandino photography and the way in which it participated in hybridity, maintaining African traditions while embracing European norms and keeping Hispanic culture at a distance. The photograph appears as part of a family collection kept partially in Barcelona (Spain) and Malabo (Equatorial Guinea) and may have been taken by Walter Dougan.

In the image, a female member of the African elite of Fernando Poo is dressed in European skirt and boots, and she stands leaning against Elizabethan columns, proudly raising her head and wearing an opulent hat. The image employs the backdrop of European fashion and architecture (and thus wealth), emphasizing Fernandina social and economic belonging as well as cultural hybridity. Fernandinos were not acritically acculturated, but instead they consciously occupied a place that made their cultural hybridity or in-betweenness explicit and visible. Since its invention, photography has been a medium that has helped to create and reinforce social binaries and hierarchies by depicting "inferiors" and "superiors" through the use of size, location, background, context, and selection. Indeed, that is the thesis of John Tagg's celebrated book *The Burden of Representation*, where he develops the fundamental idea of the photograph as a device used to criminalize and subalternize subjects. In the colonial context, such binaries were commonplace, yet this image of Mariana Dougan provides evidence that photographic manipulation could be exploited as a technology of reverse control for the African elite themselves. In this image, focused on a strong, economically independent African woman, viewers can see the representation of a "new urban subject," one that assimilated European and African visual signs in order to present Fernandinas as modern citizens of a new century.

The Fernandino photographic stance, as critics like Valenciano Mañé have successfully argued, was created with a critical use of trends stemming from Europe. Yet, while Fernandino photography challenged colonial hierarchies, as evidenced by the image of Mariana Dougan above, it clearly also buttressed class divisions, and thus it participated in the exclusion of other subaltern groups. As Alan Sekula and John Tagg point out, the visual separation of the bourgeoise from a range of subaltern groups—criminals, slaves, women, mentally-ill persons—was radical and deliberate, creating a troubling gap in the social archive. In the case of Fernandino photography, the emphasis on reinforcing class distinctions is apparent, but the images also served to connect and bind family lines, a function that follows Bourdieu's notion of photography as a medium that reinforces familial links.[11] However, while other African groups used photography (including studio photography) for familial purposes, one thing that distinguishes Fernandinos is that they were not simply depicted by "others," that is, by Europeans. Fernandinos had photographic agency, and they used it, in the above case to demonstrate the economic and cultural power of an elite African woman. Yet despite this empowering representation of Fernandinas, the photographic record also makes clear that both Europeans and Fernandinos participated in ethnic separation, trying to create and consolidate an image far from what was considered "tribal" Africa.

Knowing through Technology

Photography has its own ontology in terms of production, knowledge, and social order, and its entrance into colonial West African culture was marked by both the emerging nature of the technology and by the historical and social forces operating in the colonial period. Visual codes like poses, lighting, and framing produced photographic meaning, and social codes within the colonial context affected all of the visual coding choices. Clearly, critical readings of photographs depend on theories about the medium and the relation established between the image and its referent. As W. J. T. Mitchell describes such critique, visual culture is not only the social construction of the visual but also "the visual construction of the social" (170). Thus, the history of photography's entrance into West Africa is central to understanding these social relations and photographic meanings. Yaëlle Biro described this founding history in a curated photographic exhibition hosted at the Metropolitan Museum of New York City entitled *In and Out of the Studio: Photographic Portraits from West Africa*:

> When photography arrived on the continent, as early as the 1840s, the technology was appropriated by local communities, who adapted the new

medium according to preexisting visual codes and traditions of portraiture. By the 1880s, West-African, African-American, and European photographers had traveled along the Atlantic coast and founded temporary and permanent studios catering to the local elite. By the 1920s, significant West African urban centers had a deeply rooted photographic culture: photography had dramatically impacted notions of personhood as well as the ways in which those notions were expressed.[12]

Western notions of gender were introduced in visual forms. This was reflected in the photography of the time, which included erotized images of women, images portraying plantations, and photographs linked to particular gender norms focused on sewing, carpentry, and other occupations and work spaces. Private spaces were photographed in relation to the family as well as to the familial model imposed by Catholicism: monogamous and hetero-patriarchal, in contrast to the extended family systems common in local populations. For example, in the family portrait of Figure 13.8, on the left, viewers see a family with mother and father grouped in the center, children on the sides, and a grandparent or older family member in the background, all standing in front of what appears to be a home or perhaps a church, the women holding flowers and the father holding a cane. In the photograph on the right, a young newlywed couple sits with a similar photographic backdrop, but this time they are separated by a table, with marital witnesses and a stern-faced priest, a missionary, framed in front of the home or church. As can be seen in these two examples, even in family photographs it is difficult to term the home a private space: if the "colonial camera" was nearby or inside, an interruption in private life was taking place.

Yet the private archive (and life) of Fernandinas offers a more complicated photographic story. Although researchers have identified very few names of female land owners and planters from Fernando Poo, experts know of some, such as Amelia Barleycorn Vivour, the widow of William Vivour, who "owned the largest plantation on the island" (Sundiata 93). It is true that most family photographs center on men, who in Fernandino culture, as in European norms, were considered the head of the family, of the family business, and of political influence. Nevertheless, women clearly had important influence and relevance considering not only the role of Amelia Barleycorn Vivour but also the presence of Fernandina photographs in the John Holt archive and the marriages between Fer-

FIGURE 13.8. *María Cristina (Fernando Poo) Christian Marriage of Butukus' (chiefs) Descendants / María Cristina (Fernando Poo) Christian Marriage. A missionary.* Ramón Albanell (ca. 1915). Real Biblioteca de Palacio, Madrid. FOT/338. Inventory numbers: 10184918, 10184919. © PATRIMONIO NACIONAL.

nandinas and British officials, such as Lynslager and Beecroft. All these examples reveal that the social position of Fernandinas was high enough to warrant photographic attention, legal and economic independence, and inter-racial marriage.

As Deborah Poole brilliantly argues in her book *Vision, Race and Modernity*, the cultural and political meaning of race has much to do with the construction of images in modernity. Indeed, white women were nearly always represented surrounded by their privileges and in leisure scenes, and Poole highlights the way the dissemination of photographs became a type of social currency, a product that accrued value via dissemination. Whiteness was not simply a variable, but indeed a product of this construction. While photographic images and visual analyses from Africa have tended to focus on the black body, sometimes this act of framing and study has meant forgetting that the white body is also an invention accompanied by extremely high privileges in the social order. In that sense, references to the *bourgeois* body actually meant in most cases the *white bourgeois* body. Indeed, when John Tagg writes that from the very beginning of the technology, photography served to separate the "inferiors" from the "superiors" in Europe, what he did not explicitly state (but should have noted) is that these last were actually, and almost exclusively, white.

Indeed, Tagg's important study, while insightful in the way it critiques the power dynamics of the photographic image, still seems to avoid key historical realities in problematizing photographic history, including racial dynamics. The black body has historically been marked in photography by the lack of context, the hieratic poses (frontal and from the side), and, in most cases, by nakedness, while whiteness has appeared in a much wider range of poses, with greater context, and (tellingly) often surrounded by black servants. Spanish Guinea was not an exception regarding this visual construction of race. For example, the path for Fernandinos toward what was termed the *emancipación* (a condition that freed Africans from the tutelage of the colonial government) required passage through colonial assumptions of acculturation that insisted on the use of European clothes. That does not mean that acculturation was always forced or that the use of determined clothing was necessarily uncritical, nor that "European" or "African" have restrictive or inherent characteristics, but rather that the construction of the citizen, the subject who holds civil rights, had much to do with external and visible signs, and also that clothing choices have (and had) meaning. Fernandinos who had studied or married in the Cath-

FIGURE 13.9. *Maximiliamo C. Jones and European Farmers [sic] During the Aperitif Time.* Unknown photographer (ca. 1910–1915). © Trinidad Morgades / Laida Memba.

olic missions were portrayed very differently from people in rural areas, but more importantly, they actually were obsessively compared with other Africans in the photographic albums, underlying *visually* this distinction. The West African world was not, however, evenly portrayed, and Africans from the rural areas were the most common image in the Spanish visual media.

Fernandinas and Fernandinos depicted themselves through bourgeois visual codes, assuming not a "white role," but rather a classist discourse that linked them both with their African roots and with their British cultural ties. In Fernandino family photos such as the one above (see fig. 13.9), economic welfare is evident, but the photograph also evinces the idea of community and a sense of belonging that crosses racial divides. The image portrays a large gathering, including far more than one family group, all clustered around a long and elegantly decorated table. At least thirty-one people can be seen in the image; most appear to be black, but at least three appear to be white. The sense of connection and belonging is heightened by the close proximity of the guests, who are pictured touching shoulders or leaning against each other or the table, dressed in suits and ties and wide-brimmed hats and celebrating in a space hung with pictures. The photograph's storage location (a private Fernandino archive) and its title, "Maximiliamo C. Jones and European Farmers During the Aperitif Time," highlight not only how such racial mixing appeared commonplace within African elite culture in Fernando Poo, but also how such photographic evidence is utterly absent from the Spanish colonial archive.

The relation between technology, subjectivity, and selfhood becomes intersectional when these notions are crossed with others concerning both the ontology of photography and the political and cultural constructions of race and gender. Not all images are equally visible, even less in colonial contexts and their legacies. When one speaks of the ontology of photography (referring to both theories about it and positions toward it), two ideas are central. On the one hand, the relation of the image to its referent is not direct, but rather involves a whole set of implications of temporality. Brought into the present and through these multiple temporalities, images can make readers think about the nets of visibility they imply and the transformation of their meaning through time, including postcolonial times. On the other hand, the indexical relation between the referent and its image can also change from one moment to another. In the case of the photographs of Fernandinos, the index (one of the dimensions of a symbol related to continuity; for instance, smoke is an index of a fire) could be interpreted in terms of the civil and self-sufficient life of the African bourgeoisie. If Fernandino images indexed the African elite for Spaniards, then it is not surprising that such images were not archived in a context where Spanish sovereignty was under scrutiny and colonization was still in process. Fernandino images lacked visibility because, in the end, images are not disseminated by chance but take part in what has been called a "visual economy" or "an organization of production encompassing both the individuals and the technologies that produce images" (Poole 9). Among the three levels that Deborah Poole synthesizes in her analysis of the representation of Andean populations, she explains how the circulation of images was often perceived by the dominant European value system as fully transparent and thus could more easily produce colonial aims in colonial contexts:

> This question of circulation overlaps with the third and final level on which an economy of vision must be assessed: the cultural and discursive systems through which graphic images are appraised, interpreted, and assigned historical, scientific, and aesthetic worth. Here it becomes important to ask

not what specific images *mean* but, rather, how images accrue value. In the dominant European value system, for example, graphic images are evaluated in terms of their relationship to the reality they re-present. According to this realist discourse, the goal of all visual representation is to narrow the gap between the image and its referent. The image's value or utility is seen to reside in its ability to represent or reproduce an image of an original (or reality). Within the terms of the dominant realist discourse, this representational function of the image might therefore be thought of as its "use value." (10)

The question that is not raised, or allowed, within this dominant realist discourse is who or what power is determining value or even determining what is real? The "use value" of Fernandino images in the late nineteenth and early twentieth century was apparently quite low, since they represented a reality that Europeans, and particularly Spaniards, were not keen on seeing.

Public and Private: Press before and during the Second Spanish Republic[13]

Fernandinos started to appear in the Spanish press in the 1920s and became more visible after 1930. The likely reason for this change, following Poole's theories above, is that the "use value" of Fernandino images increased, and that led to a change in the discourse. This shift in Spanish discourse does not appear to be a recognition of Fernandino humanity or economic community, but rather a progressive instrumentalization of it. In other words, images of Fernandinos became useful for a renewed colonial discourse. The period of the Second Republic (1931–1939) was a time that, while it brought Spain economic development, laicism, and educational progress, nevertheless was deeply marked by both colonialism and racism. While African subjects during this period did gain the right to file complaints, and they experienced administrative changes as the military restructured and clericalism lost power, the basic apparatus of colonialism and racist hierarchy remained. Racial power structures shifted in form but not in substance. However, instrumentalist changes in racist power did offer openings for new technologies, and one can see that reflected in the photographic record, with many more publications focusing on educating Spaniards about the colonies and including images of Fernandinos. Indeed, reformist education, dependent as it was on the Claretian missions, was one of the main tenets of the Second Republic, and the educational system, along with photographic images, both served as effective technologies for maintaining colonial power.

This new presence of Fernandinos in the press can be seen as early as 1921, a decade prior to the official beginning of the Second Republic. For example, the wedding of Mabel Jones and Esteban Rodes, which took place in Barcelona in 1921, appeared in *La Guinea Española*, a journal edited by the Claretians. It was covered as a first-order societal event, with the couple surrounded by Spaniards and Fernandinos.[14] Nearly a decade later, but still before the official period of the Second Republic, another image appears, this time that of Jorge Dougan, with a caption calling him "el primer abogado negro de España" (the first black lawyer of Spain).[15] By the time of the Second Republic, such images had become more common. For example, Walter Dougan and Wilwardo Jones appeared in the press on behalf of a colonial commission that came to Madrid to report on the status of various issues. They appeared on October 28, 1932, in the magazine *Nuevo Mundo*, which described them as follows:

> En su cara de caoba tienen un raro brillo las pupilas del negro. Sus labios gruesos, de un rojo obscuro, pregonan, en sus dobleces carnosas, una honda sensualidad. Wilwardo Jones, este negro de Fernando Poo, viste correctamente a la europea. Su palabra es de buen tono castellano, y sus ademanes, sin estigmas aduladores, poseen cierta graciosa distinción. Su mano es carnosa y blanda, y en uno de sus dedos brilla, como una lágrima caída del cielo tropical, un brillante. Jones, de la raza bubi, es un negro acomodado de Fernando Poo. (J. R. 4)

> On its mahogany face the pupils of the black have a rare brightness. His thick lips, of a dark red, proclaim, in their fleshy folds, a deep sensuality. Wilwardo Jones, this black of Fernando Poo, dresses correctly in the European style. His speech is of good Castilian tone, and his gestures, without flattering stigmata, possess a certain graceful distinction. His hand is fleshy and soft, and on one of his fingers shines, like a tear fallen from the tropical sky, a diamond. Jones, of bubi race (sic), is a well-to-do black of Fernando Poo.

The use of sexualized and racist physical descriptions such as "thick lips," "dark red," "fleshy folds," and "deep sensuality" highlights the way photographs, even of Fernandino men, were given gendered meaning in ways common to the colonial process of feminizing the other. The article uses similarly racially charged language to describe Walter Dougan:

> Dougan se inclina ligeramente al saludar, y cuando habla tiene un dejo huidizo de desconfianza en sus gestos, como si no confiara en la buena fe del hombre blanco que se dirige a él, y temiera sus añagazas y engaños. Yo veo tras esta actitud de reserva y malicioso resabio del negro la secular tiranía ejercida por el hombre blanco durante siglos con las razas de color. (J. R. 4)

> Dougan bows slightly when he says hello, and when he speaks he has a fugitive lack of confidence in his gestures, as if he does not trust the good faith of the white man who addresses him and fears his tricks and deceptions. I see behind this attitude of reserve and malicious remorse on the part of the black the secular tyranny exercised by the white man for centuries with the races of color.

While in this second description the sexualized language is less prevalent, the attempt to feminize can be seen in terms like "lack of confidence" and "attitude of reserve," coupled with less feminized but deeply negative and violent diction like "fugitive," "tricks," "deceptions," and "malicious remorse."

Despite the social and economic position of the Fernandinos, the tone used in the above article does not differ much from the typical colonial racist discourse used to describe any Indigenous subject. For example, in the same breath that the author describes Jones as a "well-to-do black of Fernando Poo," the article identifies the photographic subject as "Jones, of bubi race (sic)." The emphasis on race is clear, yet Jones was not of "bubi race" because, for one thing, bubi does not refer to a "race." The racial focus is sadly ironic, because the visit described in the article represented an attempt by Fernandinos to gauge the metropolis's interest in colonial issues, particularly important during a political moment in which there was a new Republican form of government and that government, as evidenced by the content of the article, was attempting to argue that the Indigenous people were not being mistreated either by the Spanish authorities or by other colonial actors. The caption of the two photographs is also telling: "Walterio Dougan y Wilwardo Jones, de Fernando Poo, que han rectificado las noticias tendenciosas dadas hace algún tiempo por algunos periódicos extranjeros sobre los malos tratos que, según ellos, se le daba al proletariado negro de las plantaciones españolas. El bubi, afirman, vive en un régimen tutelar y ama a los españoles" (J. R. 4; Walter Dougan and Wilwardo Jones, of Fernando Poo, who have rectified the biased news offered some time ago by a few foreign newspapers about the mistreatment that, according to them, was given to the black proletariat in the Spanish plantations. The bubi, they say, live in a tutelary regime and love the Spaniards).

This issue of mistreatment referred to news that had appeared in the Nigerian press about the abuses of workers and the illegalities committed in Fernando Poo, and the topic had been the subject of scandal in the Lagos press. According to those reports, there was a rumor that the Fernandinos actually committed more abuse with the workers than their European counterparts because the Fernandinos owned smaller tracts of land (due to privileges in land concessions offered to the Spanish colonists), and therefore the Fernandinos were said to actually create more abusive conditions in an attempt to compete economically by making their smaller plots of land more profitable. It is curious, however, that the Spanish press would choose to refer to the Fernandinos themselves in order to defend against this denunciation. The irony becomes apparent in the final lines of the article, which seeks to emphasize that the African news stories of racial abuse were not only untrue but also outdated ("biased news given some time ago by some foreign newspapers about the mistreatment"), when in fact the article itself acknowledges that the Fernandinos had come to Spain precisely to try to confirm the absence of abuse. Thus, it becomes clear that the images of Fernandinos were objects of instrumentalized power, used to confirm a narrative of colonial absolution for racial abuse while simultaneously confirming racial and class hierarchies by feminizing the images of upper-class African men.

Fernandina Women and the Press: Education as Gender Propaganda

While media during the Second Spanish Republic offered feminized images of Fernandino men in order to further a colonial racial agenda, the photographic approach to Fernandino women instrumentalized colonial hierarchies through a discourse of education and modernization. The use of education as a tool of colonial racialized power can be seen clearly in articles from the Spanish magazine *Estampa*, which existed between 1928 and 1938 and was dedicated to societal news, attempting to incarnate the modernity that was supposed to characterize the country at that time. The appearance of Fernandinas, as well as other illustrated news of Equatorial Guinea, focused on education, and that focus began even before the official start of the Second Republic and became quite standard in its pages.[16] Discourses on education participated in the separation of labor and of the social roles of men and women and particularly emphasized gendered roles in order to spread the "advantages" of Spanish colonialism. Regarding women, two discourses were prevalent: the cultural superiority of the European lifestyle and the supposed freedom that the West offered to Indigenous women.

On February 3, 1934, *Estampa* published an article about the education and professionalization of African women, and it used the term "African" without distinguishing between ethnic belonging. The article explained how women from Spanish Guinea chose dressmaking, nursing, and typewriting as privileged professions among those appropriate for their gender. The text, signed by "F. D. R.," defended the idea that living with white women had made black women feel "anhelo de llegar a una altura cultural análoga a las mujeres del continente europeo" (a longing to reach a cultural height analogous to that of women on the European continent). The text thus emphasized that African women should conform to traditional European gender norms and that their hierarchical status could be advanced by rejecting their own traditions, depicted as obstacles to reaching the "cultural height" of the "European continent." The text also claimed that it was the wealthiest African families who had initiated this cultural process, likely referring to the Fernandinas. According to the article, after the African elite such as Fernandinas had embraced this cultural shift, other African groups followed, such as the "pamues." The images and visual codes stemming from photography printed in *Estampa* also attempted to establish a link between African women's subjugation and European discourses about modernity. For example, *Estampa* included a photograph of three black women dressed in 1920s European garb with the following caption: "Estas tres elegantes jóvenes de Guinea demuestran con su indumentaria su incorporación a la vida moderna" (These three elegant women show via their clothing their incorporation into modern life).[17] It was a condescending affirmation, focused on nothing but the clothing itself and offering the women as an educational model for assimilation of European high fashion.

Other photographs similarly posited Fernandina women as models for ideal African assimilation, sometimes using them to demonstrate pride for the Second Republic's educational endeavors, such as via an image published in *Estampa* on August 18, 1934, which suggested the importance of government scholarships: "Los indígenas de Guinea vienen a estudiar a Madrid" (37; The Indigenous of Guinea come to Madrid to study).[18] The tone of these articles and the context given for photographs of Fernandinas is overtly paternalistic and racially charged. For example, an earlier article appeared in *Estampa* on July 20, 1935, focused on two women who had come to study medicine and pharmacy in Valencia, Spain.[19] The journalist, Francisco Júcar, could hardly have been more dismissive or patronizing:

> Las dos son de igual estatura; las dos tienen la misma blancura en los dientes y las dos son negras. Si no fuera por el hecho, insólito en España, de que pretenden cursar dos carreras universitarias, las historias de las señoritas negras Victoria Ibina y Gertudis Davis, dos de las más distinguidas damas de la aristocracia de Fernando Poo, no merecerían pasar a las páginas de un periódico. Son dos historias breves y sencillas. Nacieron hace diez y seis y diez y siete años, respectivamente; crecieron, comieron plátanos, se tostaron aún más bajo el sol tropical, y un día, aconsejadas por el maestro español de Santa Isabel don Víctor Appellaniz, gran propagandista de nuestra cultura, decidieron venir a Valencia a estudiar en su Universidad. (13–14)

> They are both the same height; both have the same whiteness to their teeth, and they are both black. If it were not for the fact, unusual in Spain, that they intend to pursue two university degrees, the stories of the black ladies Victoria Ibina and Getrudis Davis, two of the most distinguished dames of Fernando Poo's aristocracy, would not deserve the pages of a newspaper. Their stories are short and simple. They are sixteen and seventeen years old, respectively; they grew up, they ate bananas, they got tanned even more under the tropical sun, and one day, advised by the Spanish teacher of Santa Isabel, Víctor Apellaniz, a great propagandist of our culture, they decided to come to Valencia to study at the university.

Júcar, in his racist depiction of the two women, emphasizes that the Fernandinas "would not deserve the pages of a newspaper" were it not for the largesse of a generous Spanish teacher who advised them to pursue a Spanish education. Yet, background knowledge on the families mentioned in the article itself belies Júcar's colonial posturing. The last name Davis was connected to a well-known Fernandino family, and Ibina to a prominent *kombe* line that had become wealthy via trade and was still important in Equatorial Guinea. The wealth and educational access represented by these dynasties had nothing to do with the actions of the Spanish government, neither before nor during the Second Republic. The access that these two women had to education came via their private initiative and efforts. The Second Republic did not substantially improve access to "Indigenous education," as it was called in those years, since it was not until decades later that ordinary Africans had access to higher education degrees. Indeed, as discussed above, Jorge Dougan, another Fernandino, had appeared in the pages of *Estampa* as "the first black lawyer in Spain" five years earlier. Thus, these two "aristocratic" women of Fernando Poo were not the first Africans to study for prestigious careers like medicine, law, or pharmacy, and their access was much more a function of economic standing than government largesse. The photographic image, however, elided the class issues and focused on representing the women as mere girls given favor by Spanish colonial benefactors. Accompanied by their white teachers and both alone, the students are represented as unimpressive, if well-educated and "integrated," girls.

As can be seen via these images of Fernandinas, photographs during the Second Spanish Republic emphasized a discourse of modernist liberal education in order to justify and frame the accomplishments of the African elite. The Spanish Second Republic saw itself as promoting laicism, demilitarization, and education among the working class, and these efforts were theoretically supposed to translate to Indigenous people, including women, in colonial Equatorial Guinea. Although societal changes were actualized in the first and second domain, republican governments proved just as racist, misogynistic, and colonialist as the previous ones. Nevertheless, the Second Republic did evince a transformation in the way Fernandinos were depicted, especially women. At the intersection of gender and race, there was a new possibility, useful as colonial propaganda: the idea that Spanish colonization had not only improved African lives but particularly African women's lives. Although in fact Fernandinos were self-sufficient thanks to their commercial background, photographs in the press told another story: finally black women could study university degrees in things like medicine and pharmacy because of Spanish Republican educators, colonial saviors of African women.

Conclusions: Racism, Sexism, and Colonialism in the Visual Realm

The community of Fernandinos was not easily integrated into the visual accounts of Spanish colonialism in Equatorial Guinea. The intersection of race and class challenged discourses focused not only on the cultural superiority of Europe but also on Spanish sovereignty over colonial subjects. Thus, despite the existence of a proliferation of images taken in Equatorial Guinea in the late nineteenth and early twentieth centuries, Fernandinos were nearly invisible in the Spanish press until the 1930s. The few exceptions found in the Spanish archives are all images of women, and the emphasis tends to be on a gendered passivity that makes the images less threatening to a colonial power structure.

When the presence of Fernandino images in the Spanish press began to change, in the late 1920s, news and publications about Fernandinos focused on emasculating men or offering educational formation for women. It seems that the African elite of Fernando Poo had finally become seen as visual tools useful for a colonial purpose: spreading the idea of the acculturation and educational formation of African subjects. As this new discourse was developing, photography became the technology used to construct visual hierarchies of difference. Photographs became one tool within a panoply of resources instrumentalized by government in order to maintain the racial and gendered power structures of colonial society. It is striking how this instrumental use of Fernandino images allowed for a rapid shift in visibility, beginning prior to the start of the Second Republic, such that Fernandinos went from near invisibility in the Spanish archive to serving as a symbol used to bolster the notion of Spain's colonial progress and prosperity. Fernandinas thus became visible to the Spanish public once the government could fit them into a convenient colonial account: one that focused on cultural assimilation via educational and gendered norms. Photography served as an effective new technology that could manipulate images of women and the colonies in order to serve the dominant discourse.

NOTES

1. I would like to thank Alba Valenciano Mañé for her help and feedback regarding this text and Benita Sampedro for her encouragement to write this text.
2. Such generic and pejorative labeling was very common in colonialist photography throughout Africa, and there are myriad examples, particularly via postcards. In the Spanish case, the website *Crónicas de la Guinea Ecuatorial* offers many samples of such photography and labeling. Rather than offering personalized portraits, individuals are referenced by racial and ethnic terms. For additional examples from other African colonies, see Geary. Similar issues are evident in most postcards of that period.
3. However, as Geary points out, even private photographs were sometimes used for postcards or illustrated books during the Spanish colonization of Equatorial Guinea (119).
4. The *carte de visite* was a photographic format that became very popular in the second half of the nineteenth century. Patented in 1854 by Eugène Disdéri, their size and price made the *cartes de visite* especially successful for studio portraits, becoming a social object that could be used as a gift or exchanged.
5. Please see *Mujer Fernandina, Santa Isabel*. Unknown photographer. Postcard. ca. 1915–1925. Source: Crónicas de la Guinea Ecuatorial. bioko.net/postal/displayimage.php?album=12&pos=32.
6. For more details on photographic typing as well as other genres employed for social control and racial formation, see John Pultz's critical analysis *The Body and the Lens: Photography 1839 to the Present*.
7. In addition to Pultz's *The Body and the Lens*, see for example Rizzo and Hayes's *Gender and Colonialism* (2012), Lindner and Dorte's *New Perspectives on the History of Gender and Empire* (2018), and Engmann's "Under Imperial Eyes, Black Bodies, Buttocks and Breasts. British Colonial Photography and Asante 'Fetish Girls'" (2012).
8. For further information on this aspect of Spanish colonialism in Guinea, see Nerín, *Guinea Ecuatorial, historia en blanco y negro. Hombres blancos y mujeres negras en Guinea Ecuatorial*.
9. Researchers can confirm the photographic activities of the Fernandino Walter Dougan because of the authorial stamp bearing his name in family photographs kept in a private archive.
10. The *emancipado* was a legal status established in Spanish Guinea in 1928 that recognized certain Indigenous people as "plenamente capacitados para realizar todos los actos de la vida civil" (Real Decreto, 17 July 1929, art. 2; completely able to achieve all the requirements of civil life). This meant they were not under the tutelage of the Spanish colonial administration. In practice, in 1928 only Fernandinos were recognized as *emancipados*.
11. For further exploration of this concept, see Pierre Bourdieu's *Un art moyen* (Medium art).
12. See the Metropolitan Museum, *In and Out of the Studio: Photographic Portraits from West Africa*, www.metmuseum.org/press/exhibitions/2015/in-and-out-of-the-studio.
13. Part of this subchapter was included in my PhD thesis. See Plasencia 2017.

14. Please see *Mabel Jones's wedding*. Unknown photographer, 1921, in *La Guinea Española* (unidentified issue).
15. Please see *Different portraits of Jorge Dougan in Spain*. Unknown photographer, ca. 1930, in Benavides pp. 13–14. Source: Hemeroteca Digital, Biblioteca Nacional de España.
16. For instance, they appear in *Estampa*, 15 May 1930.
17. Please see *Portrait of three women*, Unknown photographer, in F. D. R. "Las negras de Guinea se hacen modistas, enfermeras, mecanógrafas." Source: Hemeroteca Digital, Biblioteca Nacional de España, hemerotecadigital.bne.es/issue.vm?id=0003460319, page 23.
18. Please see *Portrait of woman*. Unknown photographer. Location: "Los indígenas de Guinea vienen a estudiar a Madrid." *Estampa*, year 7, n. 345, 18 August 1934, 37. Source: Hemeroteca Digital, Biblioteca Nacional de España.
19. Please see *Different portraits of Victoria Ibina and Gertrudis Davis in Valencia*. Unknown photographer. Location: Júcar, Francisco. "En Valencia hay dos aristócratas negras que estudian las carreras de Medicia y Farmacia." *Estampa*, year 8, n. 392, 20 July 1935, 13–14. Source: Hemeroteca Digital, Biblioteca Nacional de España.

WORKS CITED

Bhabha, Homi. *The Location of Culture*. Routledge, 1994.

Benavides, José D. "El primer abogado negro de España aspira a ser diputado de Fernando Poo." *Estampa*, vol. 3, no. 113, 11 Mar. 1930, pp. 13–14.

Biro, Yaëlle. *In and Out of the Studio: Photographic Portraits from West Africa*, 31 Aug. 2015–3 Jan. 2016, The Metropolitan Museum of Art, New York. www.metmuseum.org/press/exhibitions/2015/in-and-out-of-the-studio.

Bourdieu, Pierre. *Un art moyen*. Paris, Les Éditions de Minuit, 1965.

Crónicas de la Guinea Ecuatorial. bioko.net/postal/thumbnails.php?album=12. Accessed 23 September 2020.

Edwards, Elizabeth, and Janice Hart. *Photographs Objects Histories: On the Materiality of Images*. Routledge, 2004.

Engmann, Rachel Ama Asaa. "Under Imperial Eyes, Black Bodies, Buttocks, and Breasts: British Colonial Photography and Asante 'Fetish Girls.'" *African Arts*, vol. 45, no. 2, 2012, pp. 46–57.

F. D. R. "Las negras de Guinea se hacen modistas, enfermeras, mecanógrafas . . ." *Estampa*, vol. 7, no. 317, 3 Feb. 1934, pp. 22–23.

Foucault, Michel. *Discipline and Punish: The Birth of the Prison*. 2nd ed., Vintage, 1995.

Flusser, Vilém. *Towards a Philosophy of Photography*. Reaktion Books, 2000.

García Cantús, Dolores. *Fernando Poo: una aventura colonial española 1778–1900*. 2003. Universidad de Valencia, PhD dissertation.

Geary, Christaud. *In and Out of Focus. Images from Central Africa, 1885–1960*. Philip Wilson Publishers, 2002.

Haney, Erin. *Photography and Africa*. U of Chicago P, 2010.

"Los indígenas de Guinea vienen a estudiar a Madrid." *Estampa*, vol. 7, no. 345, 18 Aug. 1934, p. 37.

J. R. "Los negros caen víctimas de la enfermedad del sueño y las enfermedades tropicales." *Nuevo Mundo*, vol. 39, no. 2016, 28 Oct. 1932, pp. 4–5.

Júcar, Francisco. "En Valencia hay dos aristócratas negras que estudian las carreras de Medicina y Farmacia." *Estampa*, vol. 8, no. 392, 20 July 1935, pp. 13–14.

Lindner, Ulrike, and Lerp Dörte, editors. *New Perspectives on the History of Gender and Empire: Comparative and Global Approaches*. Bloomsbury Academic, 2018.

Mitchell, W. J. T. "Showing Seeing: A Critique of Visual Culture." *Journal of Visual Culture*, vol. 1, 2002, p. 165.

Nerín, Gustau. *Guinea Ecuatorial: Historia en blanco y negro. Hombres blancos y mujeres negras en Guinea Ecuatorial, 1843–1968*. Península, 1998.

Plasencia Camps, Inés. *Imagen y ciudadanía en Guinea Ecuatorial (1861–1937). Del encuentro fotográfico al orden colonial*. 2017. Universidad Autónoma de Madrid, PhD dissertation.

Poole, Deborah. *Vision, Race and Modernity. A Visual Economy of Andean Image World*. Princeton UP, 1997.

Pultz, John. *The Body and the Lens: Photography 1839 to the Present*. Abrams, 1995.

Real Decreto, 17 July 1929, art. 2.

Rizzo, Lorena, and Patricia Hayes. *Gender and Colonialism: A History of Kaoko in North-Western Namibia, 1870s-1950s*. Basler Afrika Bibliographien, 2012.

Schneider, Jürg. "Portrait Photography: A Visual Currency in the Atlantic Visualscape." *Portraitures and Photography in Africa*, edited by John Peffer and Elisabeth Cameron, Indiana UP, 2013, pp. 35–65.

Sekula, Allan. "The Body and the Archive." *October*, vol. 39, 1986, pp. 3–64.

Sundiata, Ibrahim. *From Slaving to Neoslavery: The Bight of Biafra and Fernando Po in the Era of Abolition, 1827–1930*. U of Wisconsin P, 1996.

Tagg, John. *The Disciplinary Frame: Photographic Truths and the Capture of Meaning*. U of Minnesota P, 2009.

Valenciano Mañé, Alba. "De vestidos y colonización en Guinea Ecuatorial. En busca de agencias escondidas en las narrativas coloniales (1840–1914)." *Guinea Ecuatorial: Políticas/Poéticas/Discursividades, Revista Debats*, edited by Benita Sampedro, vol. 123, no. 2, 2014, pp. 28–41.

CHAPTER 14

Punishing Narratives

The Challenges of Gender and Scientific Authority in Spanish Science Fiction Film

RAQUEL VEGA-DURÁN

Women have always been part of the scientific world in Spain, but their presence has tended to go unnoticed. Among the most renowned Spanish scientists of the twentieth century are the pharmacists Sara Borrell Ruiz (1919–1990) and Zoe Rosinach Pedrol (1894–1973), the chemist Gabriella Morreale de Castro (1930–2017), the biologist and oceanographer Josefina Castellví Piulachs (1935–), the biochemist and molecular geneticist Margarita Salas Falgueras (1938–2019), and the engineer Elena García Armada (1971–). The foremother of the first electronic book was also a Spanish woman, Ángela Ruiz Robles (1895–1975). Unlike their male counterparts, most female researchers are highlighted in articles focused solely on their work as women pioneers—that is, via an emphasis on those who were the first women to hold a particular position. Such is the case for historical figures like Fátima de Madrid (an astronomer of the tenth century) and María Andrea Casamayor (an eighteenth-century mathematician).[1] In the twentieth century, we can find similar approaches to the work of María del Carmen Martínez Sancho (mathematics), Felisa Martín Bravo (physics), Matilde Ucelay Maórtua (architecture), and Jenara Vicenta Arnal Yarza (chemistry), the first Spanish women to receive doctorate degrees in these STEM fields.[2] These women were recognized for being female pioneers, though few gained the acclaim they deserved for their actual scientific contributions.

Although women have played an important role in the development of science in Spain, it has not been until the twenty-first century that the character of the woman scientist has begun to gain visibility in Spanish film and television. A small number of dramas, comedies, and thrillers—the three most popular genres for visual media in Spain—have included women scientists as characters. Interestingly, science fiction, one of the least common cinematic genres in Spain (and one with fewer female directors), has begun in recent years to feature women scientists as protagonists. Two notable examples are the films *Eva* (Kike Maíllo, dir., 2011; Eve) and *Órbita 9* (Hatem Khraiche, dir., 2017; Orbiter 9). These two films present their audiences with a range of female characters, some as researchers and some as subjects, whose groundbreaking work is crucial for the future of humanity.[3] As a result, the films recover women's voices, eschewing the forgotten or invisible space to which women had previously been relegated in cinema about science and scientific experimentation. Both films share a similar narrative in which a woman who has become a scientific subject makes the decision to rebel against her prescribed role in the experiment. Thus, women go from being mere objects of scientific inquiry to interpreters of their own data, thereby creating their own scientific knowledge and agency. In both films, however, elements of the narrative make it difficult for the female protagonists to exert their initiative and independence in the long term, ultimately undermining their scientific authority.

Because the presence of women scientists in Spanish film has been understudied, many theoretical approaches to film scholarship, including feminist film theory and narrative structure analysis, remain underutilized in critiquing the genre.[4] The narrative structures of *Órbita 9* and *Eva* offer particularly interesting perspectives on female authority in the scientific realm. Both films follow a three-act structure (setup, conflict, and resolution) that allows the viewer to witness the absence, the authorial emergence, and the subsequent downfall of women subjected to scientific experiments. Interestingly, the endings of both films offer a brief, but significant, fourth act that proposes a new voice for women. If at first the narrative devices adopted by these films seem to subvert women's agency, both movies contain closing twists that may cause viewers to reconsider the films' fatalist endings as hopeful glimpses toward the future of women in science.

This chapter begins with an analysis of international efforts to encourage women in science and then explores how the Spanish media, particularly in television and film, have (and have not) contributed to that effort. After exploring how Spanish film relates to worldwide campaigns to increase visibility for women in science, the study focuses on a close reading of *Eva* and *Órbita 9*, analyzing the effective use of narrative structure to challenge gender roles while simultaneously critiquing how narrative structures at times undermine women's roles as arbiters of scientific authority. The chapter ends by reflecting on how these films show an attempt to fracture the exclusive categories of female data-provider and male data-collector by creating a dialogic bridge between both categories.

During the last ten years, various international campaigns have tried to illuminate the serious gender gap related to the role of women in science. On December 22, 2015, the General Assembly of the United Nations adopted a resolution to make February 11 the "International Day of Women and Girls in Science" as part of the 2030 Agenda for Sustainable Development. Via that resolution, the UN reaffirmed "the important contribution of the science and technology community to sustainable development and in promoting the empowerment, participation, and contribution of women and girls in science, technology, and innovation" and welcomed efforts "in promoting the access of women and girls to and their participation in science, technology, engineering, and mathematics education, training, and research activities at all levels" (United Nations General Assembly 70/212).

Such efforts are essential worldwide, and in Spain this is no exception. Both the Spanish government and the Spanish media have highlighted the scarcity of women in scientific careers and the low numbers of women and girls pursuing a STEM education. According to UNESCO, only around 28 percent of STEM professionals in the world are women. In Spain, 54 percent of college students are female, but only 20 percent of science majors are women ("Apenas"). Isabel Munera highlights the challenges in her popular press article "Faltan profesionales con conocimientos tecnológicos y científicos, sobre todo mujeres" (Shortage of professionals with technological and scientific knowledge, especially women"). In this instance, film and media can play important roles in highlighting the gender gap in scientific fields and in promoting the UN's empowering effort. Indeed, starting in the 1990s, film productions and documentaries, particularly in the United States and France, began making attempts to represent women as part of the scientific world.[5]

In the case of Spain, efforts to highlight the presence of women in science have been aided by a number of public and private initiatives that promote the role of women and girls in science. European initiatives with a hub in Spain include the *Hypatia Project*, a digital collection of activities for science centers and museums to "work together with schools, industries, and academics to promote gender inclusive STEM education and communication" (*Hypatia*) and *Girls in Tech Spain*, a virtual space for women to create networks through workshops, startup competitions, and collaborative projects.[6] *Inspiring Girls*, founded in the United Kingdom by the Spanish lawyer Miriam González Durántez, encourages women volunteers to talk in schools about their jobs in STEM, the obstacles they faced along the way, and the ways in which they surmounted them. Additionally, Spanish projects such as *Iniciativa 11 de febrero* (commemorating the International Day of Women and Girls in Science) organize workshops and produce pedagogical materials to make visible the presence of women in science. The main goal of initiatives such as *Inspiring Girls*, the *Hypatia Project*, *Girls in Tech*, and *Iniciativa 11 de febrero* is to provide young girls with role models in STEM fields, both as agents of fulfillment and as sources of inspiration.

The Spanish media industry, however, has found inspiration in only a few of the many important Spanish women scientists. Albert Solé's documentary *Los recuerdos de hielo* (Ice memories) records oceanographer Josefina Castellví's return to Antarctica to commemorate the twenty-fifth anniversary of the Spanish scientific base established there. Molecular biologist Margarita Salas has been the subject of several long interviews, as well as the RTVE documentary *Ellas: Científicas* (Women: Scientists).[7] Salas, who in 2019 received the European Inventor Award in the category of Popular Prize Winner, is also the only Spanish woman scientist to have a museum dedicated to her life and work, as well as the only woman to have a figure in the gallery of science at the Wax Museum in Madrid.[8]

Although many Spanish women scientists have led fascinating lives and conducted important research that could make for interesting film material, the media has not chosen to share their stories. The works of particular American, French, and Austrian female scientists have been documented in film, with examples such as *CodeGirl* (Lesley Chilcott, dir., 2015), *Marie Curie: The Courage of Knowledge* (Marie Noëlle, dir., 2016), and *Bombshell: The Hedy Lamarr Story* (Alexandra Dean, dir., 2017). There is also a visible effort in the United States to rescue the lives and works of lesser-known female scientists from oblivion. For example, the stories of the 1880s American female "human computers" have been documented not only in Dava Sobel's 2016 book *The Glass Universe: How the Ladies of the Harvard Observatory Took the Measure of the Stars* but also via Lauren Gunderson's play *Silent Sky* (first performed in 2011). Around the same time, theaters released the acclaimed *Hidden Figures* (Theodore Melfi, dir., 2016), a film about a team of women mathematicians (also known as "human computers") who worked for NASA during the Space Race.[9]

Though similar scientific work was done in Spain, Spanish women have never received similar acclaim. To take just one example, consider the women of the Observatorio de San Fernando (San Fernando Observatory) in Cádiz, Spain. According to journalist Raquel Pico, who cites Fernando J. Ballesteros, in the late nineteenth century there were Spanish women working in the San Fernando Observatory analyzing large amounts of data on stars. Their work became a crucial part of *Carte du Ciel* (Map of the Sky), an international project of the Paris Observatory, a research institution of the Université Paris Sciences et Lettres, which worked closely with the Spanish observatory. At the time, the Paris Observatory and the Harvard College Observatory were competing for the creation of the most complete star catalogue. But while the Spanish story of the San Fernando women, who played a significant role in cataloging the position of millions of stars for *Carte du Ciel*, remains untold, the work of American women working simultaneously in the Harvard Observatory on the other side of the Atlantic has been recognized on billboards. Ballesteros cannot but pessimistically point out that "the only information that survives [about the women who worked at the Spanish observatory] is a mere mention about 'having hired four young ladies.' We don't even know their names."[10] It is telling that, even though Spain has seen many important women scientists, the biggest box-office success for a Spanish film about a woman scientist has been Alejandro Amenábar's *Agora*, an English-language biopic of Hypatia of Alexandria, the fourth-century woman astronomer, mathematician, and philosopher who lived in Roman Egypt—in other words, about a woman who was not Spanish.

Women scientist characters have indeed appeared in numerous movies and television series. Doctors, nurses, and pharmacists are the three most common roles for women scientists in Spanish films, such as *Señora doctor* (Mariano Ozores, dir., 1974; Mrs. Doctor), *Las chicas de la cruz roja* (Rafael J. Salvia, dir., 1958; Red cross girls), and *Farmacia de guardia* (Antonio Mercero, dir., 1991–1995; On-call pharmacy), respectively.[11] The characters of women health professionals in Spanish film are clearly perceived as a strong presence, since they account for a considerable number of women—whether in leading, supporting, or minor roles—whose jobs are related to science. The film narratives in most of these movies, however, overshadow the scientific component of their professions, either by presenting their work within the genre of comedy or by leading the viewer to focus primarily on these characters' personal relationships to patients and customers, even when they are in leading roles.[12] In other words, although they are in scientific professions, science is not presented as a vital part of their identities or of the larger film narrative.

The importance given to scientific inquiry in cultural production depends on genre. In Spanish film, science fiction is the genre in which a clearer relationship between women and science is evidenced. Such is the case in the films *Órbita 9* and *Eva*.[13] *Órbita 9* tells the story of Helena, a young woman who has spent her entire life on a spaceship headed for planet Celeste, the only hope for human survival after the environmental collapse of Earth. Abandoned by her parents three years before due to a depletion of oxygen on the ship, Helena resides alone. Though she believes this to be her reality, she is in fact part of a private experiment, in which ten families are monitored while they live in an underground simulator, which replicates similar conditions to those in space. The goal of the experiment is to predict and troubleshoot the hardships endured during the eventual long journey to another planet in preparation for human extinction on Earth. Having arrived in the simulator when she was a baby with her parents (both scientists who consented to be part of the experiment), Helena has never agreed to become a "colono," that is, a settler on another planet. Unexpectedly, Alex, one of the engineers in charge of the project, arrives on the ship pretending to fix the oxygen depletion problem. He spends two days there and realizes that, even though he is one of the principal investigators of the project, he can no longer countenance the deception. He decides to reveal the truth to Helena, leads her to the surface, and then to the city. Alex's boss, Hugo, is then instructed to arrest them and kill Helena, since she is no longer "useful" to the experiment.[14] In order to save Helena, Alex offers Hugo a new project—Alex will stay in the spaceship with Helena, who is now pregnant with Alex's child, and the scientists will be able to observe a birth in simulated conditions to outer space. In the film's dénouement, Alex and Helena's daughter leaves the underground ship as a young girl.

Similarly, *Eva*, set in 2041, features a time in which robots are ubiquitous. The protagonist, a cybernetic engineer, also named Alex, returns like a prodigal son after ten years of absence. He becomes part of the Robotics Department in the fictional town of Santa Irene, working on the creation of a child robot under the supervision of Julia, his mentor and head of the Robotics Department. In his search for a human model for his project, he meets Eva, the ten-year-old daughter of his former colleague and ex-lover Lana.[15] When Alex asks Lana for permission to interact with Eva in order to model his child robot on her mannerisms and behavior, Lana emphatically refuses. Despite Lana's refusal, Eva starts spending time helping Alex with his SI-9 robot. After realizing that Alex still has feelings for Lana, Eva tries to get them back together. Confused, Alex decides to abandon the project, disappointing both his mentor, named Julia, and Eva. At this point, Lana visits him and confesses that she would like him to stay because he is Eva's "father." In a *mise en abyme* scene, both Eva, who is watching the couple through a window, and the audience, watching Eva on the movie screen, experience the simultaneous realization that Eva is a robot. When Lana confronts her enraged "daughter," Eva reacts by pushing Lana over a cliff, causing her death. Then Julia confesses to Alex that Eva is the finished SI-9 robot on which Lana and Alex had been working ten years ago. Even though the SI-9 robot had failed to pass safety tests, Lana had decided to keep her operational. Alex, who has come to love his "daughter" Eva, is left to deactivate his own creation.

These two films differ in their details, but they share significant structural features, particularly in relation to characters and actions. Both films have three kinds of characters: a powerful woman in charge of the experiments,[16] a male scientist (named Alex in both films, coincidentally) who becomes emotionally involved with the female object of study, and a female character, who is introduced as the research subject and is the focus of the film. In both movies, the female subjects establish a close relationship with the male scientist who helps them come to the realization that they are subjects in an experiment. When these women try to gain agency, they are denied the ability to do so because of their subject status. As each film progresses, these women become cognizant of their realities and dream of changing their preordained futures. That is, as knowledge is power, they gain power by becoming conscious (*con scientia*) with science. Yet, in a reversal of the Pygmalion myth, these women's empowerment is thwarted by the men who "created" them when these men make final decisions about their futures. Hence, if both films present women challenging the limits of bioethics, in the end their male "creators" will reduce the women to their origins. By juxtaposing similar aspects of these two films—the encounter between "creator" and "creation," the desire of the "creation" to actualize herself, and the termination of the rebellious woman—the narratives ultimately deny any possibility of the female characters' assuming participatory roles in their own futures.

Theoretically, science fiction could be a genre characterized by endless possibilities and prospects for innovation, but both film narratives place women in the traditional role of bearers of meaning. At first, both Helena and Eva are unaware that they are the objects of experimentation; Eva does not know she is an SI-9 robot, and Helena does not know she is a terrestrial being living in simulated space conditions on Earth. In *Eva*, Alex was the creative roboticist who ideated the SI-9, that is, Eva.[17] Having abandoned this failed project ten years earlier, he returns to attempt to finish it. When the "child," Eva, becomes the model for his new SI-9, he begins to observe her, and it is his male gaze that dominates the film's narrative. Alex's gaze is obsessively focused on Eva, whom he regards as an endless source of information that he can then select and process. Thus, if at first Eva seems to be the main character in most of the scenes and the one that provides information, the unrelenting focus of the camera on Alex's gaze suggests that Alex is only using Eva. To use Laura Mulvey's well-known category, this turns Eva into a "bearer of meaning," an observed object whom Alex—a "maker of meaning"—wants to understand (Mulvey 7).[18] The narrative shapes Eva as a fountain of data that Alex then processes, deciphers, and analyzes, subsequently creating knowledge. As the model chosen by Alex for his SI-9, Eva is simply a source of data, supplying Alex with the information needed to make a humanoid. Eva is indispensable for Alex, but she does not produce knowledge, as she does not get to interpret the data she provides. Significantly, while Eva's voice dominates most of the scenes in which she appears with Alex, his silent decisions are documented by the camera, which follows his viewpoint in relation to Eva, objectifying her. Thus, even though the experiment would not be possible without Eva, the film narrative constricts her role to that of being the source of information but never the producer of knowledge.

A similar situation occurs in *Órbita 9*, in which the narrative starts with Helena, convinced that her mission is to go to planet Celeste. Raised under the assumption that she is a settler escaping from a toxically polluted Earth, Helena spends her life monitoring the spaceship's controls in order to arrive safely on Celeste. In reality, however, Helena is a mere producer of large amounts of data; cameras installed in the make-believe spaceship and a bio-witness chip in her wrist continuously send information to the central headquarters in the laboratory. The images broadcasted to the headquarters may be interpreted as a metaphor of cinema itself: they gratify the scientists' (Alex among them) desire to look. As the object of a sort of scopophilia, Helena is permanently observed, and the scientists derive pleasure from scrutinizing her. Cameras are installed in "public" areas,[19] and all the scenes of observation are clinical. This scrutiny is necessary for scientific advancement, but Alex becomes a voyeur who finds stimulation in sight and pleasure by securing useful data. Thus, the woman subject becomes the embodiment of the scientific project that proves her creators' success.

Only after he interacts with Helena in person does Alex stop believing in the integrity of the project. Once Alex, refusing to continue deriving pleasure from the images on the screen, changes his perception of Helena from object to subject, he starts to experience an ethical reaction. After pretending to be an engineer who has come to fix the oxygen depletion problem on the ship, he unexpectedly returns, without the astronaut suit, and explains to her, "I know this is going to sound crazy to you, but we are not where you think. This is not a ship, Helena. It is a simulator. On the other side of that door is Earth" (*Órbita 9*). The stern expression on Helena's face makes clear she does not believe him. To prove his point, he reveals one of the hidden cameras by pulling it off the wall and showing it to her, stating, "It has been recording since you were a baby. There are two more in this corridor and another one in the decompression room" (*Órbita 9*). The moment the cameras—enablers of the surveillance and the "staging" of Helena's life—are uncovered, things change for her. Until this point, Helena has refused to believe Alex and has not moved at all; for the whole scene, she has kept her eyes wide open without blinking and her mouth half-open, unable to understand the information. Finally, she moves her head to scan the area in awe. At this moment, Helena is urged to reevaluate her identity as a settler. If for twenty years this has been her only familiar space, Alex's revelation renders it unfamiliar, encouraging her to question what was, until then, unquestionable. Still perplexed, she hears from Alex that she has a chip in her wrist monitoring her vital signs.

Disregarding Helena's refusal to have it removed—she still cannot believe that it is true—Alex rips it out of her flesh. This is a critical juncture for Helena; Alex, violently and abruptly, detaches her from her past identity as a settler and forces her to accept that her whole life had been staged. Lacking anything to anchor her to her past, she turns to Alex as her only point of reference. For Alex this is also an important juncture, since after revealing the truth, his perception of Helena changes. If before he saw her on the headquarters' monitor screens as a source of data, now he is attracted to her as a person, not just as an image.[20]

This movement of the plot can be understood in terms of Tzvetan Todorov's categories of disequilibrium and equilibrium as basic elements of narrative. In *The Poetics of Prose,* Todorov outlines his theory of the standard narrative structure: stories tend to start in a state of equilibrium, and at some point, this state is interrupted, provoking a state of disequilibrium, which in the end is finally restored to equilibrium. In *Órbita 9*, the revelation of the cameras' existence provokes a state of disequilibrium in the narrative tension, since Helena now has a desire to make sense of her own life. Until then, the story had two main protagonists: the scientist who knew everything and the guinea pig who produced useful information for the scientist. Once the cameras are revealed to Helena, she temporarily stops being a guinea pig, and both Alex and Helena can potentially reach a balance by sharing their desire to produce knowledge. After escaping from the spaceship, Helena wants to explore the new possibilities that are available to her. Concerned for Helena, as well as attempting to control her, Alex tells her what she can do to stay out of danger while he is at work. Able to exert free will for the first time, however, Helena is swayed by her desires. In her effort to understand who she is and why her whole life has been a lie, she decides to look for her parents, who now live on the surface. Her quest only leads to her persecution, since, once she finds her family, her mother decides to report her. The uninterrupted role of Helena as the object of Alex's gaze, followed by Helena's inability to evade capture by both the project supervisors and her family, contribute to the restoration of the original narrative equilibrium, which required a balance based on the existence of opposing characters. Thus, unable to fracture permanently her identity as guinea pig, Helena's "terrestrial" life ends, and she is returned to the underground "spaceship" world.

This kind of break in the balance of the narrative also happens in *Eva*, when Eva undergoes the desire to understand herself the moment she discovers she is a robot. In a role reversal, Eva then becomes a subject with the dream of being a human; unable to process this vital information about her identity, she collapses.[21] Eva's rage builds as she tries to reconcile what she now knows was her staged life as a human and her real self as a robot. The only way to regain the equilibrium in the narrative occurs once Alex disconnects her, rendering her inoperative or, in essence, dead. In both films, the moment of disequilibrium emerges when the female characters experience the desire to understand who they are. Neither of them reaches the desired goal of changing who they have been made to be (a settler and a robot), but the disequilibrium in both movies brings about an interesting outcome, since the disruption of what Eva and Helena perceive as reality causes these protagonists to reassess their roles in life.

Eva's and Helena's desire to control their own lives becomes a threat to the scientific experiments in which they unwillingly participate. This is because the narrative presented in these two films interprets the world as a set of mutually exclusive categories. Robots/humans (*Eva*) and terrestrials/settlers (*Órbita 9*) form binary patterns in the films, defining things both in terms of what they are and what they are not. That is, their equilibrium is not based on equality but on diametrically opposed characters. Eva is a robot who wants to be a human (like Alex), and Helena is a settler who wants to live on Earth (like Alex). To realize their dreams, they attempt to change the identity imposed on them by scientific research. However, the moment they empower themselves and detach from each Alex's control, they are punished.[22] Their initiative, which makes them desire to be equal entities to their "creators," implies a detachment from the experiment, which threatens the integrity of the research, rendering the female figures useless to science and, therefore, expendable.

Established with the premise that Helena and Eva are "creations" and the Alexes are the "creators," an affiliation is implied. The creator of knowledge and the data-provider creation are mutually dependent, and the disappearance of one of the identities would threaten the existence of the other. When Helena and Eva refuse to partake in the experiments, the binary interdependence fractures the narrative equilibrium, since one of the identities refuses to participate, thus voiding the other's identity. When Helena and Eva aspire to become independent subjects, they start to use the data that they themselves have been producing in order to make sense of who they are. The subjects now become scientists with first-hand knowledge driven by a desire to gain a comprehensive understanding of their own natures. Eva's and Helena's desires to be free to comprehend their own identities could also be regarded as an attempt to appropriate the "scientific gaze," a type of dominating, data-gathering gaze that until then had belonged primarily to male scientists. This role reversal elicits a temporary transformation from bearer of meaning to maker of meaning.

This state of self-awareness and ability to interpret the data, however, must cease because it is disruptive to the pre-established construct. When Eva and Helena analyze the data they gather, they break the binary structure of this narrative; in order to keep its balance, the film narrative must squelch the female characters' desire for knowledge. Therefore, in order to reestablish the equilibrium, their actions must be stopped to reinstitute the rules promoted by the male scientists. This hierarchy will not allow them to pursue their dreams of self-realization but instead restores the patriarchal binary opposition. *Órbita 9* offers a particularly clear example of this when Helena must remain a "settler" in the experiment in order to exist. Soon after she spends time on the surface, her skin, unaccustomed to the light, starts to peel, eliminating all possibility of surviving outside the underground "spaceship." Helena's body is punished, and she is urged to go underground. In this way, the narrative restores the balance terrestrial/settler by imposing an obstacle that is impossible to surmount, forcing her to choose between dying or returning underground.

The film narrative in *Órbita 9* and *Eva* is surprisingly traditional, even though science fiction is a genre of endless possibilities for experimentation, speculation, and the changing of codes. Both films have a linear narrative, which follows the same formula: established equilibrium, balance disrupted, quest to restore stability, climax, and finally, equilibrium restored. The attempt of the narrative to regain the lost equilibrium by frustrating the women's desires, however, is not successful in these films, since the moment the women attain self-awareness, they are deemed superfluous for the experiments. Their independent desires, although short-lived, have contaminated the scientific protocol, which requires the subjects' unbiased ingenuity. Their desire to make sense of their lives by themselves adulterates their controlled existence. Seduced by the "unknown," Eva and Helena eat the fruit of knowledge of what they consider their "Garden of Eden," the space where they can be conscious, and soon lose their "innocence." In both films this leads to the need to eliminate them. However, this poses a major problem because, without them, the interdependence of researcher and subject is also disrupted. Without observation, an experiment cannot proceed, invalidating scientific inquiry.

The role of women can also be inserted within the traditional film narrative, where women are the object of the controlling gaze of the male protagonists, corresponding to the gaze of science as a male domain controlled by men. The dénouements of these films, however, prove their desire to go further, making them into more subtle explorations of the interplay between scientists and their objects of study. Eva and Helena are in fact silenced at the end of the movies; Eva will hear the fatal question, "What do you see when you close your eyes?" prior to being shut down, and Helena will return to the subterranean "spaceship," literally and figuratively buried. Eva and Helena's success in breaking free from the experiments, even for a short time, suggests the possibility for a new kind of narrative logic. For a brief period, they cease to be objects and their actions manage to derail the repressive scientific experiments that control them.

In both movies, the rebellion of the female characters also forces male scientists to rethink their methods and become more self-aware. It is Eva's attempt to know herself that reveals the bioethical issues involved in building artificial intelligence that cannot be distinguished from humans. After killing her "mother," Lana, Eva asks Alex to help her be "a good girl." When she is taken to the basement where Alex has his lab, she already knows what is going to happen. She then requests that Alex read *One Thousand and One Nights* to her, adopting the role of Scheherazade avoiding being killed by Shahryār. During the movie, Eva fills this Scheherazade role, providing Alex with information and also sparking in him the desire to know more. But Alex does not want to hear any more, as he already knows the ending. Now Alex interrupts her in order to become the storyteller. The story's framing device is thus fractured, as Eva realizes when Alex takes control of the narration. "You are not going to fix me, are you?" she asks. At this moment, Alex pronounces the fatal words, "What do you see when you close your eyes?" Eva is not just a guinea pig that has been sacrificed for the sake of science; she provokes in Alex a moment of realization, making him understand that even though she was conceived as artificial intelligence, she is able to feel emotions. Eva manages to communicate to Alex her self-awareness, and her "death" brings a moment of awareness for Alex in turn. Both are subjects driven by clinical observation and human (subjective) feelings. Both belong to each part of the binary system of their narratives. Thus, Eva manages to break down the seemingly fixed narrative, proposing another way of looking at the relation and the identity of scientists and their objects of study. Eva is disconnected as a robot, but she dies as a sentient being. She has managed to acquire the human capacity of experiencing complex feelings, a power Alex is not able to take away from her.

In *Órbita 9*, though Helena will not be able to live life as she chooses, she is vindicated through her progeny. Helena agrees to go back to the underground spaceship under the condition that her daughter will be free of the experiment, to live on Earth, after a few years. The film closes with the image of the young girl coming to Earth's surface from the spaceship Órbita 9; Helena's daughter is not forced to follow in her mother's footsteps. Helena initiated a desire for knowledge that will allow her daughter to be raised producing data for the experiment as well as collecting and analyzing the data she herself produces. Thus, it is only at the very end of the film that the spectator realizes that the narrative is more irresolute than the patriarchal traditional mode it followed throughout most of the story. Helena and Eva reveal the impossibility of existing on just one side of the narrative binary, since they ultimately inhabit both sides. Eva has become a sentient robot, and Helena is both a space settler and a terrestrial. Even though they disappear from their stories, their presence will continue to influence the future. They become the Eves of a new reality, the "mothers of all life" (*New Oxford Annotated Bible*, Gen. 3.20), the origin of something different, a hybrid that is emerging, but still undefined.

The conclusions of these films suggest the birth of a new space that challenges the feasibility of the traditional narrative based on binaries. In consonance with the current climate of raising awareness and participation of women in science, these films do not propose a space interested in the opposition between women and men but one that welcomes and inspires women to see themselves on both sides of the binary. *Órbita 9* and *Eva* timidly move beyond a straightforward "women in science" paradigm, one that places women in opposition to male scientists, to propose a different possibility. These films demonstrate how the genre of science fiction can create a space to explore questions about science from the clearly subaltern position of the guinea pig. In this space, a scientific gaze can emerge and develop, a gaze that women, traditionally bearers of data, can appropriate and make their own. This new narrative suggests that the traditionally male-dominated observational mode is part of the problem, but it does not encourage the elimination of the traditional character of the male scientist. Rather, it proposes a new space of exchange where women do not take over the space traditionally reserved for men and alienate them; this would replicate the binary narrative. In this innovative narrative, women belong to a liminal space, both as providers of data and producers of a more self-conscious scientific knowledge. The closing scenes of these films encourage viewers to reconsider the relationship between the traditionally mutually exclusive categories of female data-provider and male data-collector. With a novel scientific gaze enhanced by a major degree of awareness, women manage to create a collaborative, dialogic bridge between both categories, suggesting the potential to break down old antithetical narratives. The films' spectators are left with a new narrative possibility that may become future narrative reality.

NOTES

1. Fátima de Madrid is the focus of Juan Núñez Valdés's "Did Fátima de Madrid Really Exist?" María Andrea Casamayor is featured in one of the chapters of Casado Ruiz de Lóizaga's *Las damas del laboratorio: Mujeres científicas en la historia* (The ladies of the laboratory: Scientific women in history), which also dedicates a few pages to Fátima de Madrid, and is the focus of a number of newspaper articles, such as Patricia Puyo's "La mujer que nos acercó a la ciencia" (The woman who brought us closer to science) and María Pilar Perla Mateo's "María Andrea Casamayor, la primera que publicó un libro de ciencia en España" (María Andrea Casamayor, the first woman to publish a scientific book in Spain).
2. For more information, see Núñez Valdés et al. "Primeras mujeres farmacéuticas en España" (The first female pharmacists in Spain).
3. The characters of the women in charge of the labs where the research takes place are not well developed. They have only a secondary role, and even though they have the power to decide the direction of the experiments, in both films Katherine (*Órbita 9*) and Julia (*Eva*) end up leaving the fate of their labs in the hands of the male scientists. Women as head of scientific projects still remain underdeveloped characters in Spanish film. This chapter will focus on the female characters that are treated as objects of study, that is, as guinea pigs, a well-developed character in this type of film.
4. Analysis of narrative structure has long been a centerpiece of film studies. Early scholars of narrative theory such as Claude Lévi-Strauss, Roland Barthes, Tzvetan Todorov, and Vladimir Propp tackled narrative structure from varying perspectives, including viewing them as dependent on binary oppositions (Lévi-Strauss), seeing them as enigma codes that viewers enjoy solving (Barthes), describing an equilibrium-disequilibrium-new equilibrium structure of narratives (Todorov), and identifying a sequence of narrative elements that most plots have in common (Propp). In the 1960s and 1970s, film criticism became increasingly focused on film narrative. Influenced by the field of semiotics and the viewer's relation to the text, journals such as *Screen* and *Cahiers du cinema* and scholars such as Christian Metz, Colin MacCabe, and Roland Barthes argued against classic realism and the perception of cinema as an accurate

representation of the world. At the same time, the use of psychoanalysis and cultural studies to understand film narrative became a key interest for feminist film scholars, such as Laura Mulvey, Ann Kaplan, Mary Ann Doanne, and Annette Kuhn. Feminist film and cultural studies scholars were concerned with issues of class, gender, race, and ethnicity and thus pushed the field to recognize structures of patriarchy embedded in a range of cinematic elements, including the camera, editing, and basic plot structures. Today, this broad range of narrative theories is key to film criticism, and they tend to appear hand in hand with theories of cognitivism and intertextuality to analyze both how film narratives generate meaning and how viewers actively create meaning.

5. Films such as *Outbreak* (Wolfgang Petersen, dir., 1995), *Contact* (Robert Zemeckis, dir., 1997), *Splice* (Vincenzo Natali, dir., 2010), *Contagion* (Steven Soderbergh, dir., 2011), *Gravity* (Alfonso Cuarón, dir., 2013), *Interstellar* (Christopher Nolan, dir., 2014), *CodeGirl* (Lesley Chilcott, dir., 2015), *Hidden Figures* (Theodore Melfi, dir., 2016), *Marie Curie: The Courage of Knowledge* (Marie Noëlle, dir., 2016), and *Bombshell: The Hedy Lamarr Story* (Alexandra Dean, dir., 2017), are among the growing number of films with a woman scientist in the leading role. While a number of websites cover the relationship between science and film, most tend to feature movies with male characters in the leading role; however, in recent years there has been an increase in the number of websites (particularly blogs) focused on women scientists in film. Representative blog postings are Caitlin Busch's "Annihilation and the Hard Truth about Women Scientists in Genre Films," Amy Chambers's "Rise of the Women? Screening Female Scientists," and Eduardo Angulo's "Así vemos a las científicas en el cine y la televisión."

6. More information about the *Hypatia Project* can be found at www.expecteverything.eu/hypatia and for *Girls in Tech Spain* via spain.girlsintech.org.

7. *Ellas: Científicas* is part of a series of documentaries, titled *Ellas*, produced by RTVE, showcasing women in different disciplines traditionally dominated by men. In addition to Salas, *Ellas: Científicas* focuses on the industrial engineer Elena García. Salas and García are interviewed in the documentary by actors Blanca Portillo and Antonio Resines.

8. The museum dedicated to the life and work of Margarita Salas opened in Luarca (Asturias, Spain) in 2015.

9. This film is based on the non-fiction book *Hidden Figures* by Margot Lee Shetterly.

10. The quote originally appears in Spanish (translation mine) in Raquel Pico's "Las mujeres olvidadas de la ciencia."

11. One of the first Spanish films about a female doctor is *Señora doctor*, a comedy in which Lina Morgan plays Elvira, a doctor who works in a small town and has to fight the prejudices of the villagers who cannot accept that a woman might be able to cure them. The Spanish audience would need to wait almost forty years to see another Spanish film with an active female doctor in the leading role, *Red Lights* (Rodrigo Cortés, dir., 2012). In this latter case the film narrative does emphasize the scientific work of the female doctor, Margaret Matheson (played by Sigourney Weaver). Similar to the case of *Agora*, *Red Lights* is a Spanish production in English about a woman from another country, in this case the United States. In other Spanish movies with women doctors,

the film narrative focuses on personal relations with patients, overlooking the scientific aspect of her job, as occurs in *Lejos del mar* (Imanol Uribe, dir., 2015; Far from the sea), where Elena Anaya stars as Marina, a doctor who treats the man who she believes killed her husband. It is not an exaggeration to say that Spanish female doctors have not yet found their voice in Spanish film.

Likewise, nurses have also appeared in a number of Spanish films, but there are just a few movies in which they have a leading role. Among these films are *Las chicas de la cruz roja* (Rafael J. Salvia, dir., 1958; Red cross girls), *Escuela de enfermeras* (Amando de Osorio, dir., 1968; Nursing school), the Spanish-Canadian co-production *My Life Without Me* (Isabel Coixet, dir., 2003), and *Luz de soledad* (Pablo Moreno, dir., 2016; Light of solitude). Television series have also chosen female doctors and nurses as protagonists of popular shows such as *Hospital Central* (Central hospital; a Spanish television series that follows the life and work of the staff in a fictional hospital in Madrid), and *Tiempos de guerra* (War times; a television series about high-society women turned into nurses, who in 1921 are sent to Melilla by Queen Victoria Eugenia to assist Spanish soldiers in the Rif War).

Women pharmacists have also appeared in a number of films and television shows, for example in the television series *Farmacia de guardia* (Antonio Mercero, dir., 1991–1995; On-call pharmacy), where the lead character, Lourdes Cano, is in charge of a pharmacy, but she merely appears as a seller rather than as a person connected to science. In other films, women pharmacists have played a bit part, albeit a crucial one in the film narrative. In Pedro Almodóvar's *¿Qué he hecho yo para merecer esto!* (1984; What have I done to deserve this?) and *Mujeres al borde de un ataque de nervios* (1988; Women on the verge of a nervous breakdown), the protagonists rely on female pharmacists to aid them in order to be able to control their anxiety.

12. Clear examples of how comedy undermines the relation between women and science are *Señora doctor*, *Las chicas de la cruz roja*, and *Supernova* (Juan Miñón, dir., 1992; Supernova).
13. The history of Spanish film opened with the great success of the science fiction film *El hotel eléctrico* (1908; The electric hotel), directed by Segundo de Chomón, but the genre of science fiction has never been highly popular in Spain. This does not mean Spanish film has not produced significant and influential films in the genre of science fiction. Some examples of this are: *Acción mutante* (Álex de la Iglesia, dir., 1993; Mutant action), *Abre los ojos* (Alejandro Amenábar, dir., 1997; Open your eyes), *FAQ* (Carlos Atanes, dir., 2004), *La caja Kovak* (Daniel Monzón, dir., 2006; The Kovak box), *Los cronocrímenes* (Nacho Vigalondo, dir., 2007; Timecrimes), *Tres días* (F. Javier Gutiérrez, dir., 2008; Three days), and some classics such as *La cabina* (Antonio Mercero, dir., 1972; The phone booth) and the television series *Historias para no dormir* (Narciso Ibáñez Serrador, dir., 1966–1982; Stories to keep you up at night).
14. Katherine, the head of the project, orders Hugo to instruct Alex to kill Helena. Although Katherine is very powerful, her presence is almost unnoticeable in the movie.
15. A decade earlier, Alex and Lana had been involved while working on the robot prototype SI-9, which is the same project Alex now heads up. However, things have

changed in Santa Irene as Lana is now married to Alex's brother, who lives with their daughter Eva.
16. As noted in note 3, the characters of the women in charge of the labs—Katherine and Julia, respectively—are barely developed in the films.
17. Lana worked in the laboratory with Alex in the past, but the viewers meet Lana when she is no longer part of the lab, which she has decided to abandon to focus on her family and teaching. Lana continues to be a scientist, but she does not appear in the film as such.
18. The concepts of "maker of meaning" and "bearer of meaning" are borrowed from Laura Mulvey's emblematic essay "Visual Pleasure and Narrative Cinema."
19. "Private" spaces, such as the bedroom and the bathroom, are not monitored by the cameras.
20. Data will keep arriving to the headquarters, although it will not be new data but a re-playing of past data. Alex manages to keep feeding data to the headquarters by running older video recordings of Helena, so that no one in the headquarters will realize that Helena is no longer inside Órbita 9. Thus, scientists will keep receiving pleasure from Helena for a while, through a perpetual feeding of information that won't advance their knowledge.
21. Ironically, instead of proving able to compute endless possibilities quickly, Eva collapses as a human, thus evolving toward becoming a sentient being.
22. In *Eva*, dreams are an essential part of being human; robots, on the other hand, become deactivated when they hear the question: "What do you see when you close your eyes?" We can hear a clear echo of *Blade Runner* (Ridley Scott, dir., 1982) here, a film that has certainly influenced *Eva* and was inspired by the novel *Do Androids Dream of Electric Sheep?* (Philip K. Dick, 1968).

WORKS CITED

Abre los ojos. Directed by Alejandro Amenábar, Las Producciones del Escorpión, 1997.

Acción mutante. Directed by Álex de la Iglesia, El Deseo, 1993.

Ágora. Directed by Alejandro Amenábar, Mod Producciones, 2009.

Angulo, Eduardo. "Así vemos a las científicas en el cine y la televisión." *Mujeres con ciencia*, 19 Dec. 2017, mujeresconciencia.com/2017/12/19/asi-vemos-las-cientificas-cine-la-television.

"Apenas hay mujeres en las carreras científicas: el 80% de los estudiantes son hombres." *Antena3Noticias.com*, 11 February 2019, www.antena3.com/noticias/ciencia/mujeres-carreras-cientificas-estudiantes-hombres-video_201902115c611c4b0cf2cdb074ce4a93.html.

Barthes, Roland. *S/Z: An Essay*. Hill and Wang, 1975.

Blade Runner. Directed by Ridley Scott, Warner Brothers, 1982.

Bombshell: The Hedy Lamarr Story. Directed by Alexandra Dean, Reframed Pictures, 2017.

Busch, Caitlin. "Annihilation and the Hard Truth about Women Scientists in Genre Films." *SyFyWire*, 7 Mar. 2018, www.syfy.com/syfywire/annihilation-and-the-hard-truth-about-women-scientists-in-genre-films.

Casado Ruiz de Lóizaga, María José. *Las damas del laboratorio: Mujeres científicas en la historia*. Barcelona, Debate, 2006.

Chambers, Amy C. "Rise of the Women? Screening Female Scientists." *The Science and Entertainment Laboratory*, 2 March 2015, thescienceandentertainmentlab.com/rise-of-the-women.

CodeGirl. Directed by Lesley Chilcott, H Films, 2015.

Contact. Directed by Robert Zemeckis, Warner Brothers, 1997.

Contagion. Directed by Steven Soderbergh, Warner Brothers, 2011.

Dick, Philip K. *Do Androids Dream of Electric Sheep?* Doubleday, 1968.

Ellas: Científicas. Ellas. RTVE, 7 August 2017, www.rtve.es/alacarta/videos/ellas/ellas-cientificas/4154498.

Escuela de enfermeras. Directed by Amando de Ossorio, Coperfilm, 1968.

Eva. Directed by Kike Maíllo, Escándalo Film, 2011.

FAQ. Directed by Carlos Atanes, Fort Knox Audiovisual, 2004.

Farmacia de guardia. Directed by Antonio Mercero, produced by Andrés Gandara, 1991–1995.

Girls in Tech Spain. Women, Technology, Entrepreneurship, 2015–2018, spain.girlsintech.org. Accessed 9 September 2020.

Gravity. Directed by Alfonso Cuarón, Warner Brothers, 2013.

Gunderson, Lauren. *Silent Sky*. New York, Dramatists Play Service, 2015.

Hidden Figures. Directed by Theodore Melfi, Fox 2000 Pictures, 2016.

Hospital central. Estudios Picasso, 2000–2012.

El hotel eléctrico. Directed by Segundo de Chomón, Pathé Frères, 1908.

Historias para no dormir. Directed by Narciso Ibáñez Serrador, TVE, 1966–1982.

Hypatia Project. Funded by the European Union's Horizon 2020 Framework Programme for Research and Innovation (H2020-GERI-2014). www.expecteverything.eu/hypatia. Accessed 9 September 2020.

Iniciativa 11 de febrero. 5 July 2019, 11defebrero.org.

Inspiring Girls. Fundación Inspiring Girls España. inspiring-girls.es. Accessed 8 September 2020.

Interstellar. Directed by Christopher Nolan, Paramount Pictures, 2014.

La cabina. Directed by Antonio Mercero, TVE, 1972.

La caja Kovak. Directed by Daniel Monzón, Filmax, 2006.

Las chicas de la cruz roja. Directed by Rafael J. Salvia, Asturias Films, 1958.

Lejos del mar. Directed by Imanol Uribe, Maestranza Films, 2015.

Lévi-Strauss, Claude. *Structural Anthropology*. Basic Books, 1974.

Los cronocrímenes. Directed by Nacho Vigalondo, Karbo Vantas Entertainment, 2007.

Los recuerdos de hielo. Directed by Albert Solé, Minimal Films, 2012.

Luz de soledad. Directed by Pablo Moreno, Contracorriente Producciones SLU, 2016.

Marie Curie: The Courage of Knowledge. Directed by Marie Noëlle, P'Artisan Filmproduktion GmbH, 2016.

Mujeres al borde de un ataque de nervios. Directed by Pedro Almodóvar, El Deseo, 1988.

Mulvey, Laura. "Visual Pleasure and Narrative Cinema." *Screen*, vol. 16, no. 3, 1975, pp. 6–18.

Munera, Isabel. "Faltan profesionales con conocimientos tecnológicos y científicos, sobre todo mujeres." *ElMundo.es*, 29 May 2018, www.elmundo.es/nosotras/2018/05/29/5b0c4337e2704e872f8b45be.html.

My Life Without Me. Directed by Isabel Coixet, El Deseo, 2003.

The New Oxford Annotated Bible. 3rd ed., Oxford UP, 2001.

Núñez Valdés, Juan. "Did Fátima de Madrid Really Exist?" *Review of Social Sciences*, vol. 1, no. 2, 2016, pp. 19–26.

Núñez Valdés, Juan, María Arroyo Castilleja, and Alejandro Alonso Álvarez-Rementería. "Primeras mujeres farmacéuticas en España." *Libro de Actas del II Congreso Internacional de Comunicación y Género*, edited by Juan Carlos Suárez Villegas, Rosario Lacalle Zalduendo, and José Manuel Pérez Tornero, U of Sevilla, 2014, pp. 817–31.

Órbita 9. Directed by Hatem Khraiche, Cactus Flower, 2017.

Outbreak. Directed by Wolfgang Petersen, Warner Brothers, 1995.

Perla Mateo, María Pilar. "María Andrea Casamayor, la primera que publicó un libro de ciencia en España." *Heraldo.es*, 21 February 2017, www.heraldo.es/noticias/suplementos/tercer-milenio/divulgacion/2017/02/21/la-primera-que-publico-libro-ciencia-1160215-2121028.html.

Pico, Raquel C. "Las mujeres olvidadas de la ciencia." *Revista Contexto*, no. 104, 15 February 2017, ctxt.es/es/20170215/Firmas/10860/cientificas-NASA-D%C3%ADa-Internacional-de-la-Mujer-y-la-Niña-en-la-Ciencia-machismo-mujeres-en-la-ciencia.htm.

Propp, Vladimir. *Morphology of the Folktale*. U of Texas P, 1968.

Puyo, Patricia. "La mujer que nos acercó a la ciencia." *ElPeriódicodeAragón.es*, 24 April 2009, www.elperiodicodearagon.com/noticias/aragon/mujer-acerco-ciencia_493331.html.

¿Qué he hecho yo para merecer esto? Directed by Pedro Almodóvar, Kaktus Producciones, 1984.

Red Lights. Directed by Rodrigo Cortés, Millennium Films, 2012.

Señora doctor. Directed by Mariano Ozores, Producciones Internacionales Cinematográficas Asociadas, 1974.

Shetterly, Margot Lee. *Hidden Figures: The American Dream and the Untold Story of the Black Women Who Helped Win the Space Race*. William Morrow, 2016.

Sobel, Dava. *The Glass Universe: How the Ladies of the Harvard Observatory Took the Measure of the Stars*. Viking P, 2016.

Splice. Directed by Vincenzo Natali, Gaumont, 2009.

Supernova. Directed by Juan Miñón, Aligator Producciones, 1993.

Tiempos de guerra. Atresmedia Televisión, 2017.

Tres días. Directed by F. Javier Gutiérrez, Green Moon Productions, 2008.

Todorov, Tzvetan. *The Poetics of Prose*. Cornell UP, 1977.

UNESCO. "Women in Science." 2018, uis.unesco.org/en/topic/women-science. Accessed 9 September 2020.

United Nations General Assembly. "70/212. International Day of Women and Girls in Science." *Resolution adopted by the General Assembly on its 81st plenary meeting*, 22 December 2015, undocs.org/en/A/RES/70/212.

Appendix: List of Works by Genre Addressed in This Volume

ANTHOLOGIES

Varo, Remedios, and Isabel Castell. *Cartas, sueños y otros textos*. Era, 2002.

Arnedo Soriano, Elena, coord. *El gran libro de la mujer*. Temas de Hoy, 1997.

Robles, Lola, and Teresa López Pellisa, editors. *Antología de escritoras españolas de ciencia ficción*. Vol. 1, *Distópicas*, Libros de la Ballena, 2018.

———. *Antología de escritoras españolas de ciencia ficción*. Vol. 2, *Poshumanas*, Libros de la Ballena, 2018.

ESSAYS

Acuña y Villanueva, Rosario de. "A las mujeres del siglo XIX." *Obras reunidas*. Vol. 2, *Artículos (1885–1923)*, edited by José Bolado, KRK, 2007, pp. 1225–40.

———. "La higiene en la familia obrera." *Obras reunidas*. Vol. 3, *Prosa*, edited by José Bolado, KRK, 2008, pp. 745–78.

———. "La jarca de la universidad." *Obras reunidas*. Vol. 2, *Artículos (1885–1923)*, edited by José Bolado, KRK, 2007, pp. 1609–18.

———. "La ramera." *Las Dominicales del Libre Pensamiento*, año 5, no. 234, 28 May 1887, pp. 1–2.

———. "Testamento." *Obras reunidas*. Vol. 5, *Lírica*, edited by José Bolado, KRK, 2009, pp. 635–41.

Arenal, Concepción. *La beneficencia, la filantropía y la caridad*. Imprenta del Colegio de Sordo-Mudos y de Ciegos, 1861.

———. "Carta XXIV. Delitos contra la honestidad. Artículos 358 al 362." *Cartas a los delincuentes. Obras completas*. Tomo 3, Victoriano Suárez, 1894, pp. 307–19.

———. "Centro protector de la mujer." *Obras sobre beneficencia y prisiones*. Vol. 5, *Obras completas*. Tomo 22, Victoriano Suárez, 1902, pp. 414–18.

———. *La mujer del porvenir y Artículos sobre las Conferencias dominicales para la educación de la mujer*. Eduardo Perié, 1870.

———. *La mujer del porvenir; La mujer de su casa*. Orbis, 1989.

Arnedo Soriano, Elena. *Desbordadas: La agitada vida de la elastic woman*. Temas de Hoy, 2000.

———. "Advertencia Preliminar." *El donjuanismo femenino*, edited by Elena Soriano, Península, 2000, pp. 11–16.

———. "Centro de planificación familiar Federico Rubio." *Enciclopedia Madrid, Siglo XX*, edited by Carlos Sambricio, Ayuntamiento de Madrid, 2002, pp. 64–65.

———. *La picadura del tábano: La mujer frente a los cambios de la edad*. Aguilar, 2003.

Burgos, Carmen de. "Las mujeres de ciencia." *La inferioridad mental de la mujer. (La deficiencia mental psicológica de la mujer)*. Imprenta del Pueblo, 1905.

———. "El secreto de Moebius." Madrid, Spain, 11 Apr. 1909. *El Heraldo de Madrid/*Femeninas column.

Casas, Santiago. "Sexta conferencia sobre la higiene de la mujer." *Conferencias dominicales sobre la educación de la mujer*, 28 Mar. 1869, M. Rivadeneyra, 1869.

Haraway, Donna. "A Cyborg Manifesto: Science, Technology, and Socialist-Feminism in the Late Twentieth Century." *Simians, Cyborgs and Women: The Reinvention of Nature*, Routledge, 1991, pp. 149–81.

Martín Gaite, Carmen. "La chica rara." *Desde la ventana: Enfoque femenino de la literatura española*, Espasa Calpe, 1987, pp. 101–22.

Santesmases, María Jesús. "Towards Denaturalization: Women Scientists and Academics in Twentieth-Century Spain," *Feministische Studien*, vol. 1, no. 11, 2011, pp. 52–63.

ESSAYS (BOOK-LENGTH)

Aleu y Riera, Dolores. *De la necesidad de encaminar por nueva senda la educación higiénico-moral de la mujer*. La Academia, 1883.

Castells Ballespí, Martina. *Educación de la mujer. Educación física, moral e intelectual que debe darse a la mujer para que esta contribuya en grado máximo a la perfección y la de la Humanidad. Memoria leída por en el acto de recibir la Investidura de Doctor en Medicina. Madrid, octubre 1882.* "La educación de la mujer según las primeras doctoras en medicina de la universidad española, año 1882." Consuelo Flecha García. *Dynamis*, vol. 19, 1999, pp. 241–78.

Fernández Gutiérrez, José María. *La Novela del Sábado (1953–1955). Catálogo y contexto histórico literario*. Madrid, CSIC, 2004.

Martín Gaite, Carmen. *Usos amorosos del dieciocho en España*. Anagrama, 1972.

———. *Usos amorosos de la posguerra española*. Anagrama, 1987.

Morales, María Pilar. *Mujeres (Orientación femenina)*. Editora Nacional, 1944.

Santesmases, María Jesús. *The Circulation of Penicillin in Spain: Health, Wealth, and Authority*. Palgrave, 2018.

———. *Entre Cajal y Ochoa: Ciencias biomédicas en la España de Franco, 1939–1975.* CSIC, 2001.

———. *Mujeres científicas en España (1940–1970): Profesionalización y modernización social.* Instituto de la Mujer, 2000.

Santesmases, María Jesús, and Teresa Ortiz-Gómez. *Gendered Drugs and Medicine: Historical and Socio-Cultural Perspectives.* Routledge, 2014.

FILMS AND TELEVISION

Abre los ojos. Directed by Alejandro Amenábar, Las Producciones del Escorpión, 1997.

Acción mutante. Directed by Álex de la Iglesia, El Deseo, 1993.

Ágora. Directed by Alejandro Amenábar, Mod Producciones, 2009.

Blade Runner. Directed by Ridley Scott, Warner Brothers, 1982.

Bombshell: The Hedy Lamarr Story. Directed by Alexandra Dean, Reframed Pictures, 2017.

CodeGirl. Directed by Lesley Chilcott, H Films, 2015.

Contact. Directed by Robert Zemeckis, Warner Brothers, 1997.

Contagion. Directed by Steven Soderbergh, Warner Brothers, 2011.

Ellas: Científicas. Ellas. RTVE, 7 August 2017, www.rtve.es/alacarta/videos/ellas/ellas-cientificas/4154498.

Eva. Directed by Kike Maíllo, Escándalo Film, 2011.

FAQ. Directed by Carlos Atanes, Fort Knox Audiovisual, 2004.

Farmacia de guardia. Directed by Antonio Mercero, produced by Andrés Gandara, 1991–1995.

Gravity. Directed by Alfonso Cuarón, Warner Brothers, 2013.

Hidden Figures. Directed by Theodore Melfi, Fox 2000 Pictures, 2016.

Hospital central. Estudios Picasso, 2000–2012.

El hotel eléctrico. Directed by Segundo de Chomón, Pathé Frères, 1908.

Historias para no dormir. Directed by Narciso Ibáñez Serrador, TVE, 1966–1982.

Interstellar. Directed by Christopher Nolan, Paramount Pictures, 2014.

La cabina. Directed by Antonio Mercero, TVE, 1972.

La caja Kovak. Directed by Daniel Monzón, Filmax, 2006.

Las chicas de la cruz roja. Directed by Rafael J. Salvia, Asturias Films, 1958.

Lejos del mar. Directed by Imanol Uribe, Maestranza Films, 2015.

Los cronocrímenes. Directed by Nacho Vigalondo, Karbo Vantas Entertainment, 2007.

Los recuerdos de hielo. Directed by Albert Solé, Minimal Films, 2012.

Luz de soledad. Directed by Pablo Moreno, Contracorriente Producciones SLU, 2016.

Marie Curie: The Courage of Knowledge. Directed by Marie Noëlle, P'Artisan Filmproduktion GmbH, 2016.

Mujeres al borde de un ataque de nervios. Directed by Pedro Almodóvar, El Deseo, 1988.

Órbita 9. Directed by Hatem Khraiche, Cactus Flower, 2017.

Outbreak. Directed by Wolfgang Petersen, Warner Brothers, 1995.

¿Qué he hecho yo para merecer esto? Directed by Pedro Almodóvar, Kaktus Producciones, 1984.

Red Lights. Directed by Rodrigo Cortés, Millennium Films, 2012.

Señora doctor. Directed by Mariano Ozores, Producciones Internacionales Cinematográficas Asociadas, 1974.

Splice. Directed by Vincenzo Natali, Gaumont, 2009.

Supernova. Directed by Juan Miñón, Aligator Producciones, 1993.

Tiempos de guerra. Atresmedia Televisión, 2017.

Tres días. Directed by F. Javier Gutiérrez, Green Moon Productions, 2008.

GOVERNMENTAL PUBLICATIONS

Libro Blanco. Situación de las Mujeres en la Ciencia Española. 2011, www.ciencia.gob.es/stfls/MICINN/Ministerio/FICHEROS/UMYC/LibroBlanco_Interactivo.pdf. Accessed 15 September 2020.

Consejo Superior de Investigaciones Científicas (CSIC). Mujeres y ciencia, www.csic.es/mujeres-y-ciencia. Accessed 15 September, 2020.

Ley Orgánica 3/2007 de 22 de marzo. *Boletín Oficial de Estado (BOE),* www.boe.es/buscar/pdf/2007/BOE-A-2007-6115-consolidado.pdf. Accessed 15 September 2020.

Ministerio de Economía, Industria y Competividad Gobierno de España. "Científicas en cifras: Estadísticas e indicadores de la (des)igualdad de género en la formación y profesión científica," 2015, www.ciencia.gob.es/stfls/MICINN/Ministerio/FICHEROS/Informe_Cientificas_en_Cifras_2015_con_Anexo.pdf. Accessed 15 September 2020.

UNESCO. "Women in Science." 2018, uis.unesco.org/en/topic/women-science. Accessed 15 September 2020.

United Nations General Assembly. "70/212. International Day of Women and Girls in Science." *Resolution adopted by the General Assembly on its 81st plenary meeting,* 22 December 2015, undocs.org/pdf?symbol=en/A/RES/70/212. Accessed 15 September 2020.

MAGAZINES

Estampa
Medina, Semanario Oficial de la Sección Femenina
Nuevo Mundo
La Guinea Española
Y. Revista para mujeres

NEWSPAPER ARTICLES

"La mujer y la preparación intelectual." *Haz*, no. 13, May 1939, pp. 40–41.

"Consultorio." *Medina, Semanario Oficial de la Sección Femenina*, no. 34, 11 Nov. 1941, p. 21.

Corcuera, María Gabriela. "La mujer universitaria." *Y. Revista para la mujer*, no. 50, Mar. 1942, pp. 24–25.

"Destino de la mujer falangista." *Medina, Semanario Oficial de la Sección Femenina*, no. 1, 20 Mar. 1941, p. 3.

"Diez mandamientos de la universitaria." *Medina, Semanario Oficial de la Sección Femenina*, no. 216, 6 Jun. 1945, p. 18.

"La educación de la mujer." *Medina, Semanario Oficial de la Sección Femenina*, no. 75, 23 Aug. 1942, p. 14.

"La mujer, alma del hogar, preparada por la Falange." *Y. Revista para la mujer*, no. 59, Dec. 1942, p. 37.

"La mujer y la preparación intelectual." *Haz*, no. 13, May 1939, pp. 40–41.

"Las terribles intelectuales." *Medina, Semanario Oficial de la Sección Femenina*, no. 134, 19 Oct. 1943, p. 15.

"Lo que las armas victoriosas traen, mujer." *Y. Revista para la mujer*, no. 15, Apr. 1939, pp. 12–13.

"Muchachas en la universidad." *Medina, Semanario Oficial de la Sección Femenina*, no. 64, 7 Jun. 1942, pp. 18–19.

"Retornos a las aulas." *Y. Revista para la mujer*, no. 81, Oct. 1944, p. 34.

NOVELS

Ballesteros, Mercedes (Sylvia Visconti, pseud.). *María Elena, ingeniero de caminos*. La Novela Ideal, 1940.

Barceló, Elia. *Consecuencias naturales*. Miraguano Ediciones, 1994.

Bustelo, Gabriela. *Planeta hembra*. RBA, 2001.

Burgos, Carmen de. *La inferioridad mental de la mujer. (La deficiencia mental psicológica de la mujer)*. Imprenta del Pueblo, 1905.

Janés, Clara. *El hombre de Adén*. Anagrama, 1991.

Laforet, Carmen. *Nada*. 9th ed., Destino, 1987.

Montero, Rosa. *Instrucciones para salvar el mundo*. Santillana/Punto de lectura, 2009.

_____. *Lágrimas en la lluvia*. Seix-Barral, 2011.

_____. *El peso del corazón*. Planeta, 2015.

_____. *La ridícula idea de no volver a verte*. Seix-Barral, 2013.

Pérez Galdós, Benito. *Fortunata y Jacinta*. Edited by Francisco Caudet, Cátedra, 2006. 2 vols.

PHOTOGRAPHIC ARCHIVES

Liverpool Public Record Office (Liverpool, England)
Archivo General de la Administración (Alcalá de Henares, Spain)
Real Biblioteca de Palacio (Madrid, Spain)
Trinidad Morgades-Laida Memba Archive (Private)
Crónicas de Guinea Ecuatorial (Online). bioko.net/galeriaFA.

POETRY

Janés, Clara. *Creciente fértil*. Hiperión, 1989.
_____. *Diván del ópalo de fuego (O la leyenda de Layla y Machnún)*. Regional de Murcia, 1996.
_____. *El libro de los pájaros*. Pre-Textos, 1999.
_____. *Estructuras disipativas*. Tusquets, 2017.
_____. *Fósiles*. Twenty-one poems with nine engravings by Rosa Biadiu. 2nd ed, Z.I.P., 1985, 1987.
_____. *Fractales*. Pre-textos, 2005.
_____. *La indetenible quietud*. Thirty-two poems with six engravings by Eduardo Chillida. Boza, 1998.
_____. *La tentación del paraíso*. Poética y poesía, 29 Apr. 2014, Fundación Juan March, Madrid. www.march.es/events/100043.
_____. *Lapidario*. Hiperión, 1988.
_____. *Las estrellas vencidas*. Agora, 1964.
_____. *Límite humano*. Oriens, 1973.
_____. *Los números oscuros*. Ciruela, 2006.
_____. *Movimientos insomnes*. Del Centro, 2014.
_____. *Orbes del sueño*. Vaso Roto, 2013.
_____. *Paralajes*. Tusquets, 2002.
_____. *Rosas de fuego*. Cátedra, 1996.
_____. *Variables ocultas*. Vaso Roto, 2010.

SHORT STORIES AND SHORT STORY COLLECTIONS

Barceló, Elia. "La dama dragón." *Sagrada*, Sportula, 2019, pp. 279–230.
_____. "Piel." *Sagrada*, Sportula, 2019, pp. 279–342.
_____. *Sagrada*. Ediciones B, 1989.
Mayoral, Marina. "Admirados colegas." *Querida amiga*. Alfaguara, 2002, pp. 31–55.
_____. *Querida amiga*. Alfaguara, 2002.

TEXTBOOKS (MANUALS)

Pascual de Sanjuán, Pilar. *Flora o La educación de una niña*. Faustino Paluzíe, 1889.

Pulido, Ángel Fernández. *Bosquejos médico-sociales para la mujer*. Víctor Saiz, 1876.

Solís Claras, Manuela. *Higiene del embarazo y de la primera infancia. Para las madres*. Prologue by Santiago Ramón y Cajal, Imprenta F. Vives Mora, 1907, 364 pp.

Tratado de economía y labores para uso de las niñas. Parte componente del Educador publicado por la Casa de José González. 3rd ed., Museo de la Educación, 1861.

WEB SITES

*Crónicas de Guinea Ecuatoria*l. bioko.net/galeriaFA. Accessed 24 September 2020.

Girls in Tech Spain. Women, Technology, Entrepreneurship, 2015–2018, spain.girlsintech.org. Accessed 15 September 2020.

Hypatia Project. Funded by the European Union's Horizon 2020 Framework Programme for Research and Innovation (H2020-GERI-2014), www.expecteverything.eu/hypatia. Accessed 15 September 2020.

Iniciativa 11 de febrero. 5 July 2019, 11defebrero.org. Accessed 15 September 2020.

Inspiring Girls. Fundación Inspiring Girls España, inspiring-girls.es. Accessed 15 September 2020.

UNESCO. "Women in Science." 2018, uis.unesco.org/en/topic/women-science. Accessed 15 September 2020.

United Nations. "International Day of Women and Girls in Science 11 Feb." www.un.org/en/observances/women-and-girls-in-science-day. Accessed 24 September 2020.

Valdés, Isabel. "Especial: Mujeres de la ciencia." *El País*, 8 Feb. 2018, elpais.com/especiales/2018/mujeres-de-la-ciencia. Accessed 24 September 2020.

WOMEN AND SCIENCE IN SPAIN: VOLUMES

Bauer, Martin W., and Susan Howard. *The Culture of Science in Modern Spain: An Analysis of Public Attitudes across Time, Age Cohorts and Regions*. Fundación BBVA, 2013.

Casado Ruiz de Lóizaga, María José. *Las damas del laboratorio: Mujeres científicas en la historia*. Debate, 2006.

Flecha García, Consuelo. *Las primeras universitarias en España (1872–1910)*. Narcea, 1996.

García de Cortázar y Nebreda, María Luisa, et al. *Mujeres y hombres en la ciencia española. Una investigación empírica*. Ministerio de Trabajo y Asuntos Culturales/Instituto de la Mujer, 2006.

Magallón Portolés, Carmen. *Pioneras españolas en las ciencias. Las mujeres del Instituto Nacional de Física y Química*. CSIC, 1998.

Muñoz-Páez, Adela. *Sabias. La cara oculta de la ciencia*. Penguin Random-House, 2017.

Ortiz Gómez, Teresa, and Gloria Becerra, eds. *Mujeres de ciencias: Mujeres, feminismo y ciencias naturales, experimentales y tecnológicas*. Universidad de Granada–Instituto de Estudios de la Mujer, 1996,

Osca Lluch, Julia. *La aportación de la mujer a la historia de la ciencia y de la técnica en España*. Instituto de Historia de la Medicina y de la Ciencia López Piñero, 2011.

Vázquez Ramil, Raquel. *La Institución Libre de Enseñanza y la educación de la mujer en España: La Residencia de Señoritas*. Lugami Artes Gráficas, 2001.

Contributors

SILVIA BERMÚDEZ is professor of Iberian and Latin American studies at the University of California Santa Barbara. Her critical work focuses on Spanish cultural studies, with emphasis on music studies, feminism and women's studies, Galician studies, poetic discourses, and politics. She is the author of *Las dinámicas del deseo: subjetividad y lenguaje en la poesía española contemporánea* (1997), *La esfinge de la escritura: la poesía ética de Blanca Varela* (2005), and *Rocking the Boat: Race and Migration in Contemporary Spanish Music* (2017). She co-edited the volume *From Stateless Nations to Postnational Spain/De Naciones sin estado a la España Postnacional* (2002, with Antonio Cortijo Ocaña and Timothy McGovern), *A New History of Iberian Feminisms* (2018, with Roberta Johnson), and *Cartographies of Madrid: Contesting Urban Space at the Crossroads of the Global South and Global North* (2019, with Anthony L. Geist).

MARTA DEL POZO ORTEA is assistant professor of Spanish at the University of Massachusetts-Dartmouth. Her current research addresses the emerging ecological paradigm by fostering a posthumanist dialogue that transcends the dualisms arts/sciences, mind/matter, myth/logos. She focuses on epistemologically-hybrid literary pieces, documentary, the visual arts, and different forms of performativity in Spanish and Latin American culture. Her articles have appeared in *Hispania*, *Letras Hispanas*, *The Latin Americanist*, *Hispanófila*, *International Journal of Iberian Studies*, *ALCES XXI*, *The International Journal of the Image*, and *Revista Internacional de Cultura Visual*. She has also translated into Spanish the poetry of Richard Jackson, Robert Bringhurst, Leonard Schwartz, as well as *The Dead* by James Joyce. As a poet, she received the International Poetry Prize Antonio Gala for *Hambre de imágenes*, and her book *Nigredo* was the recipient of the 2017 Iberoamerican Prize Entreversos. She is the founding editor at Quantum Prose, New York.

DEBRA FASZER-MCMAHON is professor of Spanish and dean of the School of Humanities at Seton Hill University. Her research interests include transnational poetics, immigration, cultural studies, and women writers. She has published *Cultural Encounters in Contemporary Spain: The Poetry of Clara Janés* (Bucknell UP, 2010), *African Immigrants in Contemporary Spanish Texts: Crossing the Strait* (Routledge/Ashgate Press, 2015, co-edited with Victoria L. Ketz), as well as articles in *Hispania, Afro-Hispanic Review, Anales de la Literatura Española Contemporánea, Letras Femeninas, Transmodernity: Journal of Peripheral Cultural Production of the Luso-Hispanic World*, and *Symposium: A Quarterly Journal in Modern Literatures*. She is currently working on a book-length project about La Generación de la Amistad and Saharawi poetics in Spain.

MIRLA GONZÁLEZ received her PhD from the University of Kansas in 2019. She is currently assistant director of Undergraduate Studies, Online and Professional Education, and teaching faculty of Spanish in the School of Modern Languages at the Georgia Institute of Technology. Her research addresses the intersection between science (biotechnology and robotics), gender, race, and ethics. She is particularly interested in female science fiction writers. She is the development editor of OER Acceso.

ROBERTA JOHNSON is professor emerita at the University of Kansas and adjunct professor at the University of California, Los Angeles. She is the author of *Carmen Laforet* (Twayne, 1981), *El ser y la palabra en Gabriel Miró* (Fundamentos, 1985), *Crossfire: Philosophy and the Novel in Spain 1900–1934* (U Kentucky P, 1993; Spanish translation, 1996), *Las bibliotecas de Azorín* (Caja de Ahorros del Mediterráneo, 1996), *Gender and Nation in the Spanish Modernist Novel* (Vanderbilt UP, 2003), and *Major Concepts in Spanish Feminist Theory* (SUNY, 2019). She co-edited the *Antología del pensamiento feminista español 1726–2011* (Cátedra, 2013) with Maite Zubiaurre and *A New History of Iberian Feminisms* (U Toronto P, 2018) with Silvia Bermúdez. She has received major grants and prizes from the Graves Foundation, the National Endowment for the Humanities, the Fulbright Commission, the Guggenheim Foundation, Isabel la Católica knighthood from His Majesty King Juan Carlos of Spain, and the Premio Victoria Urbano from La Asociación Internacional de Literatura y Cultura Femenina Hispánica.

VICTORIA L. KETZ is professor of Spanish, chairperson of Foreign Languages and Literatures, and director of the Central and Eastern European Studies Program at La Salle University. She has published articles and presented papers on pedagogy, mainly in the area of integrating innovative curriculum or new technologies in the classroom, as well as critical literary inquiry, focusing on recuperating historical memory, immigration, violence against women, urban studies, and twentieth-century writers such as Montero, Grandes, Ferré, Rodoreda, Ruiz Zafón, Millás, and others. In 2015, she co-edited with Debra Faszer-McMahon the collection *African Immigrants in Contemporary Spanish Texts: Crossing the Strait* (Routledge/Ashgate Press). Presently, she is preparing a special edition of Revista de Estudios de Genero y Sexualidad on disability studies with Esther Fernández. Her current theoretical book project examines the representation of violence against women and the usurpation of the female voice in contemporary Spanish writings.

MARYANNE L. LEONE is associate professor of Spanish at Assumption College. Her research interests include contemporary Spanish fiction and cultural production, women writers, ecocriticism, ecofeminism, migration, and disability studies. She has published articles in the journals *Anales de la Literatura Española Contemporánea, Ecozon@: European Journal of Literature, Culture and Environment, Letras Hispanas, Revista Canadiense de Estudios Hispánicos,* and *Revista de Estudios de Género y Sexualidades,* and in *Toward a Multicultural Configuration of Spain: Local Cities, Global Spaces* (Ana Corbalán and Ellen Mayock, co-editors. Farleigh Dickinson University Press, 2015). She is working on a co-edited book project on the environment and the more-than-human in Spanish cultural production from the medieval through the post-crisis periods.

ELLEN MAYOCK is the Ernest Williams II Professor of Spanish at Washington and Lee University and adviser to the English for Speakers of Other Languages (ESOL) program. Mayock works on twentieth- and twenty-first-century narrative and film, gender studies, and cultural studies. Her publications include the textbook *Indagaciones* (Georgetown UP, 2019; with Mary Ann Dellinger and Beatriz Trigo), the monographs *Gender Shrapnel in the Academic Workplace* (Palgrave, 2016) and *The 'Strange Girl' in Twentieth-Century Spanish Novels Written by Women* (U of the South, 2004), and the co-edited volumes *Toward a Multicultural Configuration of Spain. Local Cities, Global Spaces* (Fairleigh Dickinson UP, 2014; with Ana Corbalán), *Forging a Rewarding Career in the Humanities* (Sense/Brill, 2014; with Karla P. Zepeda), and *Feminism Activism in Academia* (McFarland, 2010; with Domnica Radulescu), along with articles and book chapters in a variety of venues. Mayock writes poetry and drama as well as maintaining a blog, Gender Shrapnel.

LESLIE ANNE MERCED is associate professor of Spanish and associate chair of the Department of English, Modern Languages and Literatures, and Fine Arts at Rockhurst University. Her research interests include gender studies in modern and contemporary Peninsular literature, with emphasis on the *fin de siglo*, turn of the century (nineteenth/twentieth), in Spain. She also researches and writes with a transnational focus, publishing work on Sor Juana Inés de la Cruz, the Generation of 1898 and the literary canon, as well as the relationship between medicine and gender. Her conference presentations deal thematically with Caribbean-Transatlantic encounters between women writers, such as Puerto Rican poet Lola Rodríguez de Tió and Spain's Concha Espina. Her studies and presentations on Carmen de Burgos's translation theory highlight her second area of expertise, German literature and culture of the nineteenth/twentieth centuries.

INÉS PLASENCIA is a cultural researcher, manager, and writer interested in the relationships between visual culture, political imagination, and citizenship, paying special attention to colonial contexts and their continuities. She holds a PhD in history and theory of art from the Autonomous University of Madrid with a thesis entitled "Image and Citizenship in Equatorial Guinea (1861–1937): From the Photographic Encounter to the Colonial Order." She is currently associate lecturer at the Autonomous University of Madrid and teaches colonialism and visual culture at Duke University in Madrid. She has also worked for different cultural institutions in Spain, such as *Tabakalera* and *Intermediae-Matadero*, and has recently been part of the research and conceptualization team of *Repensar*

Guernica, a website and digital archive project, at the Reina Sofía Museum in Madrid.

MARÍA JESÚS SANTESMASES is research professor at the Institute of Philosophy of the *Consejo Superior de Investigaciones Científicas* (CSIC) in Madrid. Her research deals with post-WWII biology and biomedicine, women scientists and gender, the practices of human cytogenetics in the atomic age, and the history of antibiotics in Spain. Her publications include *Mujeres científicas en España 1940–1970: Profesionalización y modernización social* (2000), *Entre Cajal y Ochoa: Ciencias biomédicas en la España de Franco, 1939–1975* (2001), "Towards Denaturalization: Women Scientists and Academics in Twentieth-Century Spain" (2011), and with Teresa Ortiz-Gómez *Gendered Drugs and Medicine: Historical and Socio-Cultural Perspectives* (2014). Among her recent publications are *The Circulation of Penicillin in Spain: Health, Wealth and Authority* (Palgrave, 2018) and "Circulating Biomedical Images: Bodies and Chromosomes in the Post-Eugenic Era." *History of Science*, vol. 55, 2017, pp. 395–430.

DAWN SMITH-SHERWOOD is professor of Spanish in the Department of Foreign Languages at Indiana University of Pennsylvania. Her research interests include Spanish women's literature, second-language proficiency, and high-impact practices in higher education (living-learning communities, capstone experiences, study abroad). She has published on these topics in professional journals (*Letras Femeninas, Antípodas*) and edited collections. She is currently engaged in a Scholarship of Teaching and Learning project to resolve the language-literature divide in post-secondary education and has published articles on her use of the ACTFL Integrated Performance Assessment in literature courses in the MLA's *Approaches to Teaching the Works of Miguel de Unamuno* and *MIFLC Review*.

MIGUEL SOLER GALLO holds a doctorate in Hispanic philology from the University of Salamanca, where he teaches Spanish language and culture to students from abroad. His research interests include twentieth- and twenty-first-century Spanish women's narrative with a special interest in discourse studies and analysis of political and ideological language. His publications include book chapters and journal articles that treat the ideology of the Sección Femenina, the women's group associated with the Spanish Falange, a fascist organization. He is currently editing several works by, as well as completing a biography of, the Spanish woman writer and lawyer Mercedes Formica (1913–2002), a member of the Generation of '36.

ERIKA M. SUTHERLAND is associate professor of Spanish and chair of the Department of Languages, Literatures, and Cultures at Muhlenberg College. Her research in contemporary Spanish narrative focuses primarily on the representation of disease, health, and the female body in both literary and medical texts. Her recent chapter in *MLA Approaches to Teaching Emilia Pardo Bazán* outlines this pedagogy, while articles in *La hora de Galdós. Actas del XI Congreso Internacional Galdosiano*, *Mulheres Más. Percepção e Representações da Mulher Transgressora no Mundo Luso-Hispánico*, and *Excavatio* explore different aspects of medical knowledge in nineteenth-century Spanish literature. Her forthcoming publication, *(Con)textos femeninos: Una antología de escritoras españolas* (Peter Lang, 2019), co-edited with Elizabeth Smith Rousselle, is a critical anthology of Spain's women writers, both well-known and unjustly forgotten. Her current research considers women's representation of hunger prior to and following the Spanish Civil War.

RAQUEL VEGA-DURÁN is senior lecturer at Harvard University and previously was associate professor of Spanish at Claremont McKenna College. Her research focuses on migration studies and transatlantic literature and film, while she teaches courses on Peninsular literature, Spanish art history, film analysis, women's narratives, transatlantic cultural history, and world migrations. She is the author of *Emigrant Dreams, Immigrant Borders: Migrants, Transnational Encounters, and Identity in Spain* (Bucknell UP, 2016) and has published articles about immigration, the concept of identity in Spain, visual female narratives, Mediterranean dialogues, and the repopulation of rural Spain in *Afro-Hispanic Review*, *Letras femeninas*, and *Quaderns de cine*, as well as chapters in *African Immigrants in Contemporary Spanish Texts*, *Toward a Multicultural Configuration of Spain*, and *Todos a movilizarse*. She is currently working on a book-length project about the representation of rural Spain and its depopulation in film, literature, and photography.

Index

Numbers in *italic* indicate figures or tables.

\# (hashtag), 24–25, 47, 49, 238, 243, 252, 254, 257n9
\#DistractinglySexy, 24, 48–49, 51
\#HonorThyFather, 243
\#HonorThyMother, 48, 243
\#MeToo, 24, 257n9, 309
\#Mutante, 22, 47, 243
\#Raras, 22, 47, 243
\#WomensPlace, 14, 22, 24, 33, 36–37, 47–49, 80, 243

Abir-Am, Pnina, 121, 137
abortion, 57, 269, 270, 274–75, 285, 287n11
abuse, 104, 171, 223, 242, 328, 342
acculturation, 336, 346
Acuña y Villanueva, Rosario de, 6, 15–16, 89–98, 100–101, 104–18
Adams, Rachel, 230, 232
"Admirados colegas" (Mayoral) 20, 291–93, *298*, 308
affective, 82, 167, 171, 325, 328
African, 12, 225, 319, 336, 337, 339, 348–49
  American, 35, 334
  colonialization, 224, 330, 344, 347n2
  education, 339–40, 342–46

African (continued)
  families, 329, 333, 343
  migrant workers, 328
  north, 184n7
  social classes, 12, 21, 315–20, 325–29, 332–33, 338, 342–46
  sub-Saharan, 319
  traditions, 318, 332–33
  west, 316–18, 333, 334, 347n12, 348
  women, 12, 327, 332–33, 343, 345–46
Agora (Amenábar), 354, 365n11
alchemy, 151, 166
alcoholism, 106–7, 111
Alemany, Carme, 121
Aleu Riera, Dolores, 98, 116, 210–11, 231n8
Allen, Ann Taylor, 277, 287
AMIT Association of Women Researchers and Technologists, 8, 10, 27n10
android, 240, 255, 259n29, 367n22
angel, 73, 90–91, 107
Annandale, Ellen, 65, 67
Antología de pensamiento feminista (Johnson), 3, 29
Arenal, Concepción, 6, 116–17, 231n7, 232
  and abolitionism, 115n3

Arenal, Concepción (*continued*)
  on charity and wealth, 96–98, 113
  on hygiene, 102–4
  on prostitution, 92–93
Arkinstall, Christine, 67, 91, 113, 115–16
Arnal Yarza, Jenara Vicenta, 211, 351
Arnedo Soriano, Elena, vii, 52, 66n3, 66n7, 66n9, 66n12
  women's health, 14, 16, 52–55, 57–65, 67n16, 67–69
artificial intelligence, 12, 71, 278, 362
ascientific tradition, 15, 70, 72
Asociación de Centros de Planificación Familiar de España, 54, 68
astronomy, 143, 170, 216
Augé, Marc, 35

Ballesteros, Mercedes, 18–19, 208–9, 219–20, 222–30, 231n11, 232nn12–16
Barceló, Elia, 20, 266–86, 287n6, 288–89
barriers, 4, 22, 235, 307
  breaking, 158–59
  challenging, 4
  cultural, 25n2
  institutional, 73
  social, 78, 230
Barthes, Roland, 364
Bauer, Martin W., 74, 76, 87
Bell Burnell, Jocelyn, 80, 250, 260, 308
Bermúdez, Silvia, vii, 3, 14, 28, 52, 66–68, 205
Bernárdez Rodal, Asunción, 11, 28, 66n8, 67n13, 67n14, 68
Bhabha, Homi, 330, 332, 348
Bieder, Maryellen, xiv, 3, 28
biodiversity, 12, 248
bioethics, 287, 357
biology, 130, 157, 159, 170, 196, 243, 246–47
  and biochemistry, 120, 121, 123, 127, 130
  Nobel in, 27n8
  and microbiology, 129

biology (*continued*)
  and molecular, 60, 69–70, 119, 121, 123, 127
  and physics, 76
  study of, 36, 67, 108, 122, 124, 215
biotechnology, 12, 20, 266–70, 272, 279, 286, 286n5
Biro, Yaelle, 333, 348
Blanco, Delia, 58, 60
Blasco, María, 5, 27n8
Blasco Ibáñez, Vicente, 190
Body/Politics: Women and the Discourses of Science (Jacobus), 27n12, 29, 313
Bohr, Niels, 173, 184n11, 185
Bombshell: The Hedy Lamarr Story (Dean), 354, 365n5
Borges, Jorge Luis, 179, 184n13, 185
Borrell Ruiz, Sara, 9, 33, 34, 350
botany, 99, 108, 170, 173
Bourdieu, Pierre, 75, 230n2, 232, 333, 347n11, 348
bourgeoisie, 107, 270, 333, 338
  morality and codes, 91, 104, 335, 337
  privilege, 101, 106
  readership, 95
  society, 98, 100, 102
Britain, 27n15, 48, 319, 320
  British Archives, 21, 320
  culture, 184n6, 317–18, 337
  empire, 258n17
  influence, 318, 330
  literature, 249, 265
  occupation, 317, 318
  officials, 325, 336
  scientists, 80, 134, 241
Bruna Husky, 70, 239–45, 254–56, 256n1, 257n6, 259n28
Bubis, 317–20
Burgos, Carmen de, xiv, 6, 17–18, 187–204, 204n1, 205nn5–8
Bustelo, Gabriela, 20, 265–68, 277–80, 283–84, 286, 287nn12–13, 288
Butler, Octavia, 265, 288

Cabré i Pairet, Montserrat, 55, 66n2, 68, 69
Calkins, Mary Whiton, 308
Calvo, Antonio, 134, 137
Calvo-Iglesias, Encina, 34–36, 41, 49, 50, 51
Camus, Michel, 144, 158, 160
Canopus in Argos, 265, 289
Careaga, Pilar, 212
Carrasco, Antonio, 226, 232
Carson, Rachel, 248
Carte du Ciel, 354
Casado Ruiz de Lóizaga, María José, 364
Casamayor, Amaría Andrea, 350, 364n1
Casanova, Sofía, 188
Casas, Santiago, 101–2, 109, 117
Castells Ballespí, Martina, 99, 114, 117, 231n8
Castellví Piulachs, Josefina, 9, 21, 33, 162, 350, 353
Catalan, 9, 169, 184n7
Catholic
   Catholicism, 73, 230n1, 266, 275, 334
   Church, 72, 74, 108, 207, 217, 275, 280, 285
   Isabel the, 231n9
   morality, 73, 97
   newspapers, 115n4
   precepts, 86n4, 230n1
   Spain, 5
   tradition, 267, 319
Caza menor, 55, 69
Cegarra Salcedo, María, 212
censorship, 18, 35, 55–57, 228–29, 278, 281
Centro de Planificación Familiar Federico Rubio, 58, 67
Centro Protector de la Mujer, 97, 116
Chacel, Rosa, 6
Charpentier, Emmanuelle, 8, 26n6, 312
chemistry, 15, 30, 108, 130, 170, 211, 215, 237, 351
   Nobel Prize, 8, 26n6, 78, 237, 312n19
   women in, 5, 23, 71, 81–82, 119–20

Chesler, Phyllis, 63, 68
"chica rara, La" (Martín Gaite), 35–38, 51.
   See also Martín Gaite, Carmen
chicas raras, 14, 297
child rearing, 277, 283
Chillida, Eduardo, 144, 160, 162n5, 164
cloning, 267, 277, 279–80, 284
CodeGirl (Chilcott), 354, 365n5
collaboration, 129–30, 255, 330
Colombine, 192, 205. *See also* Burgos, Carmen de
colonialism, 21
   Francoist, 13, 225, 233–34
   gender relations, 328, 347n7, 349
   problematics, 12, 315
   racial inequities, 315, 339
   Spanish, 316, 318, 343, 346, 347n8
colonization, 226, 315–16, 318–20, 338, 345, 347n3
complementarity principle, 171, 173
Comte, Auguste, 81
Consejo Superior de Investigaciones Científicas (CSIC)
   employed by, 4, 11, 16, 36, 39, 119, 122, 131
   organization, 34, 40, 74, 124
   studies done by, 10, 87
Consecuencias naturales (Barceló), 20, 266, 268–73, 275, 277, 288
contraceptives, 20, 267, 273, 277, 284
   banning of, 53, 57
   free, 59–60
   history of, 126, 270
   use of, 57, 269, 272
Corrales, Capi, 121
Cortázar y Nebrija, García de, 75, 87n6, 87
Creager, Angela, 28, 51, 125
creation, 142–43, 173, 180, 284, 330, 356, 357
   artistic, 150, 360
   institutions of, 52, 66n7, 67n14
   literary, 89, 157, 158, 246, 310n7, 303

creation (*continued*)
    meaning, 183, 193
    scientific, 248, 258n17, 259n20, 354
    See also creator; procreation
Creative Couples in the Sciences (Pycior), 121, 137n2
creator, 47, 97, 174, 294, 308, 357–58, 360
CSIC. *See* Consejo Superior de Investigaciones Científicas
Cuestiones de mujeres (Arnedo), 53, 60, 65, 68
Curie, Irène Joliot, 26n6, 46, 48, 62, 83, 241–42. *See also* Curie, Marie
Curie, Marie, 45, 80–81, 86, 237, 257n4, 258n13
    achievements, 78, 237, 239, 242, 252
    affair with Paul Langevin, 257n4
    challenges, 79, 241–42, 244, 250
    chica rara, 47, 243
    corporeality, 83, 251
    and Curie, Pierre, 238, 241, 308
    daughters, 48, 259n32
    diary, 78–79, 82–83, 85, 87n7, 237, 254
    and literature, 71, 77, 79, 81–83, 86, 251, 237, 250
    loss, 80, 82, 86, 252
    Nobel laureate, 15, 26, 45, 71, 78, 237
    radium, 83, 241–42, 249, 259n20
    widow, 49, 237, 252, 254
    women scientists, 15, 80, 237, 249, 252
    women's roles, 15, 45, 71, 78, 80, 237, 238, 251, 259n23
    writing, 15, 45–46, 49, 71
Curie, Pierre, 46–47, 78–85, 237–38, 240–41, 251–52, 254, 308
"Curso de higiene vulgar" (Ovilo), 104, 117
cybernetics, 17, 19, 356
cyborg, 13, 71, 169, 255

Dalí, Salvador, 169, 177, 179
"dama dragón, La" (Barceló) 268, 288. *See also* Barceló, Elia
dark matter, 74, 152

Derrida, Jacques, 144, 206
Desbordadas. La agitada vida de la elastic woman (Arnedo Soriano), 55, 67
Desde la ventana: Enfoque femenino de la literatura española (Martín Gaite), 35, 51
Diamond, Marian, 308
dictatorship of Franco
    censorship, 57, 123, 134, 136
    cultural productions, 12, 208
    obstacles faced by women, 41, 212
    political and social forces, 40, 123, 207, 225, 266, 275
    scientific research during, xiv, 41, 128, 135, 239
digital, 270, 353
    breaking the divide, 26, 30
    disruptions, 11
    markers, 46
discipline, x, 23, 36, 49, 112, 253, 256, 271
    academic, 43, 74–75, 215
    artistic, 215
    interdisciplinarity and, 45, 47
    scientific, 5, 42, 211, 238, 242, 267, 365n7
Discipline and Punish: The Birth of the Prison (Foucault), 270, 288, 348
discourse
    colonial, 320, 337, 339–41, 346 (*see also* colonialism)
    cultural, 3, 214
    Falangist, 213, 216, 219, 226, 229, 230, 230n4
    gendered, 13, 50, 113, 161–62, 222, 225, 230n2, 232n16, 343
    literary, 209
    mystical, 142–43, 145–46, 151, 161
    official, 91, 209, 219, 225, 229, 231n7, 244, 318, 339, 343–46
    poetics, 16–17, 141–43, 154
    political, 18, 112, 207
    public, 60, 159
    public health, 15, 90–94, 98, 102–5, 112

discourse (continued)
  scientific, 2, 27n12, 142–46, 151,
    154–55, 158–61, 196, 266, 280, 284
disequilibrium, 258n19, 359–60, 364n4
domestic hygiene, 93–95, 101, 108
donjuanismo femenino, El (Soriano), 55, 57,
    69, 372
double shift, 76, 87
Doudna, Jennifer A., 8, 26n6, 312
Dougan, Walter, 328–9, *329*, *331*, 332,
    340–42, 345, 347n9, 348n15
Duden, Barbara, 129
Durán, María Ángeles, 67n13, 133
dystopia, 13, 20, 266–67, 278–79, 283,
    287n14, 288

ecofeminist, 171, 257n6
ecology, 185, 245, 248
education
  African, 340, 343, 346
  anti-oppression, 26n5, 29, 73, 230n1, 283
  formation, ix, 24, 222
  gender implications, 28, 43, 342, 346
  health, 104–6, 114
  inequities, 208–10
  medical and scientific, 98–99, 113, 352
  Ministry of Education and Science,
    26n8, 311n13
  opportunities, 44, 97, 231n7, 339, 345
  university, 7, 101, 213–16, 230
  women and, 5, 25n2, 34, 42–44,
    98–101, 214–17, 221, 344–46
egalitarianism, 24, 57, 282, 325
Einstein, Albert, 171
  collaborators, 49, 50, 177
  Einstein Rosen Bridge, 297–98
  scientific leader, 17, 144, 146, 161,
    311n13
  Theory of Relativity, 20, 145, 150–51,
    176, *178*, 293
  Theory of Special Relativity, 20,
    292, 299
  See also relativity

Ellas: Científicas (RTVE), 353, 365n7,
    368, 373
El País, 8–9, 33–34, 49, 119, 133, 162,
    163n11, 269
Ema, Charo, 60
empowerment, 111–12, 330, 352, 357
engineering, 81, 86, 209, 221, 224–28,
    230n5, 352
  agronomic, 34
  bioengineering, 236, 245
  civil, 220, 222, 223
  genetic, 20, 267, 277, 279, 281, 284
  women in, 7, 235
England, 34, 115n3, 226
entanglement theory, 17, 173–74, 183,
    185, 302
*Entre Cajal y Ochoa: Ciencias biomédicas
    en la España de Franco, 1939–1975*
    (Santesmases), 30, 39, 51, 119,
    137, 373. See also Santesmases,
    María Jesús
environment, 19, 76, 110, 230,
    244–46, 254
  challenges, 248, 254, 256, 257n5,
    258n18, 355
  hostile, 5, 210, 225, 228, 237, 238
  issues, 12, 129, 245
  milieu, 130, 132, 300
Epps, Brad, 182, 185
Equatorial Guinea, 21, 27n13, 224–26,
    315–17, 332–46, 347n3
equilibrium, 249, 359
  narrative, 359, 360–61, 364n4
  pressure, 247
Espejismos, 55
estrangement, 267–68, 277, 287n6
ethnicity, 12–13, 279, 315, 319, 365n4
eugenics, 13, 278, 284
European Commission, 74
European Science Foundation, 126
European Union, 4, 25, 368, 377
experimentation, ix, 151, 179, 306,
    357, 361

experimentation *(continued)*
  scientific, 142, 154, 162, 297, 309, 310n7, 351
extraterrestrial, 255–56

Facebook, 11, 71, 250
Falange, 207–8, 213, 216, 231n9, 375
  ideology, 225
  Women's Section, 18, 37, 208–9, 213, 217, 231n6
family, 42, 230, 239
  definition of, 282, 284
  health, 106, 110, 251
  interactions, 38, 325, 359
  liberal, 34, 304
  portraits, 328, 329, 332–34, 337, 347n9
  structure, 107, 114, 133, 219, 221, 275–77, 280–83, 334–35
  support, 25n2, 34, 43, 214, 232n13, 251, 259
  working class, 90, 96, 108, 110
family planning, 53–54, 57, 59–60, 67n15
fantasy, 17, 166, 265, 269, 295, 305
FECYT (Spanish Foundation for Science and Technology), 77, 87, 287–88
Federico Rubio Family Planning Center, 58, 67, 372
feminism, 3, 28, 51, 57–59, 231, 256–58
  cyberfeminism, 287n12
  ecofeminism, 171, 257n6
  equality, 54, 273, 303
  Spanish, 3, 59, 67n14, 194, 201, 205n7
  waves, 23, 41, 195, 201, 204, 252, 265, 286n4
  See also feminist; *New History of Iberian Feminisms, A* (Bermúdez)
feminist
  activism, 14, 54, 57, 58, 65, 66n7, 67n13, 194
  agenda, 14, 65, 292, 302
  anti-, 225
  approach, 18, 20, 204

feminist *(continued)*
  critiques, 13, 22, 188, 193–94, 198, 200–203, 223, 277
  healthcare, 52, 58, 63, 274
  ideology, 20, 44–45, 53, 61–63, 195, 214, 217–18, 235
  movements, 13, 53, 58–60, 63, 115n3, 218, 268, 277, 283
  perspective, 55, 57, 229, 243, 252, 267–68
  projects, 2–3, 57–58, 64, 250, 285
  scholars, xiv, 53, 65, 66n7, 75, 78, 121, 133, 249, 309n2, 365n4
  science, x, 249, 253, 292
  studies, 3, 4, 27n12, 365n4
  theory, 47, 48, 59, 351
  See also feminism
Fernández Vargas, Valentina, 136, 137n4
Fernandinas, xi, 12, 20, 315, 319–28, 332–37, 343–46
Fernando Poo, 328, 330, 332, 334, 338
  colony, 317–19, 330
  commerce, 317, 320, 330
  culture, 21
  photography, 318–20, *321*, 335
  place of origin, 12, 342
  social classes, 315, 317–20, 325, 340–41, 344–46
Ferreira, Ana Paula, 57, 66–68
Fields Medal, 8
film, 21, 78, 351, 366n13, 367n22
  cultural product, 12, 373
  documentary, 244, 246, 258n14
  female scientists in, 7, 247, 352–55, 365n5, 365n9, 365n11, 367n17
  structure, 356, 357, 360–64
  studies, 4, 364n4
  terminology, 284, 356
Flecha García, Consuelo, 117, 210, 231n8, 233, 372
Flora o La educación de una niña (Pascual de Sanjuan), 94–96, 118, 377

Flusser, Vilem, 316, 348
Fortunata y Jacinta (Galdós), 89, 114, 118, 375
Foucault, Michel, 270, 287n8, 288, 316, 348
Fox Keller, Evelyn, 40, 51, 75, 286–88, 295, 310n6, 313
  theoretical framework, 3, 13–14, 22, 27n12, 28–29
framing, 20, 301, 333, 336, 362
France, 27n13, 34, 73, 78, 169, 259n19, 317
Francoism, 55, 135–36, 207, 209, 212, 219–21, 229, 230n1. *See also* dictatorship of Franco
Franklin, Rosalind, 80, 134, 137n3, 250, 308
Frente de Liberación de la Mujer (FLM), 54, 58
Freudian, 169, 177

Gaia hypothesis, 248
Gajic, Tatiana, 230n3, 233
Gallego Méndez, María Teresa, 67n13, 230n4, 231n6
García Armada, Elena, 350, 365n7
García de Cortázar y Nebreda, María Luisa, 75, 87
García Márquez, Gabriel, 282
García Nieto, María Teresa, 11, 28
Gautier, Marthe, 308
gaze, 167, 216, 325, 357, 359, 361–64
gender
  equality, 10, 57, 235, 249, 303
  and medicine, 14, 52, 65n1, 126
  roles, 11, 20, 63, 189, 266–67, 269–70, 327
  and science, 14, 27n12, 28–29, 71–73, 166, 286n4, 288, 313
Gender and Nation in the Spanish Modernist Novel (Johnson), 3, 29

Gendered Drugs and Medicine: Historical and Socio–Cultural Perspectives (Santesmases), 16, 29, 119, 127, 137, 373
genes, 13, 264
genetic engineering, 20, 129, 267, 277, 279, 281, 284
  See also engineering
genre, 2, 4, 22, 79, 219, 232n12, 243
  breaking conventions, 66n10, 67, 191, 243
  film, 21, 351, 355, 365n5
  hybrid, 36, 45, 71, 242–43
  literary, xiii, 5, 6, 7, 20, 220, 267, 290
  photographic, 325, 347n6
  science fiction, 20, 257n6, 266–69, 286, 310n7, 357, 361, 363, 366n13
  scientific, 151, 236
Gen X, 278, 287n12, 289
geology, xiii, 141. *See also* sciences
Germany, 27n13, 187, 198, 311n13
Ghez, Andrea M., 26, 312
Girls in Tech Spain (spain.girlsintech.org), 353, 365n6, 368, 377
Gleick, James, 310, 313
Góngora, Luis de, 303
González, Felipe, 123
González Ramos, Ana M., 6, 28, 30
Gould, Stephen Jay, 245–46, 248, 254, 258n13, 258n14
Graham, Helen, 275, 288
gran libro de la Mujer: Salud, psicología, sexualidad, nutrición y derechos de la mujer, El (Arnedo), 14, 53–54, 61, 64–65, 67, 371
gravity
  concept of, 151, 174, 183, 302
  force, 176, 299
  nature of, 166, 177
Grunberg-Manago, Marianne, 128
guinea pig, 21, 296, 359, 362, 363, 364n3
Guzmán y de la Cerda, Marta Isidra de, 310n5

Haraway, Donna, 4, 28, 75, 169, 185, 255–56, 259–60, 272
Harding, Sandra, 27n12, 29, 286n4, 288
hashtags, 22, 24, 37, 46–49, 51, 80, 86, 243, 257n9. *See also* # (hashtag)
Hawking, Stephen, 17, 144, 146, 176, 178, 184n12, 310n11
Hayles, N. Katherine, 294, 302, 308, 310n7, 313
Heidegger, Martin, 154, 162n8, 163
Heisenberg, Werner, 146, 161
   scientific leader, 17
   uncertainty principle, 147–49, 163, 296
herstory, 308
heterosexuality, 268, 280–81
Hidden Figures (Melfi), 35, 51, 74, 354, 365nn5–9, 368, 373
hidden third, 155–59, 164
hierarchy, 268, 274, 304, 310, 361
   racial, 327, 339
"higiene del obrero, La" (Pérez Cano) 111, 118
"higiene en la familia obrera, La" (Acuña), 90, 92, 105, 116, 371. *See also* Acuña y Villanueva, Rosario de
historical memory, 135–36
   recuperation of, 14, 36, 45, 135
historiography, 35, 135
history, 136, 272
   art, 166, 182
   of biochemistry, 120–21, 123, 127
   colonial, 21–22, 318
   cultural, 14, 65n1, 292
   exclusion of women from, 13–14, 23, 33, 40–41, 278, 292, 305, 308
   literary, 20, 35, 37, 39, 267, 292–93
   manipulation of, 278, 282, 284, 307
   photography, 318, 329, 333, 336
   political and social, 42, 58–60
   Spanish, 72, 196, 275, 291, 293, 311n15, 366
   technology, 24, 52
Hochschild, Arlie, 71, 88

homosexuality, 268, 278, 280–84
Hubbard, Ruth, 301, 313
humanities
   career in, ix, 124
   field of study, 2, 12, 18, 120, 219, 256
   vs. sciences, 43, 155, 236, 246, 251, 294
hybrid, 15–16, 71, 86, 171, 363
hybridity, 36, 45, 168, 169, 171, 330, 332
hygienist, 94–95, 105
Hypatia Project, 353, 365n6, 368, 377

Icaza, Carmen de, 37
Ignotus Award, 269
independence, 64, 317, 336, 351
infant
   health, 5, 98–99, 107
   intelligence, 94, 97
inferioridad mental de la mujer, La (Möbius), 192, 205, 375
inferiority
   intellectual, 200, 320
   mental, 18, 188, 196–97, 200–201
   Mental Inferiority of Women, The (Möbius), 18, 190, 192
   women, 189, 195–97, 200–201, 216, 241, 297
   See also Möbius, Paul Julius
innovation, 236, 272, 357
   Ministry of Science and Innovation, 10, 66n2
   scientific, 248, 255, 256, 267, 305, 352
Institute of Philosophy, 16, 36, 119, 124, 383
Instituto de la Mujer (Women's Institute), 40, 41, 66n7, 74
Instituto Nacional de Estadística (INE), 10, 75
Instrucciones para salvar el mundo (Montero), 19, 70, 86n2, 88, 236, 239, 258n12, 261, 375
interior exiles, 136
International Day of Women and Girls in Science, 9, 30, 33, 352–53, 370, 374

intertext, 15, 47, 77, 79, 83, 86, 254, 304
intertextuality, 22, 36, 45, 83, 292, 365n4

Jacobus, Mary, 27, 29, 313
Jaime, Pilar, 58, 60
Jameson, Fredric, 268, 282, 287
Janés, Clara, 16, 17, 141–62, 162nn2–5, 162–63n7–10, 375
"jarca de la Universidad, La" (Acuña) 100, 116, 371
Jiménez-Landi, Antonio, 210, 233
jobs, 62, 171, 224
  employment, 10, 120, 257n3, 355, 366
  female, 27n11, 262, 275
  professional, 227, 237, 353
Johnson, Katherine, 35
Johnson, Roberta, 3, 28–29, 68, 194, 200, 233
Jung, Carl, 166, 168–69, 177, 180, 182–83, 184n6, 185

Kammerer, Paul, 245–46, 248, 254
Kent, Victoria, 211
Kervasdoué, Anne de, 53, 60–61, 66n3
Khraiche, Hatem, 21, 351

Labanyi, Jo, 288
laboratory, 24, 26n4, 83, 166, 170, 286, 296, 300, *301*, 358
Lacan, Jacques, 144
Laforet, Carmen, 37–39, 43–44, 47, 55, 66n10, 81, 251. See also *Nada* (Laforet)
Lágrimas en la lluvia (Montero), 19, 70, 88, 236–56, 257n10–11, 259n27, 261, 375. See also Montero, Rosa
Langevin, Paul, 81, 252, 257n4. See also Curie, Marie
Latin America, 169–70, 179, 285
laws
  abortion, 275, 285
  lawyers, women, 211, 216, 219, 230, 353
  lawyers, black, 340, 345

laws (*continued*)
  legal discrimination, 11, 212, 239, 280
  rights for women, 10, 73–74, 210, 235
  scientific, 102, 146, 246–47, 260, 301, 306
Le Guin, Ursula K., 83–84, 265, 289
Le-May Sheffield, Suzanne, 73, 78, 88
Lessing, Doris, 265, 289
Lévi-Strauss, Claude, 364
LGBTQ, 239, 247, 280–81, 283–84, 289
¿liberación era esto? Mujeres, vidas y crisis, La (Sáez Buenaventura), 54, 63–64
Libro Blanco: Situación de las Mujeres en la Ciencia Española (Sánchez de Madariaga), 10, 30, 66n2, 68, 74, 88
libro de los pájaros, El (Janés), 141, 144, 146, 150, 160, 163
Linares Becerra, Concha, 37
Lindner, Ulrike, 347, 349
literary Canon, 1, 291
Lope de Vega, 20, 291, 293, 299–300, 303, 309n1, 309n4, 311–12n15
López-Sancho, M. Pilar, 10, 29, 77, 88
Lovelock, John, 245, 248, 254, 258nn17–18, 259n19, 260
Löwy, Ilana, 126

machismo, 226, 251, 269, 272
Machung, Anne, 71, 88
madness, 77, 82, 110, 128, 263
Madrid, Fátima de, 350, 364
Maeztu, María de, 212
Magallón Portolés, Carmen, 121, 211, 234
Maíllo, Kike, 21, 351
"Making a Difference: Feminist Movement and Feminist Critiques of Science" (Fox Keller), 13, 22, 28, 51
Mañeru, Ana, 122
"Manifesto for Cyborgs, A" (Haraway), 4, 28, 169, 185, 259–60
Manuel de Villena, Ernestina, 90
Marić, Mileva, 49–50
marriage, 214, 223, 226, 280, 282, 285, 325, 334–36

Martin, Emily, 294, 313
Martín Bravo, Felisa, 350
Martínez Salmeán, Javier, 60
Martínez Sancho, María del Carmen, 350
Martínez Sierra, María, xiii, 6, 26n4
Martín Gaite, Carmen, 35–41, 43–45, 47, 51, 66n10
Martín-Márquez, Susan, 232, 234
masculinity
  in intellectual tradition, 2, 6, 75, 87n6, 161, 182, 252
  in media and literature, xiii, 8, 77, 224–26, 232–34, 258n12, 282–83, 287
  women and masculinization, 19, 218, 221–22, 230n5, 272–73
maternal
  health, 98–99
  maternity, 58, 117, 272, 274, 276–77, 282
  role, 269–70
  work, 133, 200
mathematics
  awards in, 8, 72
  formulas, 149, 155, 163, 171, 173, 239
  mathematicians, 241, 245, 311, 350
  in STE(A)M, 81, 86, 143, 162n3
  women in, 18, 24, 25n2, 34–35, 88, 120, 162, 215, 217, 256, 350, 352, 354
matriarchal society, 20, 208, 278, 279, 282
Matute, Ana María, 39, 55, 66n10
Mayock, Ellen, vii, 12, 15, 22–24, 29, 50n3, 51, 70–88, 257n5, 261
Mayoral, Marina, viii, 20, 291–313
McClintock, Anne, 232, 234
McClintock, Barbara, xiii, 26
media
  news, 10–11, 71, 74, 133
  social, 24, 242–43, 250
  Spanish, 76–77, 216, 342, 351–53
  visual, 45, 320, 337
medicine
  biomedical, 39, 48, 119, 126, 249, 267
  history of, 67n15, 93, 102, 114n2, 125, 129, 130, 275

medicine (*continued*)
  medicalization, 15, 17–18, 71, 86
  preventive, 60, 240
  publications, 55, 91, 127, 187, 190
  women in, 54, 57–58, 65, 92, 98–99, 131, 210, 231n8, 241
Medina-Doménech, Rosa, 225, 232, 234
Meitner, Lise, 80, 250
memoir, 15, 45, 50, 71, 86, 204n1, 238, 242
metaphysics, 142, 242, 43, 149, 170, 243
Mexico, 23, 27n13, 125, 131, 166, 170, 184n7, 185–86
Minahan, Stella, 66n5, 68
mirada tuerta, la, ix, 23, 27n14. *See also* Roig, Montserrat
Mirzakhani, Maryam, 8
misogyny, 18, 188–90, 192–93, 196–98, 202, 254, 257n9, 271
Mitchell, W. J. T., 333, 349
Möbius, Paul Julius, viii, 17–18, 187–206
modernity, 155, 159, 163n9, 164, 221, 288, 318, 336, 349
Moi, Toril, 249, 261
monster, 18, 47, 255, 275, 313
Montero, Rosa, 235–36, 246, 249, 256–62
  Instrucciones, 238–39, 246–48, 253–54
  Lágrimas en la lluvia, 239–40, 254–56
  peso del corazón, El, 246
  See also specific titles
Morreale, Gabriela, 9, 34, 162, 350
motherhood, viii, 20, 59, 99, 195, 241, 265, 270, 277, 287
Moyano Act (1857), 93, 98–99
mujer del porvenir, La (Arenal), 92, 98, 102, 116, 231n7, 232
Mujeres, ciencia e información (García Nieto), 11, 28
Mulvey, Laura, 308, 313, 357, 365, 367n18
mundo de Yarek, El (Barceló), 269, 288
Munera, Isabel, 352
Muñoz, Emilio, 120–22
Muñoz-Páez, Adela, 211, 234

mysticism, 16–17, 141–43, 145–51, 153–54, 161–62, 170

Nada (Laforet), 35, 37–39, 47, 51, 251
Nerín, Gustau, 347, 349
neuroscience, 12, 40, 149, 180
Nevares, Marta de, 293, 304, 309nn1–4, 313
New History of Iberian Feminisms, A (Bermúdez), 3, 28, 68, 205
Newton, Isaac, 155, 161, 176, 178, 184n12, 237, 299
Nicolescu, Basarab, 17, 144, 146, 154–60, 163n9, 163–64
Nietzsche, 144, 188
Nobel Prize
  awardees, 24, 48, 50n4, 146, 159, 180, 286n5, 312n19
  Curie, Marie, 15, 45, 71, 78, 237–38, 242
  exclusions of women, 7–8, 74, 79, 86n5, 88, 250, 308, 314
  and Spain, 40, 50
  women recipients, 7–8, 24, 26n6, 27
novela rosa, 37–39, 50n2, 220, 226–27, 232n12
Núñez Puente, Adela, 232, 234
Núñez Valdés, Juan, 364

O'Byrne, Patricia, 36, 44, 49, 51
Ochoa, Severo, 30, 39–40, 50n4, 119–21, 128, 137n1, 286n5
ontology, 17, 154, 161, 171, 174, 333, 338
Orbita 9 (Khraiche), 21, 351–69
Ortiz Gómez, Teresa, 55, 68–69, 119, 122, 126, 137n1
Osca Lluch, Julia, 235, 249, 261
Oudshoorn, Nelly, 273, 289
Our Bodies, Our Selves (Boston Women's Health), 53, 68
Ovilo, Felipe, 104–5

Pacheco, Guillermina, 89–90, 113
parallax, 142, 146–49, 160, 162n6, 165

Pardo Bazán, Emilia, xiv, 6, 30, 187, 195, 231n7, 234
Pascual de Sanjuán, Pilar, 94, 118
Pérez Cano, Vicente, 111, 118
Pérez Galdós, Benito, 89–90, 114, 118
Pérez Sánchez, Gema, 280, 289
Pérez Sedeño, Eulalia, 121–22, 124
periodicals, 15, 134, 212
Perriam, Chris, 268, 289
peso del corazón, El (Montero), 19, 236–46, 254–56, 256n2, 257n10, 261
pharmacy
  history of pharmaceuticals, 126–27, 270–72, 289
  shopping at, 101
  studying, 23, 43, 344–46
  women pharmacists, 23, 33, 344–46, 350, 354–55, 364n2, 366n11
photography, viii, 12–13, 19, 20–21, 315–49
physics
  Nobel in, 24, 26n6, 78, 86n5
  and poetry, vii, 15–17
  study of, 35, 80–82, 108
  theoretical and quantum, 12, 141, 143, 145
  women in, 10, 29, 71, 74, 76, 78, 88, 108, 122
picadura del tábano: La mujer frente a los cambios de la edad, La (Arnedo Soriano), 54, 67
Piercy, Marge, 265, 289
playa de los locos, La, 55–57, 66n10–11, 69
poetry, 12, 16–17, 141–65, 248, 309n4
Poland, 27n13, 78, 237, 250, 257n3
Poniatowska, Elena, 23
Poole, Deborah, 336, 338–39, 349
posthumanism, vii, 17, 166, 169–70, 173, 183, 257n6, 260
postwar period (Spanish Civil), viii, 12, 18–19, 36, 51, 207–34
Prádanos, Luis I., 257n6
Pratt, Dale J., 257n6

Preciado, Paul B., 270–71, 287nn8–9, 289
pregnancy, 99, 203, 269–70, 273–76, 280, 287, 355
Prigogine, Ilya, 144, 146, 161
Primo de Rivera, José Antonio, 44, 207, 212, 217, 225
Primo de Rivera, Pilar, 44, 212, 231
procreation, 57, 284, 285
PSOE (Partido Socialista Obrero Español), 14, 54, 58, 66n12, 103–4, 275, 280
public health
    discourse of, vii, 15, 89, 90–93, 98, 103, 108
    policies, 65, 109, 111–14, 115n6
    promotion of, 96, 104
Pulido, Ángel, 92, 93, 108, 118

race
    class and, 116, 255, 341–42, 346
    construction of, 284, 336, 340–41
    ethnicity and, 234, 319, 330
    gender and, 234, 255, 315, 325, 328, 329, 338, 345
    human, 98, 110
radioactivity, 78, 82–83, 242
Radway, Janice, 232, 234
"ramera, La," 90–92, 108, 110–11, 114, 116
Ramón y Cajal, Santiago, 27n8, 50n4, 72, 100, 118
rape, 269, 274–75, 287n10
Real Academia Española (RAE), 141, 164, 201, 205
relativity
    concept of, viii, 20, 50, 145–46, 166, 170, 201, 302
    Einstein's theories of, 145–46, 151, 163, 176, *178*, 292
    and time, 184n12, 299, 308, 310n11
religion, 11, 73, 76, 96–97, 108, 194, 199
reproductive rights, 12, 14, 20, 52–53, 266, 275
reproductive technology, 266, 279, 284–85

Richmond, Kathleen, 230–31, 234
ridícula idea de no volver a verte, La (Montero)
    creative nonfiction, 36, 45–51
    critique of gender, vii–viii, 15–16, 236–38, 241–43, 249–53
    and science, 19, 22–23, 70–72, 76, 78–88
    see also Montero, Rosa
Robles, Lola, 286nn1–2, 289
robotics, 12, 356–57, 359–60, 363, 366n15, 367n22
Rodrigo, Antonina, 23, 29
Roig, Montserrat, ix, 23, 29
romance, 37, 50n2, 220, 223, 234, 292
Rose, Hilary, 306, 314
Rosinach Pedrol, Zoe, 350
Rossiter, Margaret, 121–22
Rubin, Vera, 74, 86n5, 88
Rubio y Galí, Federico, 57–60, 65, 67
Ruiz Robles, Angela, 350
Russ, Joanna, 265, 290
Russia, 189, 204n1, 237, 250, 257n3

Sáez Buenaventura, Carmen, 54, 62–64, 66n7, 69
Sagrada (Barceló), 268, 288
Sáinz, Milagros, 7, 30
Sáinz González, Jorge, 6, 7, 30
Salas Falgueras, Margarita, 21, 26n3, 27n8, 30, 132, 350, 353, 365
    and CSIC, 4, 128
    death, 26, 30
Salazar, Teresa, 211
Sánchez, Ana, 121
Sánchez-Conejero, Cristina, 265, 268–69, 272, 290
Sánchez de Madariaga, Inés, 10, 30
Sánchez Ron, José Manuel, 122, 311n13, 313
Sánchez Ulibarri, Juan, 109, 115n6
San Fernando Observatory, 354

Santesmases, María Jesús
  compared to literary figures, 36, 39–45, 49, 50, 50n1
  history of science, vii, 3, 5, 10, 12, 26n4, 15–16, 55
  interview with, 119–37
  works by, 29–30, 51, 66n2, 69, 372–73
Scanlon, Geraldine, 231, 234
Schleiermacher, Friedrich, 188, 190–91, 204, 206
Schrodinger, Edwin, 17, 144–46, 149, 154, 161, 165
science fiction
  authors, 83, 265, 267
  film, 21, 350–64, 366n13
  genre, viii, xiii, 7, 13, 256, 256n6, 286, 357
  Spanish, 20, 269, 277, 288–90, 293, 310n7, 351, 355
scientific inquiry
  and literature, vii, 14, 19, 73, 139, 240, 286n4, 355, 361
  and poetry, 17, 142, 156, 161
  and religion, 73, 171
  and women, 3, 16, 71, 86, 295, 351
scientific method, 292, 295
Sección Femenina de Falange, 37, 208–9, 212–13, 217, 230n6, 230n10, 233–34
Second Spanish Republic, 342, 345
secularism, 76, 341
Sekula, Alan, 333, 349
selfhood, 214, 218, 222, 236, 238–39, 290, 330, 338
Serrano, Elena, 5, 30
Sessa, Duke de, 293, 299, 301, 303–4, 306
Shakespeare, William, x, 23
Sheldrake, Rupert, 245–48, 254, 258n14, 259n25, 260
Shetterly, Margot Lee, 74, 88, 365n9
Showalter, Elaine, 249, 262
Silent Spring (Carson), 115, 248
slavery (slaves, enslavement), 194, 218, 279, 282, 317

Smit, Theresa Ann, 310, 314
Smith, C. U. M., 60, 67n16, 69
Sobel, Dava, 74, 88, 354
socialism, 66n12, 75
  party, 54, 103, 275
  El Socialista (newspaper), 103, 111–12, 117–18
Solé, Albert, 353
Solís y Claras, Manuela, 99–100, 118
Sols, Alberto, 121
Sorbonne, 237–38, 241
Soriano, Elena, vii, 14, 52–57, 59, 64–65n1, 66–67nn3–10
space-time, 296–97, 308, 310nn10–11, 311n14
space travel, 20, 184n12, 256n1, 301
Spanish Civil War, 18, 40, 123, 169, 207, 251, 266, 270
Spanish Penal Code, 59, 66n3
STEAM, ix, x, 4, 14–16, 81, 83, 133, 160–62
STEM
  chicas raras and, 33–51
  concept of, vii–viii, 1, 81–86, 163n11, 209, 245–56, 291–93
  degrees in, 7, 26n4, 29, 351–53
  fields, ix, 2–29, 160–62, 211, 214–16, 228, 230, 235–40
  stereotypes, 4, 7, 22, 28, 38, 43, 77, 291–92, 330
strange attractors, 146
Strathern, Marilyn, 300, 314
Strickland, Donna, 24, 26n6, 27
Suárez, Edna, 125
Suárez, Renée, 60
subjectivity, 2, 24, 146, 183, 270, 278, 309, 338
Subramaniam, Banu, ix, xin1, 27n14, 30
surrealism, 166, *168*, 168–69, 185
surveillance, 270–71, 358
Sussman, Herbert, 230, 234
Suvin, Darko, 267, 290
Swan Leavitt, Henrietta, 80, 250, 260

Tagg, John, 332–33, 336, 349
technohuman, 19, 236, 255–56
technology, viii, 36, 49
   in literature, 235–60
   medicine and, 3
   reproductive, 266–87
   science and, viii, 2–3, 86–87
   women in, 4–30, 119–21
   See also STEAM; STEM
television, 21, 120, 220, 278, 311n15, 351–54, 366nn11–13
thermodynamics, 146, 246
time travel, 20, 291, 296–99, 301, 308, 310nn10–11, 311n11, 313. *See also* relativity
Tipler Cylinder, 297, 311n11
Todorov, Tzvetan, 359, 364n4
Toral, María Teresa, 23, 29
Tower Sargent, Lyman, 282, 290
transdisciplinarity, 160, 164, 184n9, 186
translation, viii, 4, 17–18, 82, 160, 163n9, 187–206, 256n2
trope, 26n4, 26n5, 86n2
Tudor, Jeannette F., 63
twin paradox, 299, 311n14

Ucelay Maórtua, Matilde, 212, 351
Unamuno, Miguel de, 72
uncertainty principle, 147–49, 163, 296. *See also* Heisenberg, Werner
UNESCO, 4, 352
Unidad de Mujeres y Ciencia, 25n1, 74, 88
United Nations, 9, 30, 352
United States, 53, 131, 257n9, 258n17, 265, 353–54, 365n11
   travel to, 34, 50n4
   women and science in, 74–76, 81, 98
Urioste, Carmen de, 278, 290
utopia, 20, 266–68, 279, 282–83, 285, 287–88, 290

Valdés, Isabel, 9, 30, 33–34, 51, 163, 165
Valiente Fernández, Celia, 66, 69
vanguard, 71, 286
Varo, Remedios, vii, x, 17, 166–83, 184n6, 185
Vasquez Ramil, Raquel, 210, 234
Venuti, Lawrence, 188, 191, 193–94, 198, 204n3, 205–6
Versteeg, Margot, 5, 30
video games, 26n2, 287n7
Visconti, Sylvia, 220, 232

Walter, Susan, 5, 30
wave theory, 144–45, 148–49, 158
Wells, Susan, 98, 118
whiteness, 336, 344
women's liberation, 54, 64–65, 66n8, 68
Wonenburger, María, 24–25, 29, 34, 162
Woolf, Virginia, 1, 23–24, 30
Wyer, Mary, 5, 11, 25, 28, 30, 293, 305, 310, 313–14

xenogenesis, 265

Zambrano, María, 6, 212
Zizek, Slavoj, 149, 165
Zubiaurre, Maite, 3, 29, 67–68

CPSIA information can be obtained
at www.ICGtesting.com
Printed in the USA
LVHW030353161220
674266LV00006B/413